Quality Management in Engineering

Quality Management in Engineering

A Scientific and Systematic Approach

Authored by

Jong S. Lim

CRC Press
Taylor & Francis Group
Boca Raton London New York

CRC Press is an imprint of the
Taylor & Francis Group, an **informa** business

CRC Press
Taylor & Francis Group
6000 Broken Sound Parkway NW, Suite 300
Boca Raton, FL 33487-2742

Printed on acid-free paper

International Standard Book Number-13: 978-0-367-23008-1 (Hardback)

Library of Congress Cataloging-in-Publication Data

Names: Lim, Jong Seong, author.
Title: Quality management in engineering : a scientific and systematic approach / Jong Seong Lim.
Description: Boca Raton : Taylor & Francis, a CRC title, part of the Taylor & Francis imprint, a member of the Taylor & Francis Group, the academic division of T&F Informa, plc, 2020. | Includes bibliographical references.
Identifiers: LCCN 2019016194 | ISBN 9780367230081 (hardback : alk. paper)
Subjects: LCSH: Quality control. Classification: LCC TS156 .L56 2020 | DDC 658.4/013—dc23 LC record available at https://lccn.loc.gov/2019016194

Visit the Taylor & Francis Web site at
http://www.taylorandfrancis.com

and the CRC Press Web site at
http://www.crcpress.com

FSC www.fsc.org MIX Paper from responsible sources FSC® C013056

Printed and bound in Great Britain by
TJ International Ltd, Padstow, Cornwall

To my wife, Eunice, and my

two sons, Seth and Tim.

They are the pioneers who walk with me

and talk with me with encouragement,

new knowledge, and ideas.

Contents

Preface

Civilization has evolved with the accumulation of knowledge and idea diffusion. Knowledge is not created, but rather evolves and grows by sharing and transferring it. Knowledge is not a zero-sum economy. Knowledge creates wealth and future.

In the economy of the 21st century, sharing and transferring knowledge is pivotal to increase the momentum towards a knowledge-based economy. Knowledge is the power of the future economy. Big Data processing, artificial intelligence, smart robots, autonomous vehicles, 5G Wi-Fi technology, new energy systems, and advanced medical technologies are becoming our everyday languages. Our traditions may perish, but knowledge will spread and survive for our future civilization.

Every new technology stems from sciences and previous technologies. In order to deliver value, new technology must be transformed into the form of products, whether they be software, physical products, or services. During this transformation process, the product may lose efficiency, intention, and even the validity of the concept unless the transformation is implemented in a robust fashion, accommodating the consumers' usage conditions and environments.

In past decades, we have witnessed that just one small mistake in a product can cause hundreds of deadly accidents and shut down numerous sales opportunities. For example, in 2009 there was Toyota's unintended acceleration recall; in 2014 GM's ignition switch recall; and in 2014 Takata's airbag recall. And history repeats itself. There was VW's emissions scandal in 2015, Mitsubishi's fuel mileage cheating scandal in 2016, and BMW's engine fires in 2018. Those safety issues caused many injuries, deaths, and fears, and the massive recalls followed only after government investigations and interventions.

There are still many other examples of such catastrophic quality issues: in 2013, a battery problem with a Boeing 787 grounded the state-of-the-art airplane for more than three months; and in 2017, Samsung's flagship product, the Note 7, suffered from a disastrous battery issue that caused it to disappear from the market. More recently, in March 2019, Boeing's best-selling airplane 737 Max Jet has been grounded globally after airline crashes that killed almost 350 people in all.

When we delve deeper into such problems, they are not just unlucky or random cases. They are the inevitable fallouts of mismanagement, weak systems, and fragile engineering.

Some people would like to believe that our future technologies will guarantee failsafe products. However, theoretically that is not accurate forecasting. The more advanced the technology, the more complex the system. Therefore, new technologies require much more robust quality design and manufacturing process controls. Without the fundamentals of quality by design and process controls, new innovative technologies can be the cause of potential future crises. For example, the autonomous vehicle is an inevitable future technology already making its way into the auto industry. It relies on many advanced smart sensors, complex data fusion technology, a very accurate 3D mapping system, and an artificial intelligence system. Many auto companies tout the robustness of their technologies and systems. Unfortunately, since 2016 there have been several fatal accidents related to autonomous driving technology.

As we progress into the future, we will see more and more advanced technologies. Manufacturers will try to convince consumers that their products are safe and free from any defects. However, history tells us that advanced technology does not necessarily warrant quality. When airbag technology was released, it did prevent many fatal accidents. However, the airbag also killed and injured many people because of quality issues combined with the overconfidence in the technology by consumers.

Overconfidence in technology is becoming one of the greatest threats to society and companies. The biggest lesson to learn is that we cannot build up any new technologies without a strong basic foundation, which is quality. One small glitch either in the hardware or software in an autonomous vehicle or an intelligent robot can result in a fatal accident.

In the future of a technology-driven economy, the severity level of any quality problem will become more catastrophic, and the scale of the impact can be pandemic. Our society must become more vigilant in detecting and resolving potential quality issues of such advanced products.

Poor product quality can become a total loss to society. It can become a massive waste of natural resources, increase environmental pollution, and degrade people's quality of life. And this negative impact on consumers and society will return to the manufacturers like a boomerang.

Many company leaders believe that quality improvements need substantial investment and cost. However, this is simply not true. Quality improvements do not incur a cost increase. During a quality improvement process, the company instead has the opportunity to discover potentials for waste reduction in the product design and process controls. Cost savings brought on by quality improvement processes balance out the input cost and can lead to increased market share by attaining a reputation of high quality and earning the trust of consumers.

Every manufacturer's desire is to sell more products for more profit. Companies are in constant competition against each other. Competition is undoubtedly beneficial. Human civilization has evolved through it, and competing brings more effective value to society. We compete to be more efficient, productive, and innovative. From this perspective, manufacturers must rethink the fundamentals of manufacturing competitiveness.

Fundamental manufacturing competitiveness comes from four major factors: cost, productivity, technology, and quality. A common sense approach is to look at these four factors in a balanced and co-dependent way: cost impacts the selling price, productivity impacts the cost and delivery, technology makes the product attractive, and quality gives confidence and trust to consumers. It is difficult to have one without the other. They are all linked in direct and indirect ways.

The most significant common linkage of these four factors is *quality* even though it seems to be hidden or less visible at first glance. A poor businessman finds out the value of quality only after quality issues cause problems on the surface. Every consumer consciously or unconsciously looks for the quality value of the product when making a purchasing decision.

Quality is the philosophy, policy, principles, system, management, and engineering underlying the product. We must look at quality from a holistic viewpoint—at System and Management and Engineering—in an integral and balanced way.

Quality is conceived from a good system and executed through good management and engineering. Quality must be studied and planned scientifically. Scientific approaches require logical and mathematical thinking. A structured quality deployment should follow the scientific quality plan to make outputs. Quality is *not* a conceptual language. It should be tangible and conveyable.

In this book, the relevant scientific methods and tools are explained in the form of practical usage in the field. We will focus on manufacturing industry examples and applications, which are also applicable to the software industry. The approaches and methods explained in this book can apply to non-manufacturing industries as well. Many examples and case studies contained in this book are based on the author's thirty years of industrial experience and knowledge.

In the last part of this book, the author will discuss a new quality paradigm and software quality management methods associated with the future new technologies which form Industry 4.0. In future quality, any product design and production should utilize new technologies such as 3D printers, AR (augmented reality), smart sensors, AI (artificial intelligence), intelligent robots, 5G communication networks, etc. Real-time process control capabilities will improve quality management interactivity with equipment and machines. Future machines will not only produce advanced products but also be capable of improving quality by self-correction and adjustment. From this perspective, the scope of future quality activity should be integrated with new technology, with a new way of thinking.

Acknowledgments

Every cultivation starts by sowing a seed. I was encouraged to write this book by my students, my colleagues, and industry leaders. Most of all, it was Professor Bejan who motivated me and pushed me to start writing. Once something has been started, it is difficult to stop—a natural law of physics.

However, friction and resistance also play an inevitable role. I have received strong support from my family to get over both physical and psychological obstacles. My wife, Eunice, and my two children, Seth and Tim, supported me in overcoming the challenges during the journey.

My 30 years of industrial experience and knowledge is owed to my predecessors. They were my managers, bosses, peers, and teachers. Thanks to them, I gained a strong motivation to pass on their teachings to the next generation. My teachers are: Lee Hong Won, director of Samsung Electronics R&D Center, who taught me research and development functions; Shinichi Tsurui, Chief Inspector of Honda Motors, who gave me the opportunity to serve as engineering quality manager; Gilles Madore, Executive VP of Honda Canada Manufacturing, who was my role model in management; Jean Mari-Huritiger, CEO of Renault Samsung Motors, who supported many of my recall decisions; and Joa Kim, my colleague at Volvo, who was patient with me for challenges for the quality system and continual process improvement.

My colleagues at Seoul National University (72th Advanced Management Program) provided me with encouragement for this project. Daniel Im, CEO of AIFounded in Toronto, gave me valuable advice on innovative software technologies of the future. My university teaching assistant, Lu Jinhua, made diligent contributions to many drawings enclosed in this book.

There are industry leaders, colleagues, and students whom I owe my knowledge and experience: Fredric Pelata, Francois Provost, Fabrice Delacroix, Eric Wittersheim, Mizu Tani, Wayne Shaw, Frank Paluch, Jim Mankin, Ned Myer, Don Gallagher, Dale Quinn, Jim Ingram, Ross Grant, Marc Aston, Feng Qingfeng, Sun Da, Kim Joong Hee, and Kim Yang Soo. I was inspired by their passion for quality, technology, and engineering.

Knowledge is worth spreading. I am looking forward to even a small amount of knowledge touching industry leaders, managers, engineers, and students. They will spread new knowledge and wisdom for the future. In advance, I thank them.

Author

Jong S. Lim started his professional career as a quality engineer in the automotive industry in 1985. He received his Ph.D. in "Thermal Science" from Duke University (1991) under the supervision of Professor Bejan.

Dr. Lim progressed through engineering, management, and executive positions in internationally prestigious companies: Quality Engineer at Hyundai Motors (1985), Senior Researcher at Samsung Electronics (1991), New Model Quality Manager at Honda Motors in North America (1997), Quality VP at Renault Samsung Motors (2008), R&D VP at Geely-Volvo Motors (2013). Currently, he serves as a professor at Zhejiang University in China and CEO of QEMT—industrial consulting company (2019).

When he was in Honda, he was a new model quality manager developing many of Honda models for the North American market, including the Civic, Odyssey, and MDX.

In 2008, after he joined Renault Samsung Motors in Korea as Vice President of the Quality Division, he set up a robust quality system for new model development and strategic customer relationship management, and as a result, the brand achieved No. 1 customer satisfaction for 10 consecutive years.

In 2013, he joined the Geely Group as the Quality VP of New Platform Development in conjunction with Volvo for a global brand for the Chinese and European market.

In 2016, as a result of his distinguished contribution to quality in the Chinese manufacturing industry, he received the Best Foreign Expert Award from the Chinese government.

Currently, he is teaching industrial quality engineering and management programs in universities as well as consulting for manufacturing companies around the world in the areas of quality and technology development.

1

Fundamentals of Manufacturing and Engineering

Through the Industrial Revolution in the 18th century, humans began to use a different source of energy other than animal or human power. The machines transformed fossil energy into work energy. Such technological power was able to provide efficient methods and means to improve the productivity of the most primary industry, agriculture, by producing useful tools and machines.

This manufacturing power produced many different functional and mechanical tools, increasing their versatility, quality, and quantity. By harnessing this power, we were able to get more things for less money. Henry Ford was able to make cars much cheaper, allowing them to be available to the ordinary citizen. This democratization of the automobile was possible not because of Ford's genius, but thanks to America's accumulated manufacturing power. This accumulated manufacturing power and technology became the backbone of the second industrial revolution.

The success of manufacturing relies on engineering power that produces ideas, designs, and methods to realize a tangible product. Manufacturing and engineering share the same fundamental elements to grow and succeed against the competition: cost, productivity, technology, and quality. To develop these elements in a competitive environment requires innovative thinking based on reality and practicality.

In this chapter, we analyze and discuss the fundamental elements of the successful development of the manufacturing and engineering industry in society.

1. The Effects of Manufacturing on a Nation's Power
2. The Elements of Manufacturing Competitiveness
3. The Impact of Engineering on Manufacturing

1.1 The Effects of Manufacturing on a Nation's Power

Throughout history, countries that own better-structured power and advanced weapons have prevailed in hegemony and in becoming a global power because they were in a better position to win battles effectively.[1]

For example, in 1532, Francisco Pizarro with 168 Spanish soldiers conquered the 12 million inhabitants of Aztec and Inca: "Pizarro's military advantages lay in the Spaniards' steel swords, steel armor, guns, and horse."[2] Throughout modern history, it is clear that the nations with more advanced technology and manufacturing power came to have a stronger military and political power than other nations.

The power of manufacturing has been demonstrated through the First and Second World Wars. Germany, Japan, and the United States built trucks, vessels, tanks, planes, and weapons by maximizing their industrial manufacturing powers. During the war, they focused not only on killing enemy soldiers but also on destroying the enemy's industrial power—in particular, their manufacturing facilities and factories.

In the modern era, China opened its doors to foreign investment. When foreign companies opened manufacturing factories in China to utilize the local cheap human resources, China began to earn stable and significant income, empowering it to come out of poverty, and China soon became the world's second largest economy, mainly thanks to its manufacturing power.[3]

Despite some side effects, such as pollution by manufacturing plants, the advantage and benefits are clear and significant to the prosperity of the nations and human society.

There are four major significant effects on a nation's power created by the manufacturing industry: social, economic, industrial, and political.

1.1.1 Social Effect

Without employment, no country or society can sustain stability or provide happiness to its people. In developing countries, a reasonable amount of excess workforce can be an opportunity for economic growth of society when they are appropriately used to produce wealth.

Typically, in developing countries, automation is recognized as expensive and difficult due to capital and technological limitations, whereas manual handling and processing is still economical and flexible. Hence, the manufacturing sector can absorb the excess workforce in a relatively flexible way; if the labor cost is cheap, one can hire more human resources instead of investing in equipment or automation.

When people are hired to work, wealth is provided to them in the form of their compensation, which they can then use to support themselves and their family. This economic factor is fundamental for life quality and stability of society. *Manufacturing brings jobs and income for better life quality.*

1.1.2 Economic Effect

The purpose of economic activity is the creation of wealth. Manufacturing activity creates absolute wealth that we can see, touch, and use to improve the quality of our daily life.

> *The causes of wealth* are something totally different from *wealth itself*. A person may possess wealth, i.e., exchangeable value; if, however, he does not possess the power of producing objects of more value than he consumes, he will become poorer. A person may be poor; if he, however, possesses the power of producing a larger amount of valuable articles than he consumes, he becomes rich.[4]

From an economic viewpoint, manufacturing power creates a virtuous economic cycle; the creation of jobs and generation of income for individuals, the increase of purchasing power, and GDP growth of the nation. Such a cycle leads to the hiring of more human resources and investment in manufacturing factories.

Manufactured products can be readily tradable. The fundamental structure of global trading is based on goods, not services.[5] Manufactured goods and products are easy to ship and tradable. Since we can export and trade with other countries, we can boost the economy by bringing in foreign currency and goods needed for the nation.

1.1.3 Industry Structural Effect

To make the manufacturing process and the outputs more productive, we need better technologies or systems. For example, we need better equipment or machinery to produce highly innovative industrial goods via mass production.

Highly technological products like semiconductors, smartphones, robots, and biomedicines require innovative manufacturing processes and technologies. Many high-tech companies like Apple, Samsung, Google, and Huawei rely on advanced manufacturing technologies to manifest their innovative engineering design concepts into tangible products.

Necessity is the mother of invention. *A desire to grow motivates the companies to develop higher industrial hierarchical technologies.* The companies that have a strong foundation of engineering and manufacturing can have a chance to evolve into a more complex and advanced industry, capturing the future opportunity of making more sophisticated products. This evolution process will bring in higher profitable businesses through the company's technological and structural innovations, adopting more and more high-end technologies.

In this perspective, the countries that have a versatile manufacturing infrastructure can leverage this power to move to the production of more advanced technological products like new materials, new automated equipment, smart robots, and AI products in the coming era.

1.1.4 Political Effect

When the nation becomes mature in securing manufacturing technology and product know-how, the nation expands into the realm of science and technology more deeply. Necessity for innovative manufacturing technologies can generate stronger motivation and momentum for more advanced science and technology as we have seen throughout history.

Advanced science and technology can help develop further innovative products in return. The capability of science and technology can be transformed into political power in the form of economic and military power. Historically, the nations that dominated technology dominated other nations. In modern history, during the Second World War, Germany and the U.S.A. competed to secure new technologies and produce new weapons like tanks, submarines, rockets, and nuclear weapons.

Securing the power of manufacturing, science, and technology puts a nation in a better position in trade negotiations or the settling of political disputes. For example, the United States controls certain technological products such as advanced airplanes, nuclear technologies, and high-tech semiconductors, and can use these as a negotiation tool with other countries. China also controls the export of rare earth materials like Erbium and Gadolinium, which can be used for producing fiber-optic cables, MRI machines, rechargeable batteries, and hybrid vehicles.

Manufacturing power becomes political power.

1.2 The Elements of Manufacturing Competitiveness

Manufacturing businesses are competing against each other via their business outputs—the form of the business output is a product in its tangible form. The competitiveness of a product is decided by many complex factors and criteria. In some products, special features and functions are emphasized more than others. Even in the same product category, each manufacturer emphasizes the individual characteristics of their product in a different way. For example, Honda emphasizes its engine performance and chassis' stability more than overall NVH (noise, vibration, and harshness) performance. However, Toyota, another Japanese brand, pursues uniformly balanced performance and quality—equal emphasis of all vehicle characteristics. This same strategy applies to the smartphone industry as well—Apple and Samsung have different product strategies and each emphasizes different features and performance criteria.

However, the success of a product strategy should be judged only from the customer's viewpoint. The customer's mind and priority are not a fixed target. They change and evolve depending on their needs, their cultures, and the values recognized by their society. However, regardless of product categories, there are some fundamental common elements that the customers always look for.

Whether a product is expensive or cheap, customers are always *cost* sensitive. The *productivity* of the manufacturing process is a critical parameter

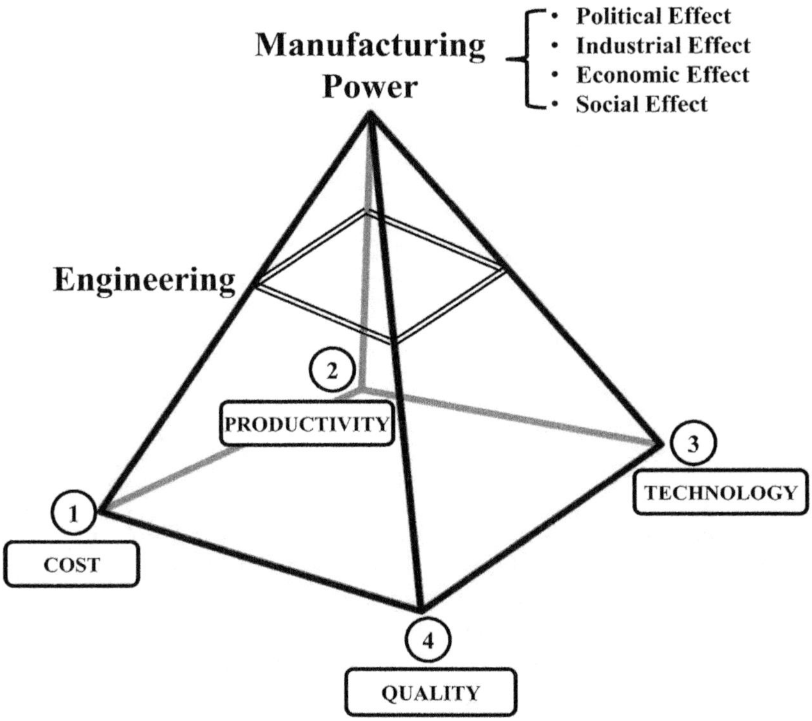

FIGURE 1.1
Elements of manufacturing competitiveness.

influencing the cost structure and delivery of the product to the market. Customers will not tolerate sloppy *quality* or unattractiveness in a product. As a product increases its complexity and value, customers tend to look for more advanced *technology* or features.

There are four elements that make up the fundamental product dimensions required for the success of any manufactured products, as graphically shown in Figure 1.1: ① Cost, ② Productivity, ③Technology, and ④ Quality.

1.2.1 Cost

In economic terms, the cost (C) and the price (P) are the fundamental factors leading to producing or purchasing a product; the cost (C) is from the manufacturer's perspective and the price (P) from the customer's perspective. The customer's final purchasing decision is based on the perceived value (V) of the product. Figure 1.2 shows a simple and clear relationship of these parameters: cost (C), price (P), and value (V).

The controlling of the cost in the first place can maximize a gap between cost vs. price and price vs. value. A clear understanding of the cost structure is the starting point to creating manufacturing competitiveness in economic terms.

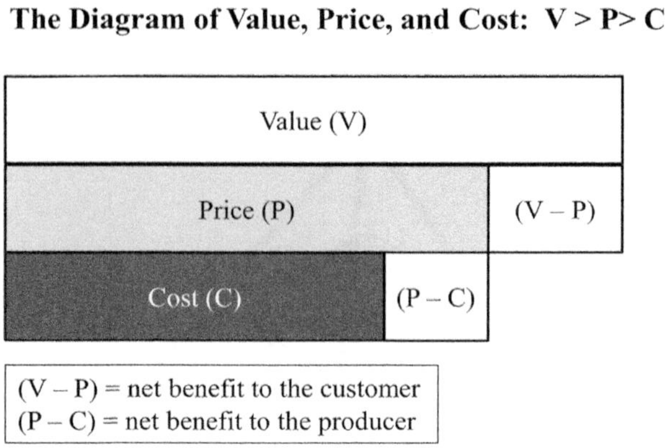

FIGURE 1.2
The diagram of value, price, and cost.

A simplified cost structure is graphically represented in Figure 1.3. The manufacturing cost is typically composed of several key factors: D&D (Design and Development) and administration costs, the material cost, labor cost, processing cost, and warranty cost (which is the failure cost charged by the customer).

In Figure 1.3, when a manufacturer can sell the product for a higher selling price (P) than the manufacturing cost (C) (see Case-A), the manufacturer can earn a profit from the sale, so they will survive in the business. Since a customer should receive more perceived value (V) than the buying price (P) of a product, *the selling price (P) is determined not by the manufacturer, but by the customers and the market. Therefore, if the selling price (P) is lower than the manufacturing cost (C) (see Case-B), the manufacturer loses money, and the business cannot be sustained.*

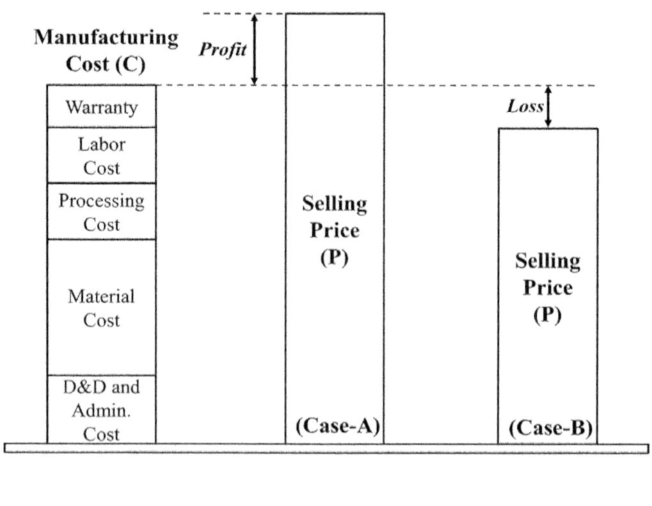

FIGURE 1.3
Manufacturing cost, selling price, and profit.

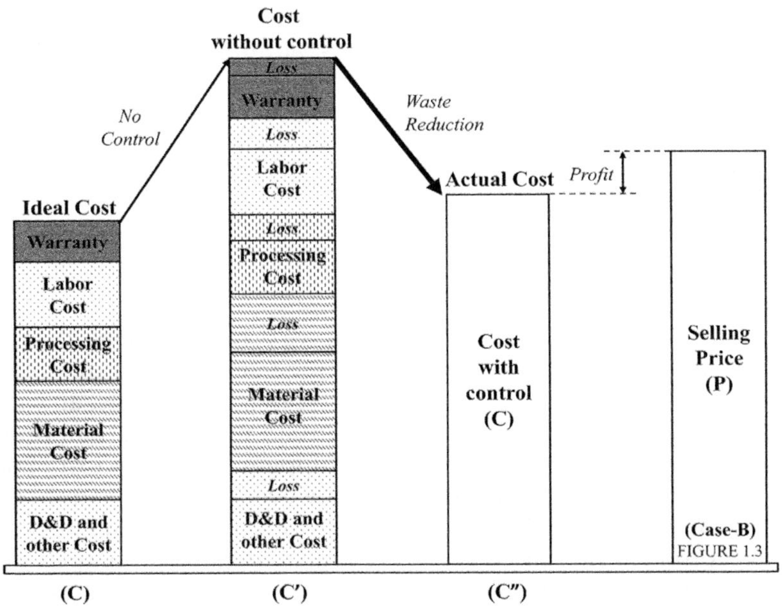

FIGURE 1.4
The impact of losses on manufacturing cost.

(C) in Figure 1.4 is an ideal cost structure to succeed in the manufacturing business. However, the reality is that there is a substantial amount of losses existing in each factor as the *result of waste*. The amount of the loss kills the competitive position of a product in the market. Generally speaking, the manufacturing cost structure of a similar product is more or less the same between competitors. However, the amount of the *"loss due to waste"* can be significantly different from each other. Therefore, identifying the loss areas and deploying waste reduction activities should be a critical management focus for cost control in manufacturing to lead to (C″) from (C′) condition at least so that it can make a profit to survive even as it passes through a (Case-B) situation in Figure 1.4.

The loss comes from waste in each process; therefore, it is the manufacturer's responsibility first to identify the waste and to remove them.

> *A newly invested machine has a working capacity of 100 hours per week. If it is used only 60 hours per week due to the production plan or the maintenance issue, it can result in a 40% loss in the processing cost. A typical number of working hours of one workforce is 8 hours a day, but if the person works only 6 hours, 25% becomes the loss in the labor cost. Also, the material input is 100 kg, but if only 80 kg becomes the product, 20 kg is the loss. The more loss areas we identify, the more opportunities we have to improve the cost structure. Our theoretical target should be "Zero Loss"—100% machine running, 0% of defects of incoming parts, and 0% of processing mistakes.*

In the manufacturing sector, the typical types of loss are: logistics loss, processing loss, equipment loss, material loss, administration loss, inventory loss, outsourcing loss, defect loss, and yield loss. According to the research done by a Japanese consulting firm, a typical amount of total loss is 20%–30% of the total manufacturing cost.[6]

We will engage in more detailed discussions of the loss due to waste in Chapter 9, "Lean Quality Management."

1.2.2 Productivity

Even a simple manufacturing process requires complex process design and management. The complexity comes from the many steps of different processes: production planning, parts order, receiving materials, and inspection; and material processing such as welding, machining, painting, and assembling; final quality inspection and shipping. The complexity and difficulty vary from industry to industry; however, systematic and efficient process control is a common requirement for successful manufacturing. Industrial products like automobiles are probably the most complex regarding scale and number of processes to control. Supply chain management alone is very complicated for these products. For example, even with highly modularized parts, the typical number of suppliers to control is between 250 and 300 per vehicle.

The right price, quantity, quality, and delivery time of a product is highly dependent on the productivity of the manufacturing. Productivity is the measurement of the efficiency of a person, machine, factory, and system in converting inputs into useful outputs. Productivity is computed by dividing average output per period by the total costs incurred or resources (i.e., capital, energy, material, personnel) consumed in that period.[7] Productivity is a critical determinant of manufacturing competitiveness.

In a given design and part, manufacturing productivity depends on the efficiency of the manufacturing system and process—see Figure 1.5. Productivity can be expressed as a function of many relevant parameters. However, in mathematical terms, productivity can be expressed merely by Eq. (1.1).

$$\text{Productivity} = \frac{\text{Total Value Delivered}}{\text{Total Cost Incurred}} \tag{1.1}$$

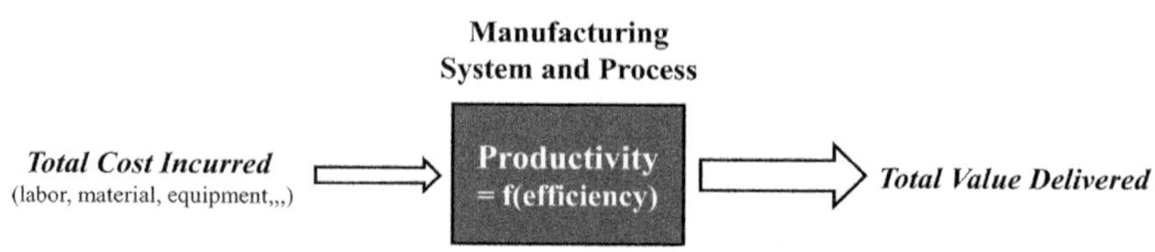

FIGURE 1.5
The definition of productivity in manufacturing system and process.

To maximize productivity (1.1), each manufacturing parameter in the system must be optimized, leading to maximum value delivered while minimizing the input cost. It is common for engineering departments to insist on investing in advanced equipment or full automation processes using convincing data and reasons. However, such high investment increases the cost of manufacturing in the form of depreciation. Other fundamental improvements should therefore be implemented before any investment decisions; begin with waste reduction, process optimization, increase in work efficiency through standardization, training, quality improvement in design, etc.

Discovering the waste and reducing it is the first job to improve productivity. In most cases, workers on the ground floor of a workplace know waste concerns better than the managers. It is the responsibility of management to create a system that motivates employees to remove waste through a bottom-up process.

> *Many Japanese firms have Kaizen programs run by the employees. In the case of Honda, "NH (New Honda) Circle" is a globally deployed program. The company motivates their employees to voluntarily participate in the program. The company compensates the employees' extra working hours for Kaizen activities and provides incentives in proportion to the outcome contributed to the company. There is an annual worldwide competition with the best-performed circles from every Honda plant in the world. The winning teams receive special medals and awards. With such a system and process in place, waste is continually removed from the work floor, minimizing losses and maximizing the productivity in the plant.*

1.2.3 Technology

Technology is the power that distinguishes humans from other species. Two-and-one-half million years ago, primitive humans developed stone tools, and 300,000 years ago they developed a technology to control fire.

Without technology, we could not have been better than animals. Modern technology does not come just from trial and error, but it comes from science and mathematics. Without the power of science, technology, and engineering, modern products have no unique appeal to customers. Securing innovative and advanced technology is becoming a more significant part of manufacturing competitiveness in the knowledge economy of the 21st century.

However, when we talk about technology we do not talk only about advanced or cutting-edge technologies. There are various technologies that can make any product and process distinctive in the customer's perception of its value. These fall into two groups: one is *product technology* such as special features and functions that can be felt, touched, and perceived by customers; the other is *manufacturing process technology* that facilitates the improvement of products in terms of cost, productivity, and quality. In most manufacturing industries, both these technologies overlap in product design and manufacturing.

In the manufacturing industry, innovative technologies have substantial advantages to attract high-tech-driven society members. Customers are willing to pay more money for unique technologies, functions, and features.

From numerous market quality surveys and reports, we have seen *immature or unfriendly technologies* bring about more trouble than benefit.[8] When new technology is thoroughly verified and validated with respect to its reliability and quality, it can be a very powerful tool in improving trustworthiness and brand image, leading to increased sales of the product.

1.2.4 Quality

Providing value to the customer is a necessary condition for the success of a business. Hence, we need to understand holistically how the customer perceives something as valuable. This viewpoint should include the economic perspective as well as the psychological value of a product or service.

If management is overly stringent in terms of quality, they may overemphasize the importance of quality to a level where the customer may not even be able to perceive a difference. This concept can be termed *"quality for quality"* or *"over-quality."* This extreme approach convinces neither business leaders nor customers to pay more money in exchange for such value and benefits.

W. Edwards Deming, a pioneer and authority in the area of quality and productivity, defines quality by the following equation:[9]

$$\text{Quality} = \frac{\text{Results of Work Effort}}{\text{Total Cost}} \tag{1.2}$$

Deming states that the customer's *definition of quality* is the only one that matters. The company's activities can be counted as valuable only when they contribute to the customer's satisfaction in the form of the *QPR (Quality Price Ratio)* as expressed in the Eq. (1.2).

In Deming's definition of quality, the meaning of "Results of Work Effort" is the amount of value recognized by the customer. In reality, not all cost invested or spent transforms into value to the customer. In order to improve quality, we have to pursue two paths at the same time: one is to minimize unnecessary costs (like rework cost due to quality defects), and the other is to maximize the output of the quality activity, which together results in maximum customer satisfaction with minimum input cost.

The definition of quality by Deming in Eq. (1.2) is similar to the general definition of "Productivity" expressed in Eq. (1.1). Deming explained the effect of quality by *"chain reaction"*: "Improvement of quality transfers wasted portion of man-hours and machine-time into the manufacturing of good products and better services. The result is a chain reaction, eventually lowering costs, improving the competitive position of the product, making customers happier, more sales, and more jobs."[10]

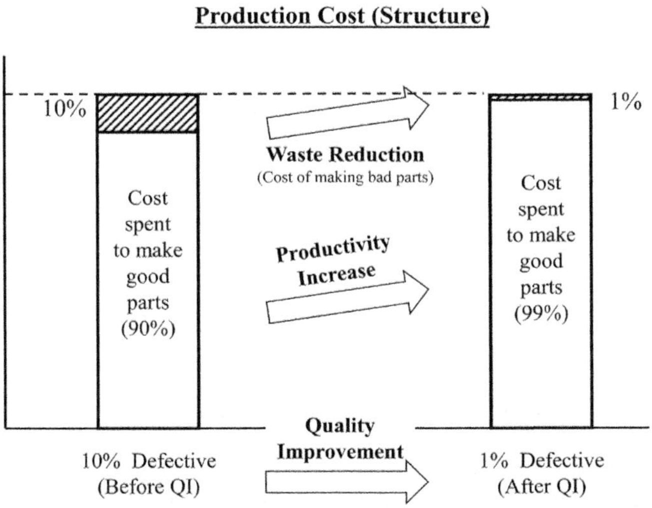

FIGURE 1.6
Illustration of productivity increase with improved quality (data).[12]

This chain reaction of the quality effect is demonstrated in Figure 1.6: when we improve quality, we also improve productivity. When the quality is up 9%, production of the good product is up 9%, capacity is up 9%, cost per unit of good product is lower, profit improves, customers are happier, and everybody is happier.[11]

Making things right the first time is a critical quality management principle from designing, engineering, supplier sourcing, and in-house processing to checking and shipping of a product. Such "quality is first" principle and management will dispel any misconception that quality improvements incur more cost for manufacturing and business operations.

Another definition of quality comes from Genichi Taguchi, an engineer and statistician who developed a methodology for applying statistics to improve the quality of manufactured goods: "Quality is the avoidance of financial loss to society after the article is shipped."[13] He described the relationship between quality and financial loss by the loss function:

$$L(x) = k(x - m)^2 \tag{1.3}$$

where L is the loss in money (for a more detailed explanation—refer to Eq. (2.1) in Chapter 2. Equation (1.3) describes a situation where, from the moment of manufacturing the product, through shipping to customers, and during the usage by customers, quality issues become a financial loss, $L(x)$, to society.

In modern history we have observed how significantly quality issues have brought calamities to society, and how the financial losses to society eventually became the financial losses to the companies who generated the quality issues in the first place.

With the loss function, quality can be measured quantitatively, and the quantified value can be used as a measure to correctly design a system, parameter, tolerance, and process for a product.

Quality in manufacturing is interrelated with productivity and cost; when quality has improved, it can also improve productivity, leading to lower cost of the product.

1.3 The Impact of Engineering on Manufacturing

Engineering is an activity to realize a human's idea, concept, and knowledge in a tangible form by using science, mathematics, and technology. Engineering activities are concerned with designing, constructing, building, and manufacturing a product, a system, or service. An engineering activity has boundary conditions like weight, cost, delivery time, and available technologies. The engineering discipline assures the qualities of safety, function, performance, user-friendliness, reliability, and durability.

Engineering plays a pivotal role in manufacturing functions and performance as shown in Figure 1.1. Manufacturing competitiveness comes from the ability of the engineering activity to deliver productivity, cost, technology, and quality from the design stage to the manufacturing stage. The core of engineering activity is *design*, whether it be to design a product, design equipment, design a system and process, design chemical formula, design software, etc.

1.3.1 Design for Productivity

As the steam engine brought about the Industrial Revolution with significant increases in efficiency in coal mining and transportation, so has engineering power brought about substantial productivity increases in modern manufacturing. The goal of manufacturing engineering is to increase productivity by an efficient process and smart equipment design.

Manufacturing engineers are often intrigued by complexity and fancy, sophisticated equipment. Some insist on overhauling old equipment with advanced new ones, guaranteeing productivity increases. However, a real revolution comes from small steps of evolutions. Deming said, "New machinery and gadgets are not the answer. We should be able to make important gains in quality and productivity by learning to use the machinery effectively on hand."[14] For example, when manufacturing process design is not optimized, introducing fully automated equipment will not be effective. There still can be lag or bottleneck effect between processes and equipment.

The parts would not flow smoothly, and WIP (work in process) inventory would build up.

Design for productivity is an engineering principle that a productivity increase should start with design. For example, new product engineering should create a design with fewer parts to assemble, and manufacturing engineering should design simpler manufacturing processes and smart transportations between processes. The most time-consuming processes in a typical manufacturing industry are processing, transportation, and inspection time. Here, we will examine actual examples and lessons that we should consider applying to similar ideas or concepts throughout engineering activities.

1.3.1.1 Smart Transportation

The time spent transporting parts is a waste, which can be reduced with the application of customized jigs and fixtures rather than full automation—typically, fully automated transportation takes a great deal of space and is very expensive. A smart approach for transportation can be a semi-automated process, i.e., human intelligence plus machines. Hence, flexible management of production and working procedures is important in meeting different requirements from model to model.

Transportation of semi-finished products to the next process is one of the main challenges in the semi-automatic manufacturing processes. Careless or inefficient handling of the parts results in defects or a waste of processing time and workforce.

Figure 1.7 shows an example of how a simple modification of an existing tool can reduce the complexity of the transportation process to minimize manual handling of the part. In the process of "(b) After," a weight is attached to the pallet with a rubber string. This simple mechanism helps the pallet with the part on it to slowly slide down the roller conveyor but not hit the end wall. Once the part is removed from the pallet, the empty pallet returns to the original loading position automatically because of the weight attached to the string. With such a simple and smart mechanism, we can reduce wasteful human workforce or motions compared to "(a) Before." This kind of manufacturing engineering concept can be applied to many manufacturing processes without increasing manufacturing cost.

1.3.1.2 Intelligent Quality Check

Another significantly time-consuming process in manufacturing is the inspection of manufactured parts. In the manufacturing process, a quality check at each process or at the end of the assembly line could be a necessary process. However, we need to keep any inspection and checking of the parts or processes to a minimum and ensure that they are purposeful and meaningful. Assuring quality through process control and eliminating inspection

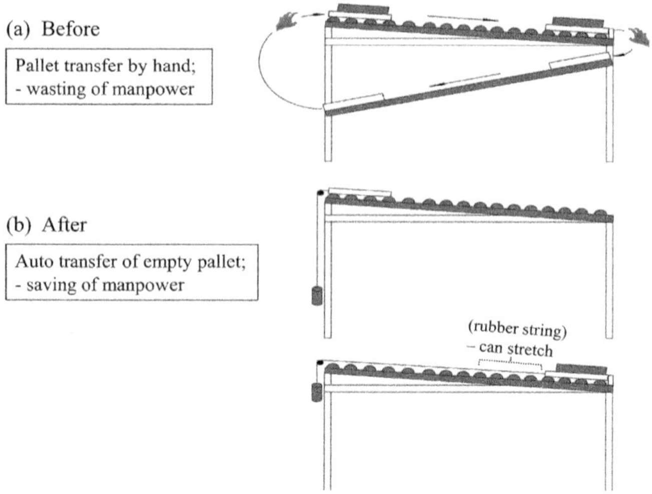

FIGURE 1.7
An example of efficient parts transportation.

and checking is the best type of quality management. However, a minimum inspection could be necessary due to limitations in process capability. In such a case, the checking process should be as efficient as possible, while still being effective. In such cases, we can also improve productivity by applying the similar concept of the smart integration of equipment and intelligent humans.

For example, a Go/No-Go gauge can identify whether processed parts are out of tolerance or not. It uses one or two pieces of a very simple jig rather than investing in expensive equipment in-line or off-line like a CMM (coordinate measuring machine).

The left-side jig of Figure 1.8 shows a plug gauge to check the hole accuracy of a part. By inserting the two different dimensional jigs, one can quickly

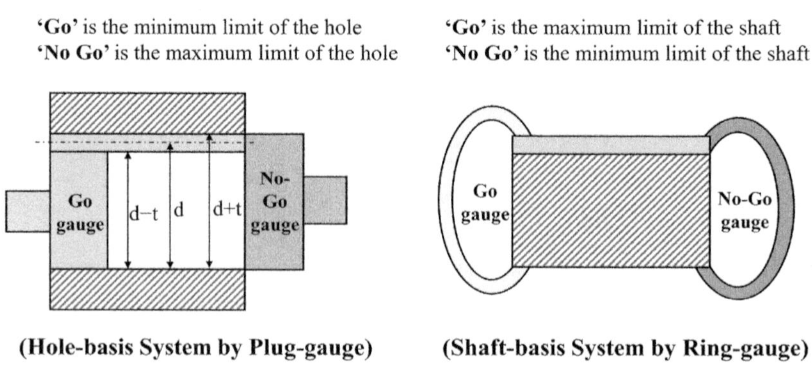

FIGURE 1.8
The concept of Go/No-Go gauge.

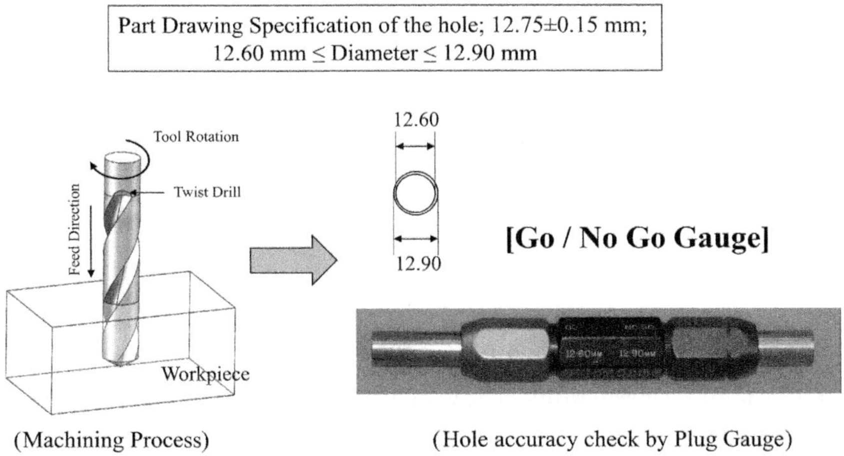

FIGURE 1.9
The effective method of part tolerance check (example: plug gauge).

confirm the accuracy of the part. In such a plug gauge, the Go gauge is designed to confirm the minimum tolerance of the hole; the nominal size of the hole (d) minus the tolerance (t), d–t; the No-Go gauge is designed to confirm the maximum tolerance of the hole, d+t. The right side of Figure 1.8 shows the reverse case: the Go gauge is the maximum tolerance of the shaft, d+t, and No-Go gauge is the minimum tolerance of the shaft, d–t. That is, the "Go gauge" should go in, while the "No-Go gauge" could not be inserted if the part is within the tolerance limit.

Figure 1.9 shows a practical example of checking the engine block hole by using a plug gauge. The gauge is composed of two sections; a Go section on one-side and a No-Go section on the other side for convenience. If the hole is within the design tolerance, the Go-jig should go in while the No-Go gauge would not go in. Such smart manufacturing engineering allows us to check the quality of a product in a straightforward and efficient way.

1.3.2 Design to Cost

The cost of a product relies on smart design from the beginning. Engineering design determines a design concept and structure, which determines the sourcing cost of the parts, the manufacturing processing cost, and the service and maintenance cost—this is a chain reaction.

The most significant cost portion of industrial products or consumer products is the parts cost. That is one of the reasons why a purchasing department works so hard during the supplier selection process. However, the purchasing department frequently complains that the design concept is expensive and in order to meet the target cost they ask engineering to "design to cost."

The engineering department typically thinks that they require a tighter tolerance of parts for better performance. However, specifying a tight tolerance is not necessarily an efficient answer because such parts are expensive and it is difficult to control the supplier's quality management. To *design to cost* is to consider cost as a design parameter starting from a product development stage. A robust design structure can assure reliable function and performance of a product through the normal manufacturing process without particular control of the parts or processes. Let's look at actual examples of *design to cost* by smart design engineering.

1.3.2.1 Tire Conicity Force Control

Tire performance clearly influences vehicle function and performance. One of the critical quality parameters to control in a tire is *conicity*. Tire conicity is the force that causes the tire to react like a cone and roll in a circular path. This force comes from the tire manufacturing process and is due to off-center tread or belt application.[15] During the engineering design process for the vehicle, engineers would like to set only small tire conicity to be allowed—for example, less than 30N force. The tire supplier would then measure the tire conicity and reject any tires higher than this specification. Naturally, such severe quality control causes more expenses to the tire suppliers, and eventually, the cost is borne by the vehicle OEM company.

However, with a robust design concept, the vehicle can be designed to be less sensitive to the amount of conicity, which allows a larger tolerance

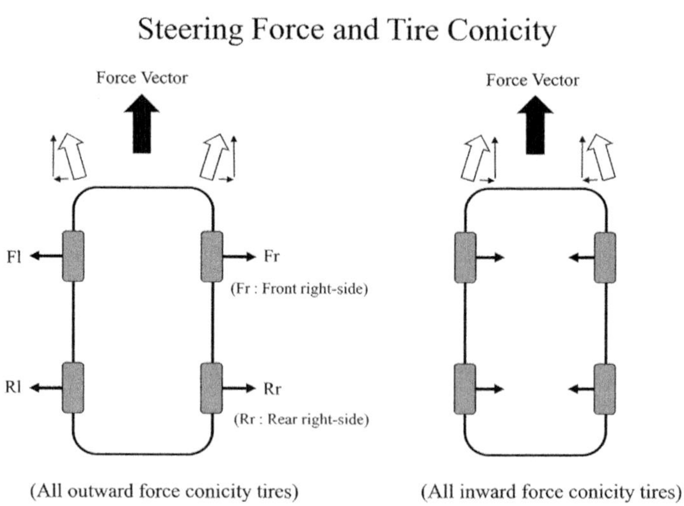

FIGURE 1.10
Vehicle design to cost—conicity control method by design.

range of the conicity specification. Another creative idea could be to ask the tire supplier to ship all four tires of either outward or inward conicity force together as a set. Figure 1.10 represents how the conicity of each tire can cancel out each other's forces, and the vehicle pulling force is minimized to help keep the driving force straight. In this case, the conicity specification can be in a larger range—as much as 60N.

1.3.2.2 Power Supply Voltage Control

A power supply is used in many electrical products. The critical quality factor is to provide a stable voltage to the serving equipment—if the voltage swings in a large variance it can shorten the life of the equipment. Therefore, circuit design to keep the voltage within a narrow tolerance range is very important.

The graph in Figure 1.11 shows a characteristic curve of a normal transistor gain (h_{FE}). A typical way to control the output voltage relies on the selection of a particular gain transistor (h_{FE}). In Figure 1.11, considering the characteristic curve of the transistor, an ideal design seems to use a transistor (20 h_{FE}) to guarantee 115 voltage output, which is the target voltage. However, since every transistor has its tolerance, it is difficult and *expensive* to select transistors with exact gain (20 h_{FE}). If we try to narrow the tolerance, it will be very expensive—a normal cost transistor has +/– 30% tolerance. With this low-cost transistor, the voltage would swing between 115 and 121 volt at (A) point—the variance is 6V, which is too high to accept.

The goal of engineering work in such a case is to design a smart circuit of the power supply that uses a transistor with a reasonable cost. As the curve shows, the *transistor characteristic curve plateaus at point (B).* At this point, the tolerance of the transistor is still +/–30%; however, the voltage output is not as sensitive as in the case of (A). The voltage would swing only between 123 and 125V—the variance range is 2V instead of 6V as in the case of (A). At this point, the voltage output is 124V instead of 115V.

We can re-engineer this problem by adjusting the value of another component in the circuit. The current limiting resistor recommended for the original circuit is 40 K ohms. However, we will modify this recommendation in our new design to 60 K ohms. By making this change, we are able to move the entire voltage vs. gain to meet our target voltage, 115V.[16]

This is the power of engineering for a *"design to cost."*

1.3.3 Design with Technology

Design engineering has the power to bring advanced technologies into a new product through design structure and concept. Engineering should have the capability and experience to realize new technologies in a new

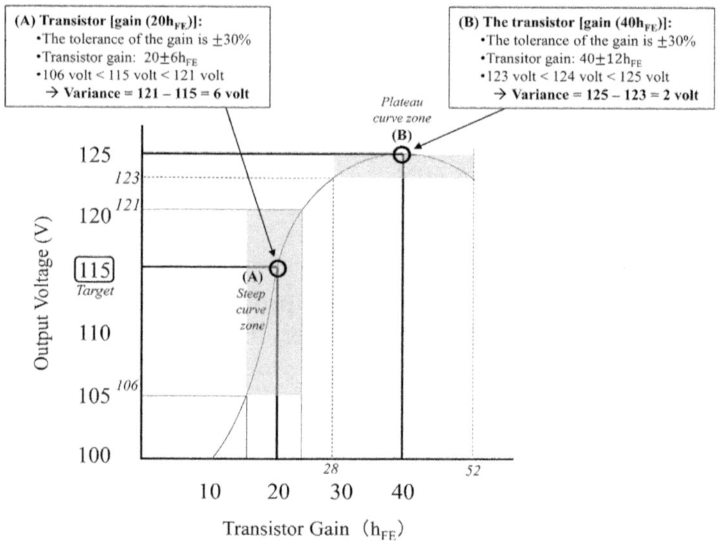

FIGURE 1.11
Power supply circuit design to cost (Thomas).[17] Copyright (© 1990) from *Engineering Quality by Design* by Thomas Barker. Reproduced with permission of Taylor & Francis Group, LLC.[18]

product. A caveat is that the new technology should be one already confirmed through verification and validation, and with engineering's empirical data and knowledge.

Design with technology brings about breakthroughs in products like semiconductors, AI software, intelligent robots, and autonomous vehicle technologies. Figure 1.12 shows an example how design with technology brought about a breakthrough in the history of a semiconductor to make Moore's law possible—the law that states that the number of transistors on a chip doubles every two years. In this example, a DRAM (dynamic random-access memory) is the most regular type of integrated circuit; the higher the density, the higher the memory capacity. The technology breakthrough requires smaller pitch design between circuit lines on a wafer; the thinner the pitch, the denser the chip, and the bigger the memory size. Forming a thin pitch is the critical technology during a photolithographic and chemical etching process.

1.3.4 Design for Quality

Quality defects can occur in various ways. Most defects arise from the incoming parts from the supplier chain, parts handling, and processing in the facility. Once defects are found in-house or in the market, a root cause analysis starts, and a suitable countermeasure is applied to remove the cause.

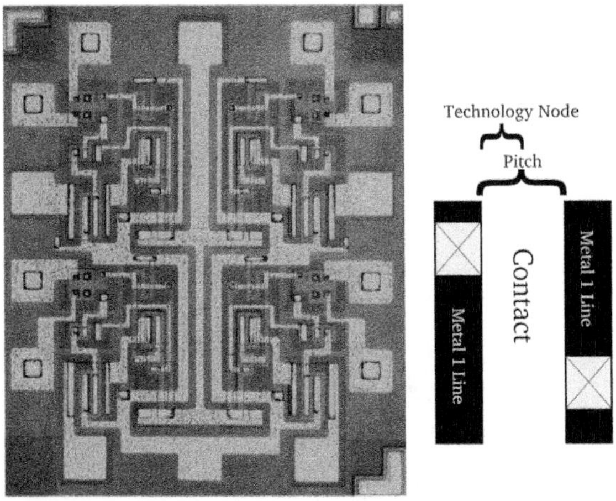

Half Node	Year	DRAM
800 nm	– 1989	4M
600 nm	– 1994	16M
350 nm	– 1995	64M
250 nm	– 1997	256M
180 nm	– 1999	1G
130 nm	– 2001	2G
90 nm	– 2004	4G
65 nm	– 2006	8G
45 nm	– 2008	16G
32 nm	– 2010	32G
22 nm	– 2012	64G
14 nm	– 2014	128G
10 nm	– 2017	256G
7 nm	~2019	
5 nm	~2021	

FIGURE 1.12
Semiconductor design technology and chip density (data).[19,20,21] (Image credit: licensed under the Creative Commons.)[22]

If the action stops here, this does not bring about a robust design process for quality. A robust design process is intended to reflect the previously known concerns and potential concerns and transform this information into new design standards and specifications as represented in the robust design process flow in Figure 1.13. Whenever a new project starts, engineering design

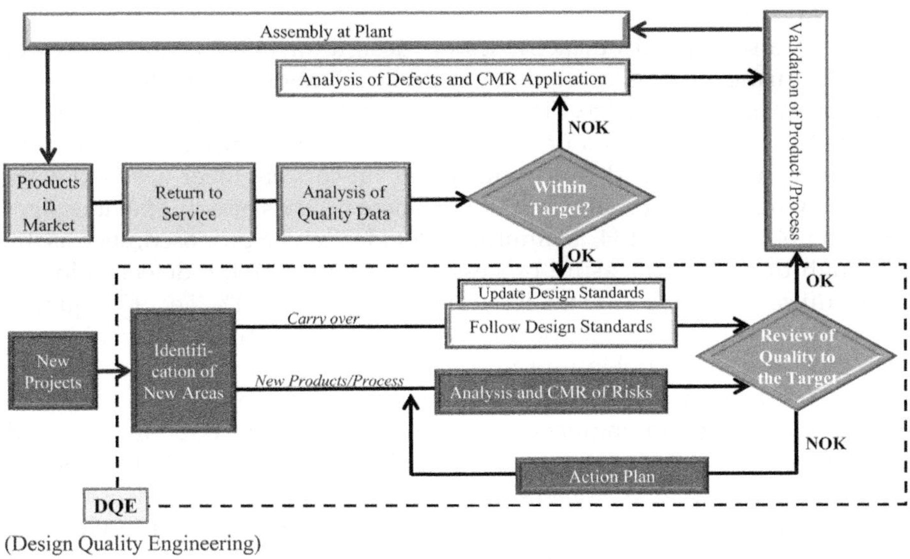

FIGURE 1.13
Robust design process for quality.

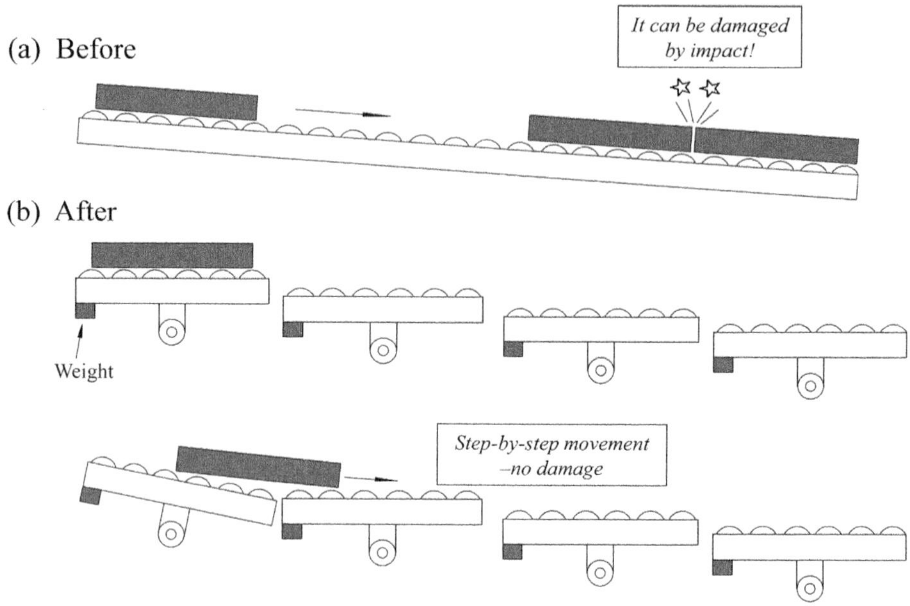

FIGURE 1.14
Robust process design for quality by manufacturing engineering.

should identify the robustness of the design concept, standards, specification, manufacturing process, and so on. This kind of approach is a "design for quality" principle.

We can also expand our activities of "design for quality" to the area of manufacturing process or equipment design. Design for quality relies on technology and knowledge. Figure 1.14 illustrates an example of a potential cause of parts damage in a production process of (a). Once a potential process concern is identified, a manufacturing process engineer needs to work on removing the concern by developing a new process or equipment design. For example, in Figure 1.14, an improvement idea can be proposed, such as (b). Through the countermeasure process by applying (b), the defect ratio can be significantly reduced from 10% down to less than 1%, i.e., 90% quality improvement. Then the quality effect increases productivity as a chain reaction as we discussed in Figure 1.6.

Cost, productivity, technology, and quality are all closely interrelated, as is manufacturing and engineering. The most significant issue in advanced management is how we manage these fundamental elements to provide the best quality value to customers through quality-focused engineering and manufacturing. In the coming chapters, we will discuss comprehensive quality management methods in engineering through a scientific and systematic approach.

Notes

1. Russell, Bertrand. *The Impact of Science on Society*, Unwin Hyman, 1976. Chapter 5, "Science and War," pp. 84–89.
2. Diamond, Jared. *Guns, Germs and Steel*, Vintage, 2005. Chapter 3, "Collision at Cajamarca."
3. West, Darrell M. and Christian Lansang, July 10, 2018, "Global manufacturing scorecard: How the US compares to 18 other nations," www.brookings.edu/research/global-manufacturing-scorecard-how-the-us-compares-to-18-other-nations/. In 2015, China's Manufacturing output was $2.00T vs. the United States' $1.87T (accessed February 2, 2019).
4. List, Friedrich. *The National System of Political Economy*, J. B. Lippincott & Co, 1856, Cosimo Classics, 2011.
5. Six Reasons Manufacturing is Central to the Economy, Roosevelt Institute, May 23, 2011, www.rooseveltinstitute.org/six-reasons-manufacturing-central-economy/ (accessed January 10, 2019).
6. Zero Cost Management, APPLIWARE, Ver. 4.4, 2009.
7. Productivity, www.businessdictionary.com/definition/productivity.html (accessed January 10, 2019).
8. "Car Dependability Ratings—2018 VDS," J. D. Power, www.jdpower.com/press-releases/jd-power-2018-us-vehicle-dependability-study (accessed June 15, 2018).
9. Witzel, Morgen and Malcolm Warner eds. *The Oxford Handbook of Management Theorists*. Oxford University Press, 2013. Chapter 11, pp. 196–218.
10. Deming, W. Edwards. *Out of the Crisis*, The MIT Press, 2000.
11. Deming, W. Edwards. *Quality, Productivity, and Competitive Position*, The MIT Press, 1982.
12. Deming, W. Edwards. *Quality, Productivity, and Competitive Position*, The MIT Press, 1982. p. 3.
13. Barker, Thomas B. and Marcel Dekker, "Engineering Quality by Design," INC, 1990.
14. Deming, W. Edwards. *Quality, Productivity, and Competitive Position*, The MIT Press, 1982.
15. Bhadauria, Gaurav and Prof. Ajay Bangar, Analysis of "Variation in Non-Uniformity with Runout Temperature of Tyre" Using Taguchi Parametric Optimization Technique, *International Journal of Engineering Research and Applications (IJERA)* ISSN: 2248-9622, Vol. 3, Issue 4, July-August 2013, pp. 2104–2108.
16. Barker, B. Thomas. *Engineering Quality By Design*, New York: Marcel Dekker, Inc., 1990.
17. Barker, B. Thomas. *Engineering Quality By Design*, New York: Marcel Dekker, Inc., 1990.
18. License Number: 4507810278785, License date: Jan 14, 2019, Licensed content publisher: Taylor & Francis Group LLC Books, Licensed content title: Engineering quality by design : interpreting the Taguchi approach.
19. "14 nanometer," https://en.wikipedia.org/wiki/14_nanometer (last modified January 14, 2019).

20. Data: www.wikiwand.com/en/Semiconductor_Industry_Association (accessed February 3, 2019).
21. Data: www.theinquirer.net/inquirer/news/2436699/samsung-is-mass-producing-through-silicon-via-ddr4-memory-in-128gb-modules (accessed February 3, 2019).

2

Fundamentals of Quality in Manufacturing

"Quality" is an everyday word in our daily lives. It sounds like a conventional, established word, but at the same time, it seems to be a struggling word in the modern, high-technology world.

Conceptually, "quality" is a very broad term, which therefore needs clarification from the customer's and manufacturer's standpoints. *Quality can be a tangible product only when the customers perceive its value.* Maximizing the quality effect is every manufacturer's goal with respect to cost and productivity improvement, which is a natural path of *Quality Management and Engineering.*

Since we are used to advanced technologies everywhere in the 21st century, we presume quality is seamlessly built into every high-technology product. In other words, quality is taken for granted in every aspect of products or services. However, this is not possible without intentional efforts by technology firms.

We have many questions about quality. Sometimes we are confused about the proper definition of quality. In this chapter and the coming chapters, we answer those questions by discussing the fundamental factors of quality: what is quality, how do we seek customer satisfaction, what is the value of quality, what is the cost of quality failures, how do we measure quality, how do we deploy quality throughout the organization, what are the benefits of premium quality, how do we pursue quality, and how do we integrate craftsmanship into the modern manufacturing. After answering those questions, "quality" will not be a vague word anymore. We will have a clearer, holistic picture of our business operations, enabling us to map quality onto product designs and processes.

Why are some organizations so fervent about quality? What kinds of business outputs do they expect from quality? Every business talks about customer satisfaction, but do they truly respect customers' value and operate the business according to what they say to the customers? If we disappoint customers, what kinds of business costs do we incur? Many companies seem to have a good system on paper, but do they truly respect the system and processes in place throughout the organization? Some companies are enthusiastic about the details and refinement of their products, but do they reap the benefits from their labor?

There may not be one beautiful and straightforward equation to answer all these questions; however, by utilizing the *"scientific and systematic approach to quality management in engineering,"* we will be able to address these issues in a rational, economic, scientific, and mathematical way.

1. Meaning of Quality
2. Value of Quality to Society
3. Cost of Quality
4. Deployment of Quality
5. Measurement of Quality
6. Premium Quality for Premium Brand
7. Craftsmanship for Quality

2.1 Meaning of Quality

Quality is the perception of the products and services that the customer experiences and appreciates. The critical point here is that the perception comes from *customer's standpoint*. How the customers recognize that something has "value" is a big question that every manufacturer should ask itself first. For example, when an engineer decides to tighten the tolerance of a product or process, the outcome of the decision should benefit the customer's experience or perception, not just the engineer's own peace of mind. At the same time, the decision should benefit the manufacturer's business as well. They should be able to sell more products or reduce the overall cost in the manufacturing process or the market.

According to Joseph Juran, "Quality defined from the user's standpoint is anything that enhances customer's satisfaction."[1] Juran, an influential quality management consultant, first requires that manufacturers develop qualitative approaches to understand quality from the customer's viewpoints and then quantify it by weighting the values of the product's key factors. For example, some customers are very concerned about features and functions, while others are more concerned about design and usability.

The first step is to define the customers. Which countries are the target markets? Into which age groups, income groups, kinds of jobs, and education levels do customers fall who are interested in the different types of our company's products? This process can be done at a very early stage of product planning through *a market survey*. Through the market survey, many quantitative values can be measured as well. Appropriate statistical tools and scientific methods can be used for a determination of the weight values of each design and the product's unique features and parameters from the customer's standpoint.

Businesses in operation for a long time would be able to explain the relationship between customers and products and make a product plan by using the large datasets available already. However, the question is whether we understand our current customers. We should not assume

"Top 10" Purchase Reasons

1.	Well-Made Quality	23%
2.	Passenger Capacity	14%
3.	Value for the Money	9%
4.	Brand Reputation	7%
5.	Reliability	6%
6.	Safety	6%
7.	Expert Reviews	6%
8.	Comfort	4%
9.	Previous Model Experience	4%
10.	Interior Roominess	3%

FIGURE 2.1
Market survey of current customers. (Example: H-Company SUV Model.)

that the products sell well because of superior quality, features, price, or service. We should ask exactly why they sell. Start with the question "*Why?*"[2] Figure 2.1 shows an example survey of H-Company customers in the United States in 2004.

According to research[3], acquiring a new customer is about five times more expensive than retaining an existing one. The study by Bain & Company in 2001[4] shows that increasing customer retention rates by 5% increases profits by 25% to 95%.

Society changes, and so do customers. Potential customers are like a moving target. Trend analysis of the market and the technology are a must to guide the company's efforts to increase the customer's perception of the product's quality. One of the tools that identifies potential customers and their needs is to analyze market trends by using a perceptual mapping tool.

Perceptual mapping is a diagrammatic technique used by asset marketers that attempts to understand the perceptions of customers or potential customers. Typically, the position of a company's product, product line, or brand is displayed relative to their competition. Perceptual maps, also known as market maps, usually have two dimensions but can be multi-dimensional; they can be used to identify gaps with competitors as well as potential opportunities in the market by clarifying perceptual problems with the company's product.[5]

Figure 2.2 explains where our product is compared with other competitors by using perceptual mapping. Our current production is in the region of "sport" and "personal." We can make our future product plan with a good balance of "family" and "comfort." Such a product strategy is the desired planned choice based on the future market trend.

There are other scientific tools available for quality planning such as *QFD (Quality Function Deployment)*—the details of QFD will be discussed in a later chapter.

Quality can be defined as "on target with minimum variance." However, in nature, everything tends to go in a disordered way, increasing entropy. The

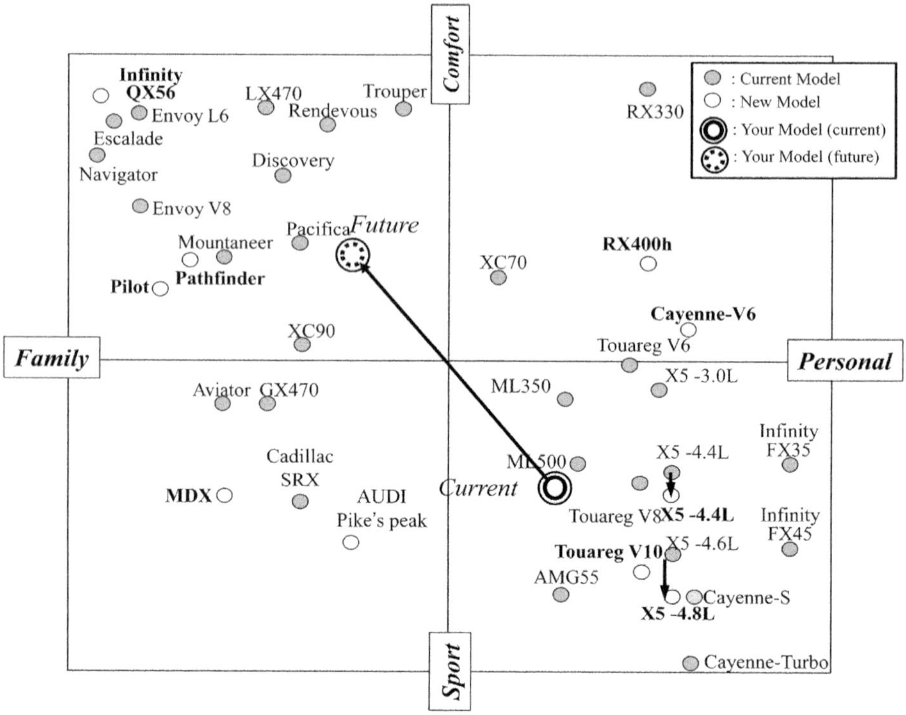

FIGURE 2.2
Perceptual mapping for future model. (Trend of SUV market, 2018 Detroit Motor Show.)

scale of the variance tends to increase over time, and the causes and the types of variance tend to increase as well. This variance causes problems over time if it is not controlled.

There are three types of variances: design variance, process variance, and usage variance. Design variance can result in process variance and usage variance. Process variance results in a waste of materials, slowing of the processes, and idling of machines and workforce, etc. If process variance is not controlled, variance is transferred to the final shipped products and results in failures in function and performance in the market. Variance is transferable from one process to the next. Variance is an infectious disease, which destroys the internal and external customers' trust and eventually breaks down sales and the business as a whole.

Fortunately, there are good quality and business leaders who were aware of these detrimental effects and performed significant amounts of research and corrective business practices. Walter Shewhart at Bell Labs in 1924 developed a compelling variation theory and a controlled method named Statistical Quality Control (SQC). Jack Welch, the former CEO of GE, was an advocate of controlling variances in business by the extensive deployment of Six-Sigma in GE's organization.[6]

2.2 Value of Quality to Society

2.2.1 Benefits to Customers and Business

We cannot separate the business from the cost and the profits, neither can we separate it from the quality. Quality must be understood from two perspectives: increasing customer satisfaction as well as increased benefits to a corporation. In other words, treating the parameters of quality and cost together in a comparable way is a more holistic approach for the business.

W. Edwards Deming defines quality as the output value versus the input cost in the sense of economic value through a system. Figure 2.3.[7]

From the quality value creation process of Figure 2.3 we defined the quality by QPR (Quality Price Ratio) in Eq. (1.2) in Chapter 1:

$$\text{Quality} = \frac{\text{Results of Work Effort}}{\text{Total Cost}}$$

When making a purchasing decision, a customer asks whether the benefits they gain is worth the price they pay. *When the judgment is done by QPR, the manufacturer's and customer's viewpoint become the same.* When people and organizations primarily focus on the quality defined by QPR, quality tends to increase while the cost falls over time. When quality improves, it can reduce rework, scrap, warranty, and litigation costs, resulting in reduction of the total cost.

Figure 2.4 is QPR graph showing how manufacturers and customers perceive quality satisfaction. The QPR value increases from region ① Pain zone to ④ Happy zone, where high quality with low cost makes both customers and manufacturers happy.[8] The result of quality improvement increases brand image, customer loyalty, price, and sales—this is a *chain reaction.*

2.2.2 Avoidance of Financial Loss to Society

Earlier we defined quality as "on target with minimum variance," and this definition can be linked with Genichi Taguchi's loss function, which is a fundamental tool for quality engineering.

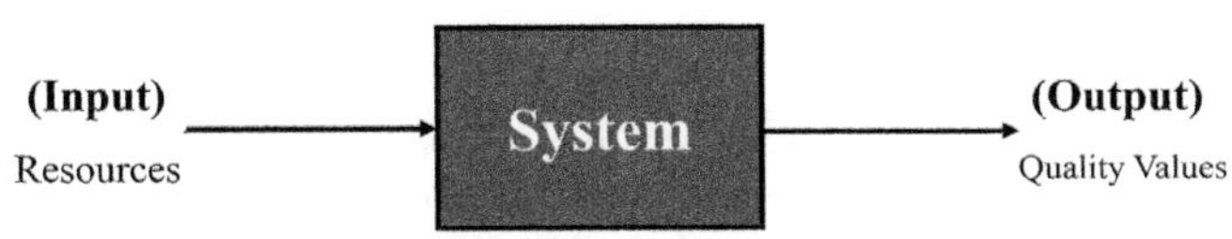

FIGURE 2.3
Quality value creation process through the company operation system.

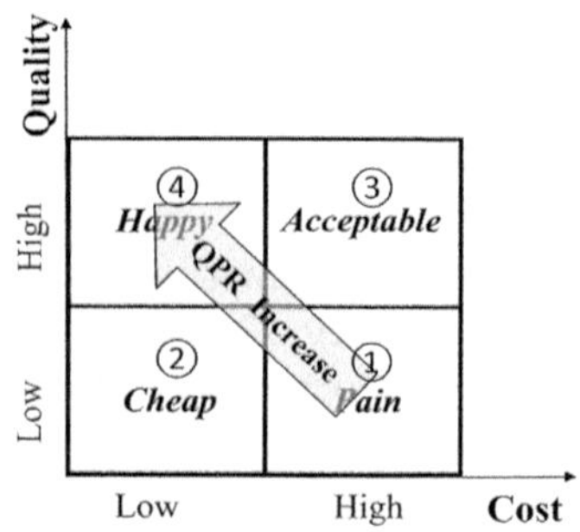

FIGURE 2.4
Quality satisfaction by QPR (quality price ratio).

Taguchi's approach to quality is much more rigidly defined in an economic sense: "Quality is the avoidance of financial loss to society after the article is shipped." A truly high-quality product will have minimal loss to society. Here the loss can take the form of malfunctions or poor performance resulting in returns, warranty, product replacement, rework, scrap, inspection, pollution, and overall deterioration of societal productivity.

The implication is that the loss to society measured in dollars is proportional to the deviation-from-target-value.

Taguchi used the loss function to describe quality:[9]

$$L(x) = k(x - m)^2 \tag{2.1}$$

where L is the loss in money, m is the point at which the characteristic (thing) should be set, x is where the characteristic actually is set, and k is a constant that depends on the magnitude of characteristic (thing) and the monetary unit involved. The continuous loss function illustrates that merely making a product within the specification limits does not necessarily mean that the product is of good quality since good quality is now defined as *keeping the product characteristic on target with low variation.*

Figure 2.5 shows a clear meaning of quality regarding the economic benefits. The simple interpretation is that the closer the characteristic value (x) of the product is to the ideal center value (m), the better the product is for function and performance meeting the customer's expectations. Here the characteristic values can be dimensional, e.g., weight, temperature, or material characteristic.

A power supply unit (or PSU) converts input AC power to low-voltage regulated DC power for the operation of a computer. To assure the perfect function of the digital circuit board in the computer, the built-in PSU is supposed to provide 5 Volt standby (5VSB) voltage output so that the standby functions on the computer and certain peripherals are adequately powered. The typical PSU unit has 5% tolerance

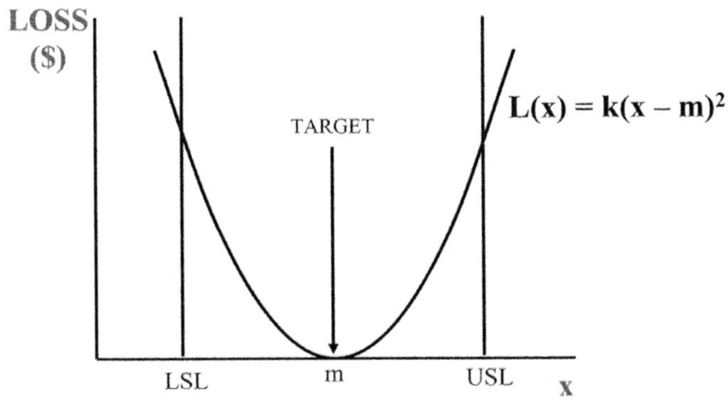

FIGURE 2.5
Quality and the loss function.

of the output, i.e., 5.0 ± 0.25V. The function of the computer should be excellent within the voltage from 4.75V to 5.25V. However, there could be a long-term durability effect between the computers receiving the accurate 5V and the other computers receiving the swing voltage, i.e., 5 ± 0.25V. As a result, some computers equipped with the PSU of high voltage variance (5 ± 0.25V) will fail earlier than the computers with PSU of accurate voltage (5V), resulting in customer dissatisfaction. In other words, the quality output of the two different PSUs is not the same even though the output voltage is still within the functional tolerance limit, 5 ± 0.25V.

As demonstrated by the loss function above, the benefit of quality is apparent. When variance becomes minimized, quality improves. When quality improves, it brings less economic loss to society (i.e., customers or producers). Such a chain reaction converts to a cost-saving effect, leading to more QPR—maximum quality value with minimum cost, benefiting both customers and producers.

2.2.3 Value to Society

As illustrated by the quality definitions included thus far, quality can be defined as "*value*" that customers appreciate in tangible or intangible form. *Value* can manifest in three major categories: *Physiological & Psychological Value, Economic Value, and Social Value.* Mathematically speaking, quality value is a function of these three categories as expressed in Eq. (2.2). This value paradigm is also relevant to Abraham Maslow's hierarchy of needs, as shown in Figure 2.6.

$$Quality\ Value = f(Physiological\ \&\ Psychological, Economic,\ and\ Social\ Value) \quad (2.2)$$

① *Physiological and Psychological Value* is related to the comfort of body and feeling of safety. For instance, a car with good suspension and handling gives the customer a feeling of safety and less tiredness. Also with premium service, customers feel they are treated in a prestigious way.

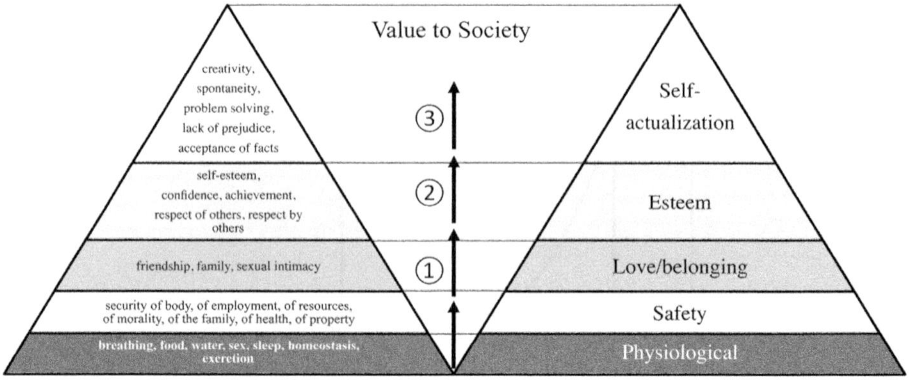

FIGURE 2.6
Value of quality to society corresponding to Maslow's hierarchy of needs. (Licensed under GNU Free Documentation License.)[10]

② *Economic Value* is much more tangible, which can be converted into monetary value. If a product is very durable, the customer can enjoy it with low maintenance costs and high resale value. Productivity of the customer will be high because the product will always work trouble-free, and the person will save the waste of time visiting service stations. This viewpoint is aligned with Taguchi's loss function.

③ *Social Value* is about how the customer's social status is recognized by others based on the products and services provided to him or her. Usually, such social value is recognized by the brand value. People are willing to pay more money for prestigious quality brands. However, history, design, and image alone cannot build a premium luxury brand. The premium luxury brand comes with superior quality and trust of the products.

These quality values are not measured by an absolute ruler. The amount of quality value is relative and variable depending on the *customer's needs, perceptions, and satisfaction of the quality.* A customer is anyone who is impacted by the product or process. A customer can be external or internal. *Internal customers* are the people working in the same organization. For instance, in an assembly line, the downstream working station is an internal customer to the upstream workstations. *External customers* are the people outside the organization who receive the goods and services of the company.

Therefore, the quality value of Eq. (2.2) can be converted into Eq. (2.3) and be proportional to the following function;

$$\textit{Quality Value} \propto f(\textit{Customer's needs, perception, and satisfaction on the quality}) \quad (2.3)$$

2.2.4 No Safety, No Quality

Safety is a human being's basic instinct and fundamental requirement for survival. This requirement is the first stage of the need in Maslow's Hierarchy of Needs[11] as shown in Figure 2.6. Safety and quality are closely interrelated. As the design of advanced technology products becomes complicated, so does the manufacturing process. These kinds of products inherently contain many potential safety risks. The technology itself does not warrant safety. There are risks of fire, malfunction of the innovative technologies, and failure of the safety features. For instance, an autonomous vehicle may reduce human driving mistakes; however, it will also have a severer impact when any component or software has even small glitches.

In recent history, there were many pandemic levels of safety issues due to quality defects. In 2009, there was Toyota's unintended acceleration recall; in 2014, GM's ignition switch recall; in 2014, Takata's airbag recall; and in 2018, BMW's engine fire recall. Those safety issues caused many injuries, deaths, and psychological worries, and the massive recalls followed only after government or media investigations and interventions.

When safety incidents occur in the internal processes or the market, it causes the shutdown of the company, plunge of the stock price, large-scale litigation, and a detrimental impact on the company's public image. In 2013, a battery problem in the Boeing 787 grounded the fleet of state-of-the-art airplanes for more than six months. In 2017, Samsung's flagship product, Note 7, was completely eliminated from the market due to a battery issue.

People do not trust a company that ignores the safety of customers or its employees. In the same way, people do not trust the brand that does not provide safety requirements.

Safety is a necessary condition for quality. Not fixing potential safety issues is like leaving a ticking bomb in the system and the market. Behind the historical safety recalls, there was the company's moral hazard issue as a common cause of the problems. Fixing the potential safety issues from the design to the manufacturing could be a pre-condition to the company's quality operation. *Safety and quality go together.*

There are two aspects of safety issues: one is *the producer or service provider's internal safety,* and the other one is *the customer's safety.* Both aspects should be looked at with equal priority. The company should assure the safety of its employees by providing a safe working environment, and the employees should care about the safety of their customers as a moral obligation. Potential safety issues should be a must-report item and a top indicator in the performance of the company's operations and its management.[12,13]

In 1986, when the Challenger spacecraft exploded, causing the deaths of seven astronauts, NASA's space shuttle program was grounded for nearly three years. During the launch, the primary O-ring blow-by (the escape of hot gas) occurred because the O-ring hardened and did not fully seal at low temperatures. Management made the decision to launch in the low temperatures, despite the fact that the vehicle was never certified to operate in

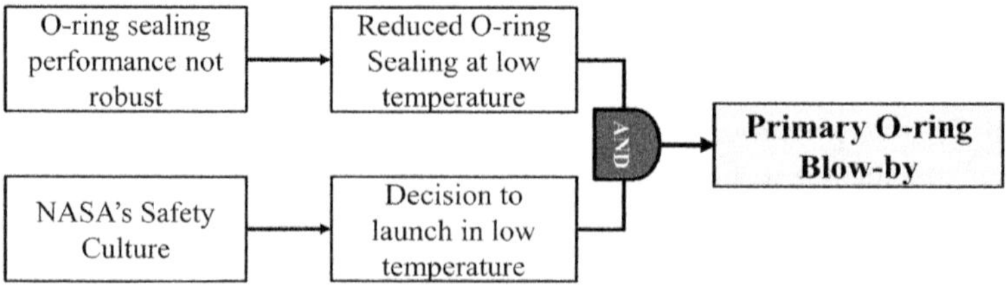

FIGURE 2.7
Chain reaction of safety issue. (Challenger Spacecraft, 1986.)[15]

temperatures that low. There were both technical and management problems that caused the disaster as illustrated in Figure 2.7.

The O-ring design used in the Solid Rocket Booster was the technical cause[14], but more importantly, NASA's mismanagement of the potential safety issues was the primary cause behind the disaster. When both the technical and the management problems were combined, the disaster became inevitable.

This type of failure—a combination of technical and management issues—is the most common cause of safety and quality failures leading to disasters. When a product has weaknesses in its technical aspect of design and process, it is like a potential bomb. When management makes a poor decision, it ignites the potential risk, inevitably resulting in a disaster.

Engineering's quality concerns exist at every stage of activities—product design, technical specifications, and validations—and potential technical concerns are embedded in the product. Management directions and decisions are also at every stage of such activities. In the case of the Challenger disaster, the Rogers Commission report pointed out that NASA's safety's culture was the primary cause of the problem.[16]

Unfortunately, sometimes we only learn from mistakes and failures: "NASA administrators tried for years to improve the agency's safety habits, but those efforts were unsuccessful until the space shuttle *Challenger* exploded in 1986. In the wake of that tragedy, the organization was able to overhaul how it enforced quality standard."[17]

History repeats, in 2019, Boeing's 737 MAX was grounded worldwide after the two fatal crashes that killed 346 people onboard. Many airlines cancelled the order of this model.

As we discussed regarding Maslow's Hierarchy of Needs in Figure 2.6, when humans' basic needs crumble, everything else crumbles. *In order to bring out the value of quality, the company should warrant safety first.* Such a priority creates a virtuous cycle: safety warranted → stable process → less variance → stable quality → value to customers.

We cannot build up quality without a basic foundation—safety. *"Safety first"* should be the company's culture and fundamental principle for business operation, engineering design, manufacturing process, and daily management.

2.3 Cost of Quality

In the previous section, we discussed the value of quality that we can get through the quality value streams. Everybody wants excellent quality value, and we know the quality value is not created without input cost. The producer's viewpoint is to create a maximum value of quality with minimum input cost through efficient business operations.

The cost of quality is the cost that comes from poor quality of the product and preventive quality activities. Poor quality represents the cost for quality failures internally or externally, and *the preventive cost is the cost that we spend on preventive quality including appraisal activities.*

In the case of the manufacturing business model, a typical cost to the company is through the processes in Figure 2.8. Each major function block is responsible for the cost incurred in its group. The responsibility of some cost items is shared with other functional groups. For example, in the case of warranty cost, it is shared with the suppliers in the form of charge-back.

Cost of quality sounds like a mystery, and each department of a company may have a different understanding or interpretation. For example, accounting people only care about direct quality costs that are shown in the company's ledger. However, salespeople are concerned about indirect quality costs as well.

From the company leader's standpoint, a proper interpretation of quality should be aligned with the customers' standpoint. If we see the graphical representation of quality cost as in Figure 2.9, quality cost becomes unequivocally clear.

The cost of quality can be expressed in the following two equations:

$$\textit{Cost of Quality} = \textit{Direct Quality Cost} + \textit{Indirect Quality Cost} \tag{2.4}$$

$$\textit{Cost of Quality} = \textit{Preventive Quality Cost} + \textit{Poor Quality Cost} \tag{2.5}$$

In Eq. (2.4), direct quality cost is the cost shown in the company's ledger, while indirect quality cost is not shown; however, it still has an impact on the company's

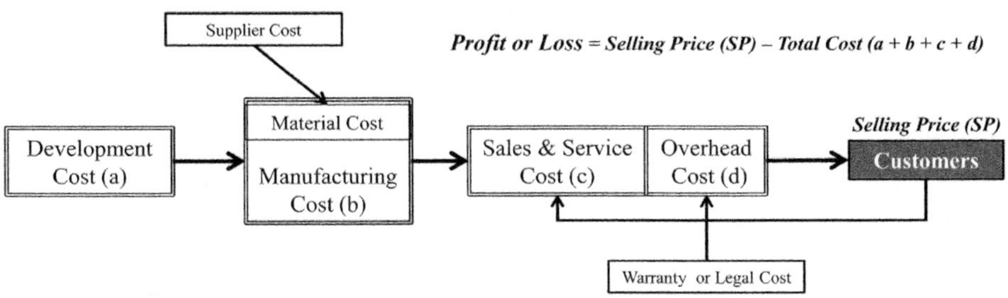

FIGURE 2.8
Total cost incurring process to a company.

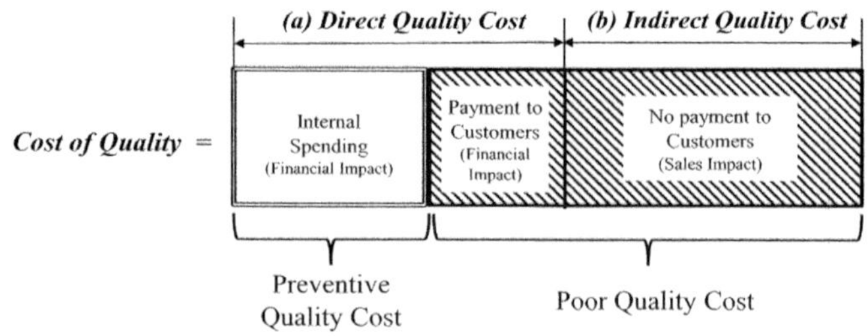

FIGURE 2.9
Structure of quality cost.

sales or revenue in the future. In Eq. (2.5), preventive quality cost is the cost of activities undertaken to prevent potential poor quality internally or externally, and poor-quality cost represents the internal or external poor quality cost.

A clear explanation is as follows:

(a) Direct quality cost is incurred in each process directly and shown in the company's financial statements as in Figure 2.8 and Figure 2.9. However, *(b) indirect quality cost* is not visible in the company's ledger or P&L statement because the customers absorbed the cost without claiming it either because the warranty does not cover the whole cost or the warranty is expired. The indirect costs manifest in the form of customers' losses with respect to their time, expense, and stresses, which are not shown in the company's financial statements. However, this indirect cost becomes the company's cost because it will be eventually be charged to the company by the customers' revenge—no longer buying the company's product or service.

2.3.1 Direct Quality Cost

Direct quality cost is the direct financial loss to the company shown in the company's ledger. Direct quality cost to the company comes in two forms. One is the *direct poor-quality cost* due to the poor quality of the company's product not meeting the internal or external customer's expectations. The other is the *preventive quality cost* spent by the company to prevent potential concerns; if such concerns are not alleviated, it could result in direct poor-quality cost. The direct cost is composed of two quality costs, which is expressed in Eq. (2.6).

$$\textit{Direct Quality Cost} = \textit{Direct Poor Quality Cost} + \textit{Preventive Quality Cost} \quad (2.6)$$

Figure 2.10 shows the structure of direct cost in two main categories: *direct poor-quality cost* and *preventive quality cost*. Direct poor-quality cost is the recovering activity cost due to quality failures internally or externally.[18] Preventive quality cost is any cost that is spent on prevention activities *including appraisal costs* for potentially poor-quality issues.

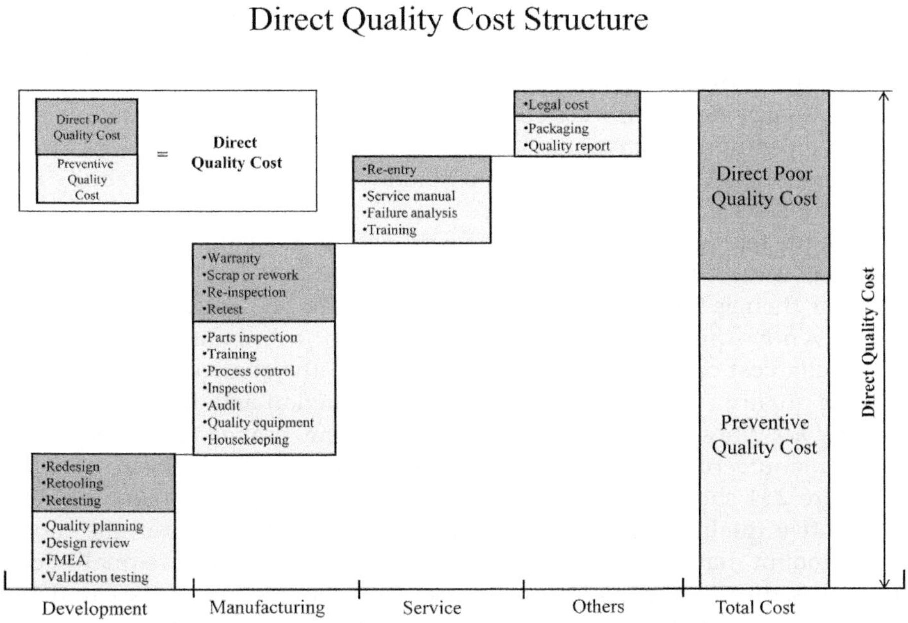

FIGURE 2.10
Structure of direct quality cost.

Direct poor-quality costs are incurred when unacceptable products and services are delivered to the next operation process or the end customer. Direct poor-quality cost is composed of internal and external poor-quality costs.

- *Internal poor-quality cost:* costs as a result of quality failures detected *before* the product is delivered to the end customers.
- *External poor-quality cost:* costs paid by the company as a result of quality failures experienced by the end customers.

These two types of *direct poor-quality costs* are incurred due to poor systems, poor design, poor suppliers, and/or poor manufacturing. Typical examples of such costs are re-design, re-tooling, re-testing, rework, scrap, warranty, and legal cost as shown in Figure 2.10.

> *When the instrument panel does not fit the car body, it will result in a process delay or stop the assembly line. Such a failure causes internal poor-quality cost. When the defects are delivered to the customers, it becomes external poor-quality cost.*

Preventive quality costs are direct costs to ensure that only acceptable products and services reach the end customers. This cost is embedded in each product creation and manufacturing process. *Preventive cost* occurs throughout each quality activity in the form of development quality cost, manufacturing quality cost, supplier quality cost, and service quality cost. Examples of each quality cost are described in Figure 2.10, such as quality-planning cost, robust quality design process cost by FMEA, design review cost, quality confirmation equipment cost, etc.

For instance, with respect to a smartphone's quality, impact durability testing is an essential quality assurance process to ensure there will be no failures in the market well after the warranty period. Such quality confirmation costs fall under *preventive quality cost.*

What, then, is the optimum cost balance that the company needs to spend on preventive quality activity? First, we need to think about a virtuous cycle of quality cost control and management. When the company improves the overall quality system, engineering, and production activities by the right quality activities, such as a robust design review, proper validation, and a stable manufacturing process, *direct poor-quality cost will decrease.*

Figure 2.11 shows the relationship between direct poor-quality cost and preventive quality cost—the summation of these two costs is the monetary amount that the company will pay directly. As preventive quality cost increases, direct poor-quality cost decreases. From the graph, we can determine the best interim operation point to keep the preventive quality cost at *P*. As fundamental quality improves, the poor-quality cost significantly decreases from the (*R*) to (*R′*) curve. With this new poor-quality cost, the best interim operating point moves from *T* to *T′* position, which creates an opportunity to reduce preventive quality cost from *P* to *P′* point.

The lesson here is that when the overall quality improves thanks to good preventive quality activities at the beginning, poor-quality cost decreases, which creates an opportunity to reduce the preventive quality cost, causing a

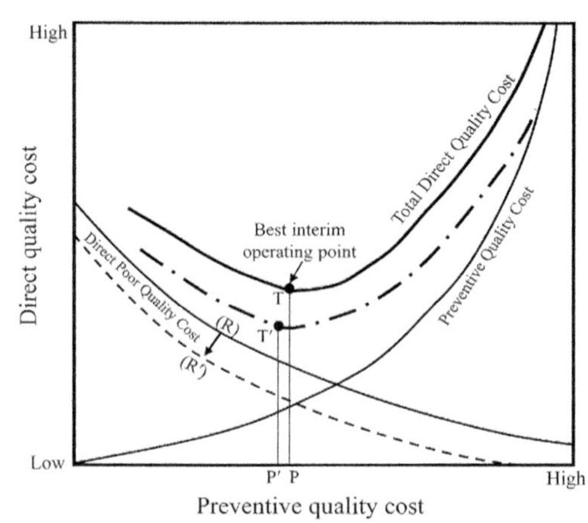

FIGURE 2.11
The effect of preventive quality activity on the direct quality cost improvement.

virtuous cycle. *Therefore, it is better to start with robust quality processes and eliminate unnecessary processes or activities in later stages when the processes become stable and mature.*

> *There are many field examples of such quality improvement effects leading to lower direct poor-quality costs. The engine plant of Honda Motors used to do a 100% engine firing testing in the manufacturing line as a method of final quality assurance. This preventive quality cost was very high in the beginning. However, with quality improvements such as robust engine design and a stable assembly process, there were few poor-quality incidents internally and externally. Then they were able to reduce the engine firing testing down to 1% sampling of the total production. Today, with a 6-sigma level of quality, the hot firing test was replaced with the cold test.*[19] *There is virtually no firing testing anymore except audit. Such a result saved a vast amount of preventive quality cost to Honda Motors.*

The lesson here is obvious—with quality improvements of the product design and manufacturing processes through robust preventive quality activities, we can minimize both the poor-quality cost and the preventive quality cost through a virtuous cycle.

2.3.2 Indirect Quality Cost

Indirect quality cost is the cost that a company does not pay directly. Taguchi explained that any form of poor-quality results in *financial loss*. Some financial losses to the customers are paid by the company, but some are not; in such cases, customers must bear the cost and pain.

Some portion of the financial loss to the customers is not shown in the direct cost of the company's ledger because the company does not pay from the company's financial coffer. The indirect cost of poor quality is *the portion of the loss experienced by customers, but which is not paid by the company.* Typically, a company pays only the direct cost for the warranty including the parts and labor cost. For instance, even in recall cases, the company does not pay the customers' traveling expenses or for taking a day of vacation to bring the product to the service networks of the company.

The indirect poor-quality cost borne by the customers comes in two forms: one is the visible and calculable cost, and the other is the invisible cost, which is difficult to quantify.

- The visible financial loss to the customer = Customer-paid direct cost
- The invisible financial loss to the customer = Customer-suffered productivity-loss cost

Either case will increase the customers' dissatisfaction so that it will discourage future purchasing or support of the product. Such indirect costs borne by the customer will come back to the producer as a boomerang effect eventually.

1. *Customer-paid direct cost:* The cost is spent by customers when the product does not meet the customer's expectation because of malfunction, poor performance, or poor perceived quality. The typical cost types in monetary terms spent by the customer are:
   - Traveling costs and time spent to visit a repair service station
   - Repair costs after the warranty period
   - Backup product costs when the product failed during or after the warranty period
2. *Customer-suffered productivity-loss cost:* Some costs paid by customers are not in the form of money, but indirect forms of expenses or loss to customers. The costs still impact the customers in indirect ways such as:
   - Productivity loss due to the product downtime
   - Productivity loss due to the physical and psychological stress caused by the poor quality of the product

Table 2.1 shows the effect of the indirect quality cost borne by the customers. The customers' dissatisfaction with the product due to its poor quality triggers chain reactions such as: talking about the bad experience to their friends, losing trust in the brand, not purchasing the brand in the future, and not being willing to pay the right price for the brand. Such retributive actions by customers result in a decrease of the company's sales revenue.

This indirect cost can be like cancer. The harmful effects grow and spread. The "White House Office of Consumer Affairs found that a dissatisfied customer will tell between 9–15 people about their experience … which leads to damage of reputation, loss of potential customers and revenue."[20] Generally, it is difficult to detect or measure the impact of indirect costs compared to the direct costs. When it is detected, it is often too late or too difficult to correct the problems. It costs a huge amount of money, time, and effort to reverse the course. There is a delay loss effect

TABLE 2.1
Effect of the Indirect Quality Cost

Result of Poor Quality	Action by Customers	Loss to Company
Dissatisfaction with the product	No future purchasing of the product	Sales volume decreases → *Lower sales revenue*
	No trust of the product or the brand	Lower value of the brand → *Lower sales revenue*

until the company recognizes the severity of the problem. The loss costs rise as the sales revenues fall, though the indirect cost does not directly appear in the company's ledger.[21]

The costs and nuisances experienced by customers have a substantial adverse effect on the product and the brand, which is paid for by the company in the form of *indirect cost*. Therefore, there should be a mechanism to monitor and measure such *indirect cost*. A general methodology is to measure (1) Customer-incurred direct cost and (2) Customer-incurred productivity-loss cost *by studying the major poor-quality cases in terms of money value.*

> *K-Company has a robust and flexible warranty system to minimize the impact of the indirect quality cost items. For example, when a product fails just after the warranty period, the company should consider a proper compensation like warranty extension or partial coverage of the repair expenses.*

2.3.3 The Impact Scale of Poor-Quality Cost

Many business leaders have a misconception about quality, thinking that the investment for quality improvement is more than the benefit of quality improvement. In some cases, they think fixing the problems later could be cheaper than spending money today. Such a viewpoint may be valid if the quality impact remains the same scale downstream. However, we know that is not the case according to our experience and knowledge.

As we discussed earlier, if the quality is good, lower amounts of poor-quality cost will occur. With the improvement of quality, we can also decrease some portion of the preventive quality cost. According to the research, the cost of building quality into the product is 5% of sales, while the cost of nonconformance is 20%[22]—the scale is four times.

The question from management is how much, in cost terms, do we need to perform the quality activities as discussed in Figure 2.11. To better understand such a real-world question, we should look at the *total poor-quality cost including direct and indirect poor-quality costs,* as represented in Figure 2.9. The total poor-quality cost is the combination of direct and indirect poor-quality costs as expressed in Eq. (2.7).

$$\textit{Total Poor Quality Cost} = \textit{Direct Poor Quality Cost} + \textit{Indirect Poor Quality Cost} \quad (2.7)$$

Every direct or indirect poor cost should be interpreted as a quantitative, objective number in dollar values to everybody in the company, including the accountant, salespeople, and company leaders.

When product design is not robust enough, extensive validation tests and quality confirmation activities are necessary to prevent or detect potential concerns in later stages. Such activities increase the preventive quality cost. However, if such preventive actions are not taken appropriately, the flawed product will flow into the manufacturing process, resulting in internal failure correction costs via scraps and repairs. This kind of outflow can generate more troubles and cost than the preventive quality costs.

> *Repairing defects requires many extra administrative activities. For instance, a special team needs to be formed for specific repair tasks, and there are still risks of side effects from the repairs. Special logistics needs to be arranged to receive good parts, and the repair guidelines need to be prepared. Later in the process, additional tests or inspections need to be done to assure its function and performance. Holding products for repair may cause an increase in the customers' complaints on late delivery or missed sales opportunities. However, such administrative and external risk costs are frequently hidden, not controlled, and not shown in the company's ledger.*

In other words, when we include such hidden costs, the potential downstream recovery cost is much higher than the upstream cost shown in the company's ledger. Considering all-inclusive hidden costs and risks, we can expect the scale of the downstream cost to be much greater than the scale of the upstream countermeasure cost.

Unhappy customers share their bad experiences of the product with more than ten other people, and study by Ebert[23] shows a scale of the downstream impact is 10 times.[24] Many empirical observations and data in the industry shows that the total poor-quality cost needs to be looked at from the perspective of *"10 times cost rule."*[25] In other words, the failure of the upstream process causes *"ten times more cost"* in the downstream process as graphically shown in Figure 2.12. For instance, if the preventive cost, which is mainly design and appraisal cost, is "1," the internal failure cost is "10 times of the prevention cost." We need to note the cost impact is *not only a direct cost, but it also includes the indirect cost that the downstream customers pay or suffer.* This *"ten times cost rule"* is the principal rule we need to apply when we calculate or decide the quality cost.

FIGURE 2.12
Cost impact (10-times scale rule).

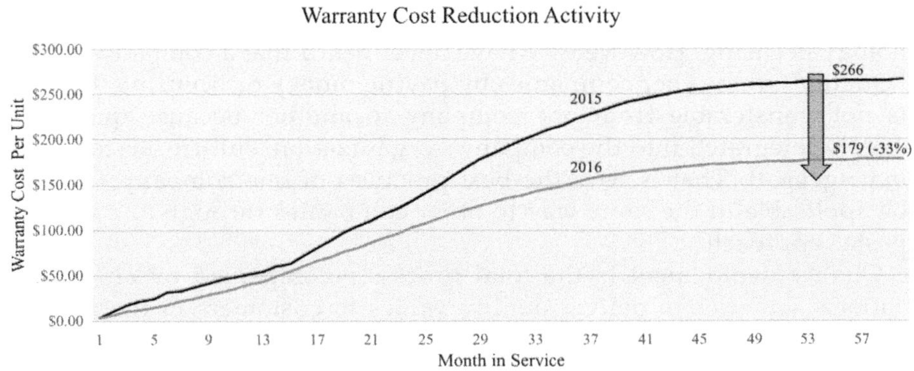

FIGURE 2.13
The effect of preventive quality activity. (Example: R-company SUV model.)

As such, we can formulate the downstream recovery cost as approximately 10 times the scale of upstream countermeasure cost as shown in Eq. (2.8):

$$\textit{Downstream recovery cost} \sim 10 \times (\textit{Upstream countermeasure cost}) \tag{2.8}$$

2.3.4 Management of Quality Cost

Figure 2.13 represents an *actual example of the effect of preventive quality activity.* When proper quality activities were deployed after the first year model had been launched, 33% of the external direct cost (warranty) was reduced; $266 – $179 = $87/unit was saved. However, the company actually spent less than $8/unit on preventive quality activity. Most of the warranty cost savings came from the corrective actions on the mistakes made for the first year model (2015 Year SUV).

Typical global top-quality automobile companies spend a warranty cost of less than 1% of total sales revenue. However, when we include indirect poor-quality costs, the impact of the total poor-quality cost can be ~10% of the total sales revenue by using the ~10 times scale of direct poor-quality cost impact.

Such a demonstration illustrates that when we include the indirect poor-quality cost, the investment in the preventive quality activities becomes significantly more justifiable. The prevention of product failures upstream is much cheaper than the fixing-cost downstream. *The amount of preventive quality cost can be justified as having 10 times the cost-savings effect in the downstream.*

2.4 Deployment of Quality

Quality starts from more attitude, principle, policy, philosophy, and methodology rather than technology.

Quality and technology are different in the aspect of their development, maturation, and continual improvement. The company may be able to buy technology from other companies through technical assistance or technology licensing. However, we have never heard that a company bought "Quality" from other companies by paying money or licensing. Quality is not transferable from one company to another because quality is deeply integrated into the company's organization, culture, system, and management. That is why the best practices of one company may not be applicable in the same way to other companies through a "copy and paste" approach.

Quality deployment is the total process accompanied by an organization's activities to deliver quality values to customers in the form of products or services. Quality deployment is not just the quality department's mission or responsibility. Ultimately, those responsible for quality are the owner(s) of the company, including the board members, and the employees.

An ideal model of quality deployment is a combination of top-down, bottom-up, and cross-functional approaches (see Figure 2.14). It should start from the top and move to the bottom of the organization, building up the quality culture.

2.4.1 Start from the Top Management

Top management is responsible for setting up *Quality Policy and Management Principles*. Every company says "customers are first." However, if actual decisions from the top do not practice the company's policy or principles, quality

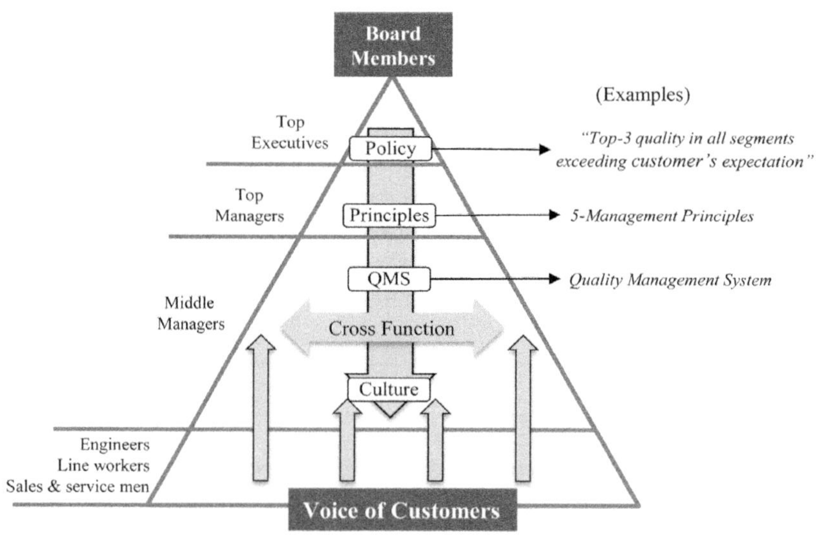

FIGURE 2.14
Quality deployment process.

can be only lip service, and quality actions will not be deployed throughout the organization. The members of the organization are watching not what the top management is saying, but what they are doing.

That is why Deming said, "Quality starts in the boardroom."[26] The boardroom is the highest decision-making committee of a company. The boardroom's policy and principles flow down to top management, and top management allocates the required resources to each downstream activity followed by the evaluation of each function's outputs. During this downstream process, top management's intention and direction is deployed to the organization by the systems and processes of the company. Deming said that a worker in actual workplace, *Genba in Japanese terminology,* cannot deliver better quality than the system allows, and managers are responsible for making the system. If top management makes short-term profit decisions by neglecting the long-term interests of customers, such harmful practices spread like a disease to managers and line workers.

Customer-oriented *decisions and practices* by the top management are a necessary condition to cause actual customer-oriented cultures to be deployed and spread to the bottom of the company hierarchy that actually interfaces with the customers. When the customer-oriented culture builds up, a *bottom-up process* is expected to occur. Then the top management can hear the real voices of the customers communicating the right decisions for the interests of the customers. This sound practice initiates a virtuous quality cycle for the long-term prosperity of the company.

2.4.2 A Bottom-up Quality Deployment

In a typical company organization, the bottom hierarchy interfaces with the customers. Quality is delivered from the engineering, manufacturing line, and sales and service networks to the customers. The bottom of the company organization responds to the customers by listening to their complaints, needs, and desires. Many global companies build up a vast IT system to monitor the quality at the end of the manufacturing line and in the market. However, a monitoring system that is not followed by the right actions cannot cause the right quality to be delivered to the customers. A good quality system and deployment require employees' actions for quality improvement in *Genba,* which is the bottom-up process.

There are two major elements for the bottom-up process: one is a *quality culture,* and the other is a *quality system.*

Quality culture is like a healthy bloodstream of a human body. When quality culture has matured throughout the organization, self-motivated behavior by employees for customers occurs spontaneously. The effort of top management can cultivate the quality culture through their customer-focused practices, including education and training of the employees. When quality culture has spread and matured, the employees respect the internal quality systems and processes to serve customers.

Quality system is a collection of standard processes prescribing how each business unit should run its operations to deliver the best outputs to downstream processes, finally reaching customers. When making standard procedure and work instructions, employees of each operation should be the maker and owner of these standards; then they become *a practical mechanism for* the customer-oriented activities in *Genba.*

The advantage of a bottom-up process is speed and accuracy, as employees react spontaneously to customers' voices, and the right information effectively reaches the top management.

2.4.3 A Cross-Functional Quality Deployment

Hierarchical management alone cannot effectively deploy quality. In hierarchical management, issues go to the boss, and the boss makes decisions or coordinates actions as needed. In today's fast-moving quality business environment, speedy, concurrent, and dynamic management is inevitable. Most quality activities require multi-functional communication and actions cross-functionally and simultaneously as a necessary management method. For instance, a countermeasure decision for even a small quality issue may require many inputs from different functional groups like R&D, suppliers, manufacturing, and sales and service departments.

Considering the complexities of today's products, a typical common root cause of the quality failures in the market is poor or ineffective cooperation between functional groups. For example, a supplier's poor-quality issues can be traced back to excessive cost-cutting by the PU (purchasing) department, poor engineering activity by the SQE (Supplier Quality Engineering) department, or faulty design specifications defined by R&D.

To promote fluid cross-functional communication and cooperation, customer-oriented quality activities should be deployed in the following way: *a scientific breakdown of the R&R* (role and responsibility), and *a promotion of teamwork-based activities.*

- *A scientific process for R&R breakdown* should focus on the customers' viewpoints and the proper cross-functional management method. Originally, QFD (Quality Function Deployment) was developed to define and prioritize engineering parameters with respect to customers' requirements in the market. A similar concept to QFD can be used to define each department's actions to best match customers' requirements in the market—see Figure 2.15.
- A *promotion of teamwork-based activities.* Complex issues are difficult to solve by one functional group or one team. We can expect better outputs when we form a joint team for complex tasks by sharing the same goals and responsibilities. For example, for a supplier's quality

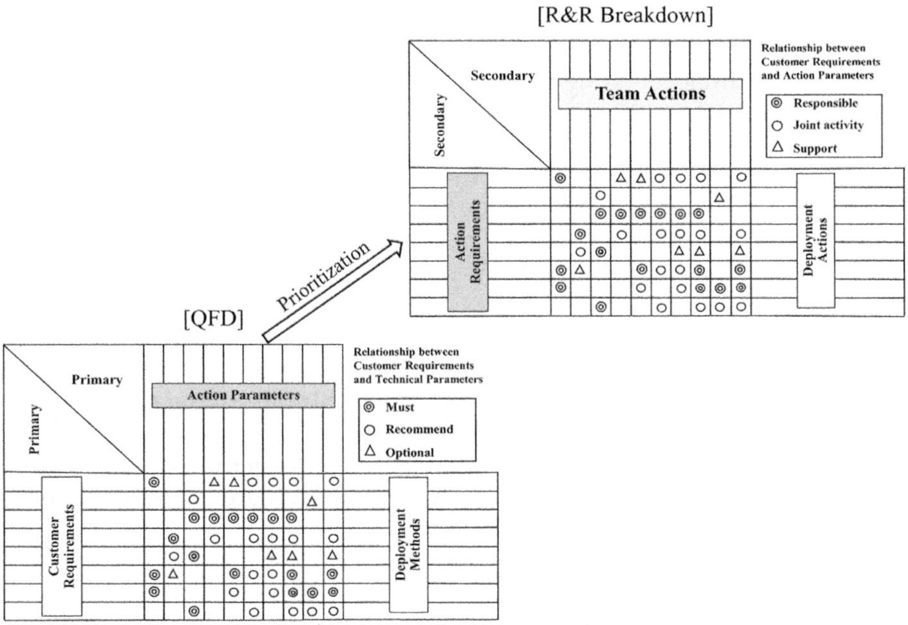

FIGURE 2.15
Cross-functional quality deployment.

capability assessment of a new technology part, a joint team of R&D, Purchasing and Supplier Engineering Quality should visit suppliers to check the current state of the part and to identify improvement points to incorporate into design or manufacturing processes.

2.5 Measurement of Quality

In the 21st-century knowledge economy, knowledge power is ever more significant. Knowledge comes with data, information, and predictive capability. With the help of IT technology, gathering big data can be done conveniently, and predicting future patterns is becoming easier with the advent of AI (artificial intelligence). The remaining issue is the human factors: what do we need to measure and what actions do we need to take?

During a quality deployment activity, it is natural to check whether our actions resulted in meaningful outputs to the internal and external customers. Deming said, "If you do not measure, you cannot control." The starting point should be what kinds of data and information we need to collect, measure, and analyze.

As for any measurement, the purpose and analysis method should be clarified from the beginning. The ultimate goal of any measurement should link to understanding the *customer's satisfaction on the quality*—market quality. Since the market quality data is absolute and objective information, it has the power to convince and motivate an organization to do proactive quality activities. There are many forms of market quality data in different types like warranty data, VOC (voice of customers), market surveys, and others.

One of the significant concerns of market quality data is that there is a substantial time lag effect, Δt_{lag}, between today's production quality and the expected quality in the market tomorrow. In other words, today's market data is only the representation of yesterday's production quality; to put it another way, it is not real-time data. Many products are already in the supply or in-house manufacturing pipelines with yesterday's quality contents. Therefore, it is a big challenge to drive today's action based on yesterday's data.

Such a situation is represented in the diagram of Figure 2.16, and it can be formulated in the following form, Eq. (2.9):

$$\begin{aligned} &In-house\ Quality\ (Product\ Planning,\ Design,\ Manufacturing,\ Sales\ at\ t) \\ &\quad = Market\ Quality\ (t + \Delta t_{lag}),\ where\ \Delta t_{lag}\ is\ a\ time\ lag \end{aligned} \tag{2.9}$$

Considering such a lag effect in the market quality data, we need to appropriately correlate *today's in-house quality data to the future market quality data.*

Making a precise correlation between today's and tomorrow's data requires scientific methods, experience, and knowledge. The first step is to set up key quality parameters at each in-house quality process that will have the most sensitive impact on the market quality. These key parameters are from the analysis of the current data and the experience of the company.

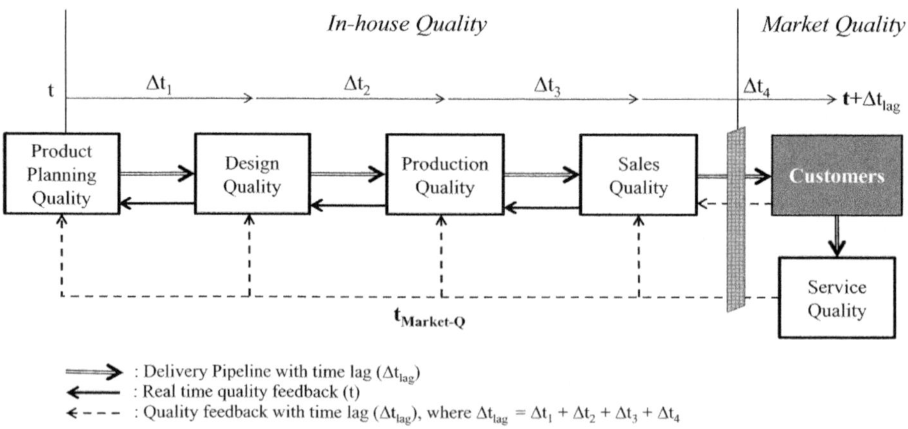

FIGURE 2.16
Quality delivery process and time lag effect of quality data.

Usually, the market warranty data and VOC are the most objective market data. The analysis of the quality data in the market produces substantial information, which is linked with the producer's design, suppliers' parts, and in-house processes.

Today's high-end cars are equipped with an air cleaner installed in the front IP (instrument panel) area or on the back seat area. Any abnormal noise of the air cleaner can damage the high-end brand image of the vehicle. Most of the time, the cause of such noise is traced back to the part quality of the fan supplier. Therefore, once the design specification has been confirmed to be satisfactory, the supplier's capability of the assembly tolerance of the motor components needs to be measured and analyzed. During the process, we may find the amount of interference between the DC brush and the commutator is on the higher side than the specifications. Then, the interference parameter becomes the primary measurement and control point in the supplier's assembly process as a temporary solution, and a permanent countermeasure by a new design should be followed such that it is no longer necessary to control during the assembly process.

Each process of the product design and manufacturing requires different critical parameters to control and measure. With the accumulated experience and knowledge of the company, an initial checklist for each process needs be made and then be regularly correlated with market data. The quality checklist needs to be dynamically updated to capture potential outflows into the market. In other words, if the in-house defects rate is extremely low, but many defects occur in the market, the measurement parameters need to be investigated and adjusted. If there are durability problems only found in the market, accelerated laboratory testing or field testing should be done to identify the exact causes, and necessary countermeasures should be applied through design changes or suppliers' parts correction processes.

Table 2.2 shows the examples of the typical critical parameters to control and measure at each quality flow.

Market quality can be expressed as a function of quality flow parameters as Eq. (2.10):

$$\begin{aligned}&\textit{Market Quality (customer satisfaction)} = \textit{Quality of (Product Planning}\\&\times \textit{Design} \times \textit{Manufacturing} \times \textit{Sales} \times \textit{Service)}\end{aligned} \tag{2.10}$$

Equation (2.10) shows that the market quality (customer satisfaction) is not the summation but the *multiplication* of quality inputs from each quality flow. If any quality parameter fails in the quality-delivery-chain to the customers, the total outcome, customer satisfaction, becomes zero.

Each quality input should be measured and correlated with the market quality data (customer satisfaction). Scientific methods should be deployed to link market data to in-house data measured by using Table 2.2. Once the risks are identified, necessary countermeasure actions need to be taken through the quality system and management.

TABLE 2.2

Key Quality Parameters to Measure at Each Process

Quality Flow	In-House Quality Measurement (today)	Market Quality Data (tomorrow)	Remarks
Product Planning	- Features and performance position with benchmarking - Styling and design competitiveness	- Initial market reaction to new products (Sales feedback, VOC)	• VOC: Voice Of Customers - data from the call center, website, or a survey by a 3rd party company • Durability: check customer's long-term experience of the products by warranty data and survey • Warranty: 3 MIS (3 Months in Service)
Design	- Performance and attributes competitiveness against benchmarking models - Functional assurance	- Feedback from sales - Warranty ratio (claims/unit, cost/unit) - Durability Quality (6~12 MIS warranty, survey)	
Manufacturing	- Incoming parts quality - Straight shipping ratio - In-house repair ratio	- Warranty ratio (claims/unit, cost/unit); - Initial Quality (3 MIS warranty, survey)	
Sales	- Contract success ratio - On-time delivery ratio	- Sales satisfaction (VOC, survey)	
Service	- Reentry ratio	- Service satisfaction (VOC, survey)	

2.6 Premium Quality for Premium Brand

Every day many similar products and services are gushing out into the market. The price, features, performance, technologies, and quality levels are more or less the same. They are not much distinguishable from one to the other. The competition in the low or medium segment is more severe than in the premium segment.

In the post-TV age, only remarkable products can survive in the market. People do not place their trust TV advertisements; instead, they heavily rely on the opinion of experts and their peer groups. With information technology, experts' opinions and comments spread quickly and heavily influence the average customers' minds and their final purchasing decisions. What has been accepted as remarkable by innovators or early adaptors can easily penetrate the early and late majority customers.[27]

Figure 2.17 shows how a premium-quality product can be distinguishable from an average-quality product. The *quality cost* and *quality value* perceived by the customer is *not linear*. There is a tipping point where $\Delta Q_{pv}/\Delta Q_{ac}$ *is*

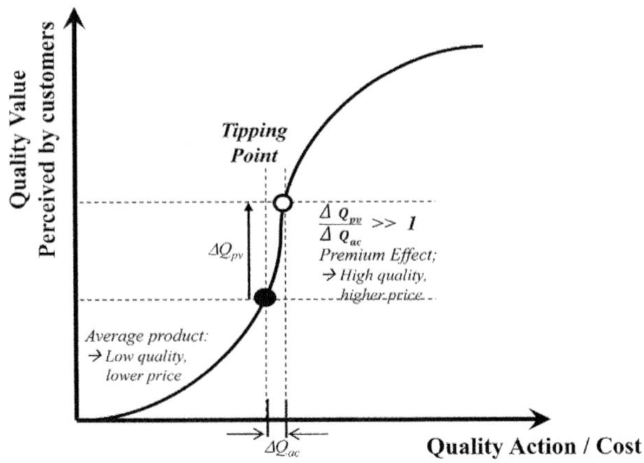

FIGURE 2.17
Premium-quality effect.

bigger than 1, where with a little more quality action and cost, (ΔQ_{ac}), higher perceived value (ΔQ_{pv}) can be attained. Better quality parts and materials, better process, and more skilled people make a difference in quality value perceived by customers. It also induces *less loss* to customers due to fewer poor-quality defects and premium perception in the market as discussed in the section 2.2.3.

In the segment of premium quality, the brand is much more distinguishable compared to a mediocre-quality brand, given its unique quality characteristics. With respect to this higher-quality segment, the major positive effects are summarized in Table 2.3.

Premium quality can enjoy a better price position and higher profit ratio. Such premium effects have been demonstrated through many quality surveys by the J.D. Power survey and Consumer Reports. Price positioning is not linearly proportional to the quality improvement curve. There is a tipping

TABLE 2.3
Positive Effects of Premium Quality and Business Benefits

	Input of Premium Quality	Business Benefits to the Company
1	Fewer internal and external quality failures (e.g., fewer repairs)	• Reduction of poor-quality cost → reduction of financial cost to the company
2	Higher perceived quality value of the product	• Higher selling price → bigger profit margin • Sales volume increase → bigger profit volume
3	Better brand value	• Higher price → more profits ratio per unit sale • Bigger market share with the better market positioning → stock price increases → market capitalization increase

point, where a little decrease of quality defects increases the price position nonlinearly—as seen in a J.D. Power survey in 2013.[28] At this tipping point, *quality value (Physiological & Psychological + Economic + Social) perceived by the customer increases nonlinearly.*

Another benefit is that a better brand quality brings about a higher market value for the company.

Numerous studies conducted in economics, marketing, and management indicate that premium quality leads to high performance of the stock with higher expectations of a bigger market share acquisition by the company. [29,30]

As a summary, the amount of the increased input cost for higher quality can be offset by the higher perceived values in the market, which eventually create a virtuous cycle, resulting in more profits for the company.[31]

2.7 Craftsmanship for Quality

Humans began to make tools 2.5 million years ago by using stones. Since then, endless and persistent endeavors have taken place to make the tools sharper, more user-friendly, and more durable. These skills have been inherited and improved from generation to generation throughout human history along with the advent of new materials from stones to bronze and steel.

With the creative minds of Homo sapiens, more useful life goods have been developed, such as clothes, shoes, and wares, which were only possible with more in-depth knowledge and experience in handling materials and processing skills. Each region and culture has developed its own unique products based on its needs, environment, and culture. The availability of materials and processing skills were the main factors in making specific products that were unique to the region.

For example, the first porcelain was made in Zhejiang province in China as far back as 1000 BC. Porcelain manufacturing required good quality clay materials like kaolin, high-temperature control skills, and hand skills to form proper shapes. These factors were readily available in the region.[32]

Once specifically purposed products are made, the techniques are inherited by the next generation. Rough manufacturing skills evolve from generation to generation, and it eventually becomes *craftsmanship*. This evolution is the common characteristics behind every form of craftsmanship. Craftsmanship is an instinct existing in human minds.

Craftsmanship is the humans' persistent mind and effort to make refined and unique artifacts. The culmination of true craftsmanship is through the *combination of science and art*. There are fundamental characteristics behind every kind of craftsmanship.

2.7.1 Fundamentals of Craftsmanship

- *Mental Sharpness:* Humans can augment their work outputs by intentionally focusing and exerting effort. It is a cultural and mind-driven force. When we have a strong will and desire, we focus and seek different ideas, and we also source different materials and processes. This effort comes from mental sharpness, which becomes the culture's meme and social mind to ignite every craftsmanship.
- *Deep Knowledge:* Without a proper understanding of the product technology and manufacturing process, we cannot improve things. Having the fundamental knowledge of a product, we also need to perform deep studying, learning, and researching. Behind every successful product there are excellent people who have in-depth knowledge and ideas of the product. The know-how is transferred from generation to generation, and the next generation adds more knowledge to the one previously known.
- *Inventive and Creative Mind:* Just copying things cannot make a product with true craftsmanship. We have to improve the design, materials, and processes with a creative mind. This inventive and creative mind is the driving force to make the product better, perfect, and unique.
- *Training and Practices:* Craftsmanship is a skill, and excellent skills are taught by one person to another through apprenticeship. The learned skills should go through numerous practices to make them a part of a person's own body and mind. *Good* practice makes perfect.
- *Building the Uniqueness:* Once specific skills are mastered, people like to seek their soul products. When people embed their character and personality into the product, the products become unique in design, appearance, and taste, truly appealing to the customers.
- *Quality Everywhere:* Inside every successful product there is visible, tangible, and hidden value. The most significant common denominator is the quality value. The craftsman builds quality everywhere and in every process: from material sourcing to designing, processing, packaging, and shipping. Then this craftsmanship becomes the face value people are willing to pay.
- *Continual Improvement:* Nothing is perfect, but we can endeavor to make it continually better. This continuous endeavor is a journey of evolutionary improvement. The journey of craftsmanship continues systematically as a loop; Plan-Do-Check-Action. There is no end.

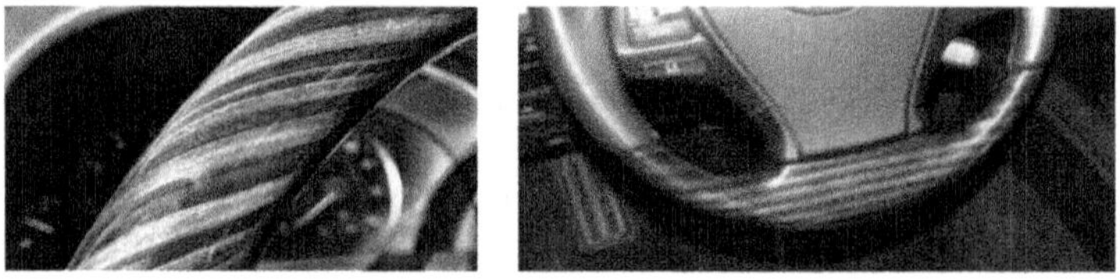

FIGURE 2.18
Craftsmanship in modern manufacturing. (Example: Wood processing.)

2.7.2 Craftsmanship in Modern Manufacturing

The traditional meaning of craftsmanship in the dictionary is "The quality of design and work shown in something made *by hand*."[33] However, with the advent of modern technologies, we no longer rely only on human hands.

As technology spreads, opportunities for *craftsmanship* grow in unlikely places in the mass production environment. Craftsmanship evolves as workmanship in the modern manufacturing process. The underlying spirit and mind of craftsmanship remains the same. The workmanship fully utilizes modern technologies, equipment, and tools.

> *Wood processing technology is one of the oldest technologies. However, it still evolves depending on designs and applications. In an automotive application of wood trim, T-Company is proud of its unique craftsmanship on its wood trim—the wood grain runs horizontally in alternating layers of dark and light wood veneers—see Figure 2.18.*[34] *The special wood processing technology was developed by the supplier. This process takes 67 manufacturing steps over the course of 38 days.*[35] *It produces high natural attractive wood feeling and appearance by modern craftsmanship.*
>
> *Another example of modern craftsmanship, i.e., workmanship, is the superb body accuracy of a vehicle. In the past, it was common practice for workers in the final inspection area to use a rubber mallet to adjust the fitness between the door and body. Nowadays, this practice is extinct, as there are better body fit designs, better parts, and statistical process control. During the manufacturing process, smart processing equipment, automatic body accuracy measurement machines, and advanced manufacturing technologies like laser cutting and robot welding are used. The outcome of this kind of workmanship produces a "zero touch body" such that it does not require any manual adjustment process in the assembly or the final inspection process.*

Modern manufacturing technologies have advanced significantly. However, the spirit of craftsmanship that desires to make things perfect remains the same in the modern manufacturing industry.

Notes

1. Juran, J. M. Juran on Quality by Design. The Free Press, A Division of Simon & Schuster, Inc., 1992.
2. Sinek, Simon. *Start with Why*, Portfolio / Penguin, 2009.
3. Gallo, Amy. "The Value of Keeping the Right Customers," *Harvard Business Review*, October 29, 2014.
4. Reichheid, Fred. "Prescription for Cutting Costs -Loyal Relationship," Bain & Company, 2001.
5. "Perceptual Mapping," https://en.wikipedia.org/wiki/Perceptual_mapping (last modified March 15, 2019).
6. "GE and Jack Welch Leadership—Case study approach," www.academia.edu/11670513/GE_and_Jack_Welch_Leadership_-_Case_study_approach, (accessed January 30, 2019). Welch was chairman and CEO of General Electric between 1981 and 2001. During his tenure at GE the company's value rose 4,000%.
7. Deming, W. Edwards. https://en.wikipedia.org/wiki/W._Edwards_Deming (last modified March 24, 2019).
8. Witzel, Morgen and Malcolm Warner, eds. *The Oxford Handbook of Management Theorists,* Oxford University Press, 2013. Chapter 11, pp. 196–218.
9. Barker, B. Thomas. *Engineering Quality By Design,* Marcel Dekker, Inc., 1990. The derivation of the quadratic loss function: from the statement of identity, $L(x) = L(x) \rightarrow L(x) = L[(x-m) + m]$, and by a Taylor series of expansion

   $$L(x) = L(m) + \frac{L'(m)}{1!}(x-m) + \frac{L''(m)}{2!}(x-m)^2 + \cdots$$

   Since the loss is zero at m (our aim value), and the first derivative is also zero at m, then, $L(x) = \frac{L''(m)}{2!}(x-m)^2 \rightarrow L(x) = k(x-m)^2$, where $k = \frac{L''(m)}{2!}$.
10. "Maslow's hierarchy of needs," https://en.wikipedia.org/wiki/Maslow%27s_hierarchy_of_needs (last modified December 1, 2018). https://commons.wikimedia.org/wiki/Commons:GNU_Free_Documentation_License_version_1.2 (last modified June 6, 2019).
11. Maslow, Abraham. "A Theory of Human Motivation," *Psychological Review*, Vol 50(4), Jul 1943, 370-396.
12. Baer, Drake. "How Changing One Habit Helped Quintuple Alcoa's Income," www.businessinsider.com/how-changing-one-habit-quintupled-alcoas-income-2014-4 (accessed January 12, 2019).
13. DUHIGG, Charles. *The Power of Habit*, Random House, 2012.
14. ThinkReliability, www.thinkreliability.com/case_studies/root-cause-analysis-challenger-explosion/(accessed August 10, 2018).
15. Feynman, Richard. www.feynman.com/science/the-challenger-disaster/ (accessed January 30, 2019). The rubber used to seal the solid rocket booster joints using O-rings failed to expand when the temperature was at or below 32 degrees F (0 degrees C). The temperature at the time of the Challenger liftoff was 32 degrees F.
16. Rogers Commission Report, https://en.wikipedia.org/wiki/Rogers_Commission_Report (last modified February 22, 2019).
17. Duhigg, Charles. *The Power of Habit*, Random House, 2012.

18. Harrington, James. *Poor Quality Cost*, ASQC Quality Press, 1987.
19. An engine cold test is done by moving the crank shaft with a motor while checking for any abnormal noises coming from internal moving parts.
20. Shulzhenko, Mary. November 1, 2017, "5 Major Reasons Why Unhappy Customers Don't Complain," www.providesupport.com/blog/5-reasons-unhappy-customers-dont-complain/ (accessed January 31, 2019).
21. Harrington, James. *Poor Quality Cost*, ASQC Quality Press, 1987.
22. Harrington, James. *Poor Quality Cost*, ASQC Quality Press, 1987.
23. Ebert, C. and R. Dumke. *Software Measurement*. Springer, Heidelberg, New York, 2007.
24. White House Office of Consumer Affairs, in a bulletin on consumers' attitudes, 1982.
25. Syed, Ferhan, February 25, 2009, What is 1-10-100 Rule?, https://totalquality-management.wordpress.com/2009/02/25/what-is-1-10-100-rule/ (accessed, January 12, 2019).
26. Nilsson, Joyce, *The Essential DEMING*, McGraw-Hill, 2013.
27. Godin, Seth. *Purple Cow: Transform Your Business by Being Remarkable*. Portfolio, 2003.
28. J.D. Power, IQS Execuive Report in Chinese Market, 2013.
29. Tellis, Gerard J. and Joseph Johnson, "The Value of Quality," *Marketing Science*, Vol. 26, No. 6, November–December 2007, pp. 758–773.
30. Shaked, Avner. and John Sutton, "Relaxing Price Competition Through Product Differentiation," *Review of Economic Studies*, 1982, XLIX, 3-13.
31. "Portugal's Shoemakers Bet on Quality" *The Wall Street Journal*, Mar. 9, 2014. As the nation's rescue lenders pushed wage and price cuts, its shoemakers did the opposite. They have defied IMF and EU cost-cutting recommendations by investing in quality, now they are thriving.
    - Shoe industry's quest for quality is export strategy in high-end market.
    - Focus on high quality by new machinery and new methods to treat leather.
    - The shoes are sold as much as $600 a pair
    - Employment has risen 10% with $2.6Billion revenue in 2013
32. Chinese ceramics, https://en.wikipedia.org/wiki/Chinese_ceramics (last modified May 10, 2019).
33. Craftsmanship, https://en.oxforddictionaries.com/definition/craftsmanship (accessed January 14, 2019).
34. Lexus LS Flagship Luxury Sedans Redesigned For 2013 Model Year, July 30, 2012, https://pressroom.lexus.com/releases/lexus+ls+flagship+luxury+sedans+redesigned+2013+model+year.htm (accessed January 14, 2019).
35. Tendo Woodwork Corporation, "Tendo is proud of this steering wheel with the wood grain running horizontally," www.fujitsu.com/downloads/EDG/binary/pdf/find/31-1e/1.pdf (accessed January 14, 2019).

3

Quality System, Management, and Engineering

It is difficult to build a house using a single material or a single component. It is difficult to operate a company using a single method. Likewise, it is difficult to build quality using only one single approach. To build a quality house for a company, a plant, and a product, an adequately balanced approach is required, by using *System, Management, and Engineering. These three elements are the three Quality Pillars.*

For a robust quality house, the three quality pillars must interact vertically and horizontally. Figure 3.1 represents their interrelationship in an intertwined way—they need to be connected, interrelated, and integrated.

This interrelationship is one of the reasons why one company's best quality system and management practices do not fit into another company. For example, in the 1990s, TPS (Toyota Production System) had a strong influence on Western industry with respect to lean engineering and manufacturing. TPS's JIT (Just In Time) has impressed many Western companies. However, the TPS system cannot simply be copied and pasted, because the system is deeply rooted in Toyota's culture, supply chain management, and engineering activities. For example, if we actually undertake the nearly zero inventory approach without the supporting infrastructure, we may jeopardize our manufacturing processes, leading to a possible shut down. Since many of Toyota's supply chains are within 100 miles of Toyota's assembly plant, most parts can be shipped from nearby suppliers' plants without much unexpected risk of traffic.[1] Even Toyota had to make some adjustments on the requirements of suppliers' stock management after taking a huge supply chain hit from the March 2011 earthquake and tsunami in Japan. Toyota stated that the plans would involve Toyota asking those suppliers to either spread production to multiple locations or hold extra inventory buffers with more dual sourcing of parts.[2]

The father of TPS, Taiichi Ohno, said "if you are going to do TPS you must do it all the way. You also need to change the way you think. You need to change how you look at things."[3]

For example, Hyundai motors did not merely copy Toyota's system. They further developed their own system, JIS (Just in Sequence). Most of the parts are pre-module assembled by Hyundai Tier-1 suppliers and provided in sequence. For example, the cockpit module is comprised of many components like the instrument panel, steering column, center fascia, glove box, etc.

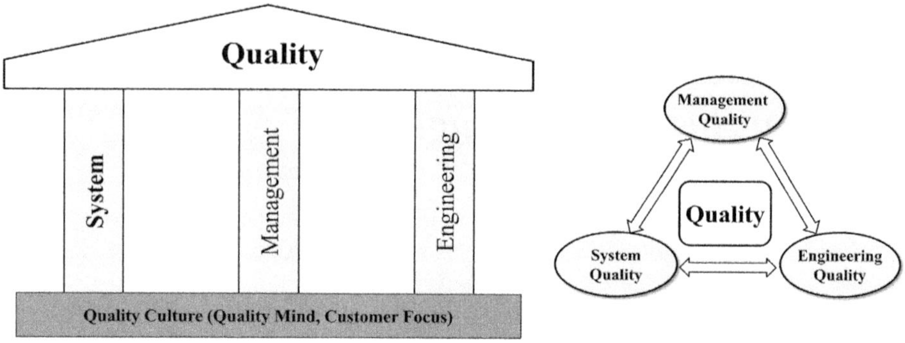

FIGURE 3.1
Quality pillars.

Furthermore, the front module is comprised of many parts like the hood latch, horn, power steering cooler tube, headlamp, etc. Their module ratio is more than 50% with an aggressive approach like a complete chassis module.[4]

> Southwest Airlines is a Dallas-based carrier that started a low-cost airline business in 1971. In the 1970s, air travel was expensive. Southwest came up with a low-cost and straightforward business model—two price categories: nights/weekends and daytime. The company grew substantially to 55,000 employees in 2017, and it has never lost money even in the aftermath of 9/11. The company recorded its 44th consecutive year of positive net income in 2017.
>
> Southwest has been a significant inspiration to other big carriers like United Airlines and Delta Air Lines. They copied the low-cost business model of Southwest Airlines, but they failed. Why did the same strategy not work for others? The strength of Southwest does not come from a single strategy or a system. Their competitive power combines the high level of employee and aircraft productivity.[5,6]

In 2000, when Robert Nardelli from GE came to Home Depot, he brought GE's famous Six Sigma with many GE managers. In the beginning, his program seemed to be very successful with profits improved thanks to the cost cuts made by decreasing the workforce. However, this was only a short-term effect. Customers were no longer able to find help on the store floor and began to go elsewhere. Home Depot began to lose loyal customers and lose market share.[7] *The "magic" copied from GE's Six Sigma did not work at Home Depot.*

From the case studies above, the lesson is clear that a "one size fits all" approach is not effective, and it can even be dangerous. One company's strategy, system, or management method cannot be merely copied by others. The company's business operation and outputs are *intertwined with people, culture, system, management, design, engineering, sales, and service.* So are the company's quality operations and outputs. We need to take a holistic approach to pursuing harmonized interactions between each element of the quality pillars: quality system, quality management, and quality engineering.

1. Quality System
2. Quality Management
3. Quality Engineering

3.1 Quality System

Operating a business and organization requires standardized systems and processes. The effectiveness of standards is observed everywhere in nature and human society. For instance, bees and ants build their territories using a specific system and method. There are more than 20,000 different types of bees on the earth;[8] however, the way they build honeycombs is the same—it is always in hexagonal shapes. Science and mathematics support that shape: a hexagon is the most efficient way to make walls with the minimum amount of wax and still maintain structural integrity.[9]

A quality system is a collection of principles and processes prescribing how the organization should perform their business operations in the most effective and efficient way to enhance customer satisfaction. An effective and efficient quality system is a necessary condition for effective quality management.

Alfred Chandler, an American business historian, said, *"structure follows strategy"*[10] in his research of four American conglomerates that dominated their industry from the 1920s onward. The system is the organization, and each organization should have a clear structure, role, and responsibility—*system follows structure.* The system comprises procedures and principles that management and employees should respect and support. The system links all the functions and processes of the organization. The systems should be designed in a way *to achieve the long-term strategy* of the company as a whole.

A system is like a black box—the inputs are converted into different forms of output through the system. For example, when we input materials and labor, the output becomes a final saleable product like a car or a phone. In the same way, quality values are created through the system—it is a magic box. Throughout the organization's systems and operations, quality is being made. *Management should have clear control of the system so that it is not a black box anymore. It should become a clear box.*

Figure 3.2 represents Deming's famous "Red Bead Experiment" in the 1980s. This experiment demonstrates the importance of the system, for which the management of a company is responsible. The box represents a system, the white beads are the good parts, and the red beads (here, shown in black) are the bad parts. The workers of the company are asked to pick up

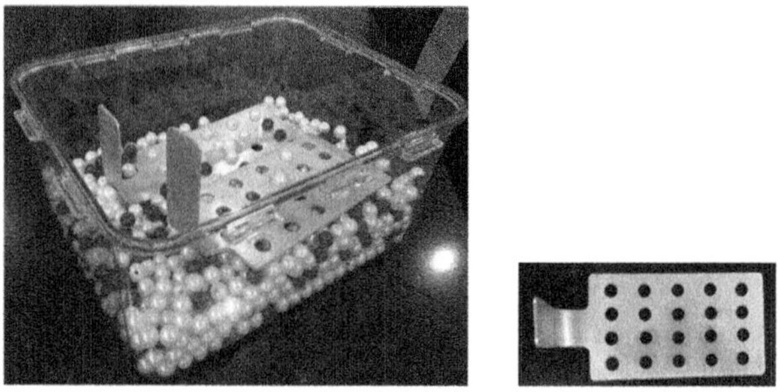

FIGURE 3.2
Deming's red bead experiment.

only the good parts (white beads) by using the paddle on the right side of Figure 3.2.[11] However, in this demonstration, picking up the only the white beads is very difficult. Why? Because the system also contains red beads—the problems, the bad parts. This demonstration clearly shows the relationship between the quality system and the expected output. That is why, from the very beginning, a good quality system is essential for the effective quality management of a company.

3.1.1 Structure of a Quality System

Business operations require effective and efficient management methods to provide the greatest amount of value to customers. *Management follows the system, but also, management is responsible for making system.* Therefore, quality systems should be aligned with the organization's goals, strategies, policies, management principles, and the collection of management processes at a corporate and department level, as in Figure 3.3.

3.1.2 Closed-Loop Quality Systems

When we design a quality system, we should have a vision that the system should be able to support quality business activities effectively and efficiently to enhance customer satisfaction. Quality systems should not be rigid, but must be robust and evolutionary—they should have a mechanism of continual improvement through a closed P-D-C-A loop.

The concept of the closed-loop system can be explained figuratively like an engineering system, as in Figure 3.4, where the output is automatically compared with the target value, and flexible adjustments of the system operate to improve the output quality. The system is not fixed, but evolves and morphs, adapting to changes in the market to accommodate the customers' needs. Therefore, this kind of system is *robust and evolutionary.*

FIGURE 3.3
The structure of quality system.

Since a quality system is linked to an organization's structure, quality policies, processes, and standards, those constituent parts must be changed and updated appropriately. For example, the quality team structure in each division, warranty policy, engineering standards, and countermeasure application processes need to be flexible and continually updated to serve the customers better.

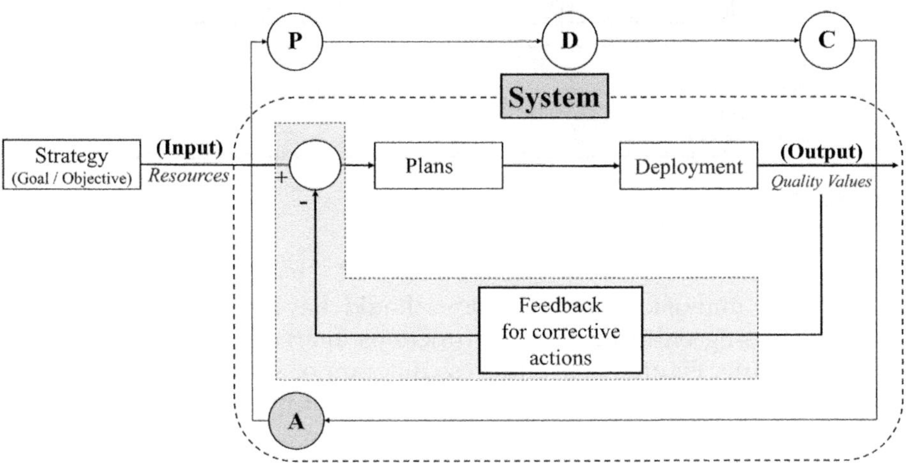

FIGURE 3.4
Closed-loop quality management system. (Example: Automatic control system of engineering.)

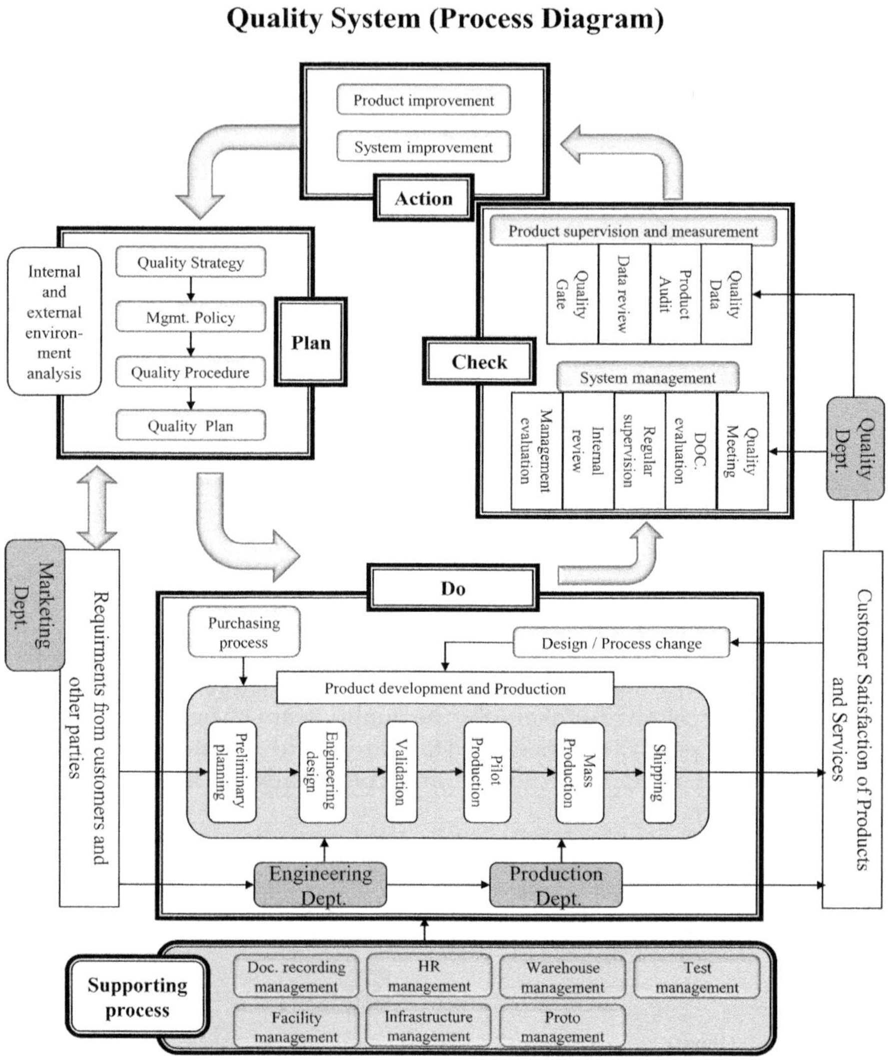

FIGURE 3.5
Process diagram of quality system.

A typical manufacturing company should have a closed-loop quality system focusing on three primary functions: marketing, engineering, and manufacturing. Figure 3.5 is a process diagram of such fundamental functions highlighting the engineering activities that most influence customer satisfaction.

First, the marketing department has to communicate the exact needs and expectations of the customers to the engineering department. The engineering department then has the responsibility of transforming the customer requirements into product design. Lastly, the manufacturing department has to design manufacturing processes, install equipment, and train people to produce good quality products to meet the design and customer requirements.

Many multinational companies have multiple markets and diverse customers with varying needs. As a starting point, they can apply their headquarters' standard systems. However, during the execution, they have to modify and adjust their quality system to adapt to the different markets and business environments—*this is an evolutionary quality system.*

> *Poor-quality incoming parts that do not meet the manufacturer's requirements generate all kinds of problems in the plant. Therefore, each manufacturer should build a robust quality system to actively manage their supplier quality depending on the level of suppliers. Supplier quality management systems employed in Japan, the United States, and China need to be different. In Japan, in most cases, the quality of suppliers' incoming parts can be trusted, and the data is stored in a system. In the United States, datasheets accompanying parts from suppliers are to be checked by the receiving plant staff. However, in China, accompanied datasheets often do not match the real quality of shipped parts. Therefore, the supplier quality management system needs to be adjusted to require actual confirmation of incoming parts from local suppliers.*

Another example is the final quality inspection checklist on the manufacturing line. Many companies treat their quality checklists like the Bible, without changing or updating it for long periods. However, the market and the customers change, and the products of the company have variance in quality over time—either improving or degrading. Therefore, the items and viewpoints of checklists should be dynamically updated to reflect actual market data and feedback from customers.

A quality system must be integrated into the information management system so that all procedures, instructions, forms, and quality checklists can be conveniently accessible online. With a quality system integrated with the IT system, necessary data and reports can be updated and available in real time, being able to drive appropriate quality actions promptly.

3.1.3 ISO 9001

ISO 9001 is the most universally used Quality Management System (QMS) worldwide. In ISO 9001, a QMS is defined as "a way of defining how an organization can meet the requirements of its customers and other stakeholders affected by its work." The first ISO 9001 standard was launched in 1987. It immediately became popular globally. In 2013 alone, over one million certificates to the standard were issued across 187 countries, and many other companies and organizations have used the standard without seeking certification.[12] One of the primary reasons for such expansion is that many global companies

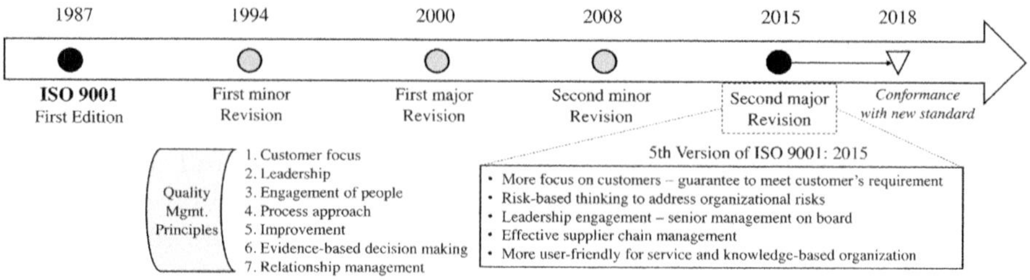

FIGURE 3.6
History of ISO 9001 and the updated content in ISO 9001: 2015.

require their suppliers to have the ISO 9001 certificate as a precondition to doing business with them.

Updated ISO 9001 was launched in 2015 (ISO 9001:2015), replacing the previous version (ISO 9001:2008). Figure 3.6 shows the history of ISO 9001 and the major updated contents in the fifth version in 2015. The standard itself is changing, updating, and evolving to adapt to the marketplace, different challenges in businesses, and increased globalization. The updated standard emphasizes a number of quality management principles including a *strong customer focus, the motivation and implication of top management, the process approach, and continual improvement*—refer to the seven quality management principles ISO 9001:2015 in Figure 3.6.[13]

Table 3.1 explains the general requirements by ISO 9001:2015. ISO 9001 does not specify what needs to be done to increase customer satisfaction but requires organizations to define objectives themselves and continually improve their processes in order to achieve them.[14] Using ISO 9001:2015 can help ensure that customers receive consistent, good quality products and services, which in turn brings about many business benefits.

ISO 9001 is certified by accreditation bodies operating under ISO/IEC 17011. Many countries have formed accreditation bodies to authorize the ISO 9001 certification.[15] Unfortunately, some companies seek ISO certification for *marketing purposes only; in such a case, it is a waste of resources and money.*

Figure 3.7 shows the reality in observing the standards, which was surveyed among 23 Japanese companies by a Japanese consulting firm, APPLIWAR, in 2012. *This was not an audit but a survey,* because honest answers could be gathered more easily in a survey rather than hidden as in the case of an audit. The result shows that the reality was not as good as we could expect. In the industry, most of the time, a company's standard is not clear, not efficient, or not updated to their actual working procedures; therefore, there is a significant gap between the formalized standards and the actual practices on the work floor.

The real benefit of ISO 9001 comes from creating a standardized system and practicing quality management according to the defined standard

TABLE 3.1

Requirements of ISO 9001:2015

Category	Requirements
Context of the organization Determine external and internal issues, the needs and expectations of interested parties, quality management system scope and its processes.	• Understanding the organization and its context • Understanding the needs and expectations of interested parties • Establishing the scope of the quality management system • Determining quality management system and its processes
Leadership Top management to demonstrate leadership and commitment, establish and communicate a quality policy, and ensure responsibilities and authorities are assigned, communicated and understood.	• Leadership and commitment to implement quality in the organization • Policy to deploy quality throughout the organization • Organizational roles, responsibilities, and authorities
Planning Organizational Quality Management System Planning to address organizational risks, opportunities, changes and quality objectives.	• Actions to address risks and opportunities • Quality objectives and planning to achieve them • Planning of actions when changes to the QMS are made
Support Provide resource needs, ensure employees are competent and aware, and include documented information to support QMS.	• Defining and providing resources required • Competence measurement and improvement • Awareness of QMS • Communication process and methods of QMS • Documented information
Operation Plan and control processes needed to meet the requirements for products and services (Design and development, external providers, production and service provision, release of products and services, nonconforming outputs)	• Operational planning and control • Requirements for products and services • Design and development of products and services • Control of externally provided processes, products, and services • Production and service provision • Release of products and services • Control of nonconforming outputs
Performance evaluation Monitor, measure, analyze, and evaluate QMS.	• Monitoring, measurement, analysis, and evaluation • Internal audit • Management review
Improvement Select opportunities for improvement, take action against nonconformities, implement corrective actions as necessary, and continually improve QMS.	• General • Nonconformity and corrective action • Continual improvement

Reality in Observation of Standard Process

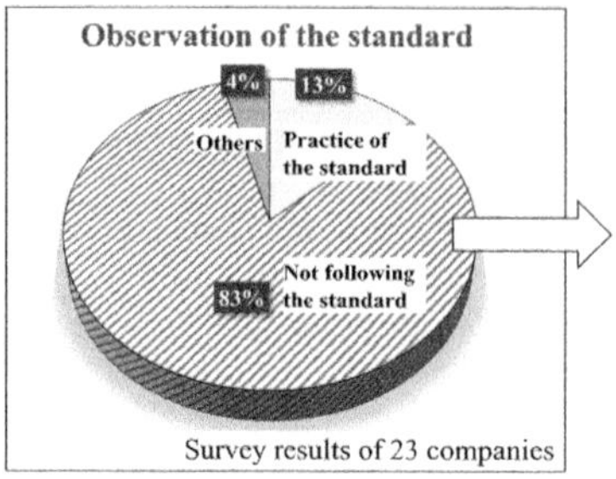

Why do you not follow the standard?

1. The standard not matching reality: 9 cases (7%)
2. The standard is too strict / not efficient: 57 cases (43%)
3. The standard is not clear: 68 cases (50%)
 (no trouble by not observing the standard)

FIGURE 3.7
Reality of observation of standard process.[16]

system and procedures so that it can deliver maximized values to the customers and the companies.

> *ISO 9001 is not the only answer for a quality management system. For instance, Toyota does not have an ISO 9001 certification. Instead, Toyota has TPS, which is benchmarked by many companies worldwide. It is crucial to build a quality management system that aligns with the company's business category, strategy, culture, and history more than following generic processes for marketing purpose only.*
>
> *During the ISO audit, some companies focus on treating auditors well, looking for positive comments. However, a compliance or renewal audit is a good opportunity to find out the weaknesses of the quality system through a third parties' eyes and expertise. Ask for keen observations and acute comments instead of sweet words.*

3.2 Quality Management

3.2.1 The Scope of Quality Management

The ultimate objective of quality and related activities is CS *(customer satisfaction)*. The purpose of CS is to increase the customer's satisfaction with a product or service, leading to an increase in the customer's loyalty and repurchase of a company's products or services.

Quality Management is "all scopes of quality management activities inside the organization to increase customer satisfaction." In this perspective, the scope of quality management activities should include QP, QC, QA, and QI as follows:

$$\text{QM (Quality Management)} = \text{QP} + \text{QC} + \text{QA} + \text{QI} \tag{3.1}$$

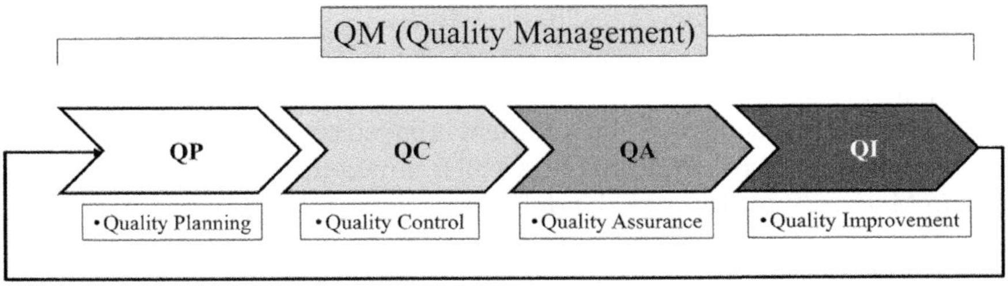

FIGURE 3.8
Deployment process of quality management.

where

QP = Quality Planning
QC = Quality Control
QA = Quality Assurance
QI = Quality Improvement

The deployment process is schematically explained in Figure 3.8, which is a step-by-step approach.

- QP is a systematic quality planning process through which quality activities are planned from product development to the sales and service stages.
- QC is quality control activities and methods to ship only qualified products to customers. QC is primitive management focusing on inspection, collection of data, and statistical process control.
- QA is quality assurance where quality is assured through systems and processes so that only qualified products are made from the beginning. QA is a more advanced stage of management to maintain competitiveness in the market.
- QI is quality improvement activities to continually improve quality in a system, process, product design, manufacturing, and sales and service. Through continual quality improvement, a manufacturer can increase competitiveness in the market, increasing customer satisfaction.

For the successful outputs of quality management, the management needs CS-minded skills and critical actions.

3.2.2 CS-Minded Quality Management Skills

Management follows a system, and the management of a company also designs the system. Without management, a system is a just a hard shell that cannot be adapted, evolved, or morphed. That is why an excellent management structure and highly skilled managers are essential for the high performance of a company.

The first step of successful management is based on multi-angle approaches. Jo Owen explained the multi-aspects MQ (management quotient) in his book *How to Manage?*[17] In the book, he defined:

$$MQ = IQ + EQ + PQ \tag{3.2}$$

Here IQ (Intelligence Quotient) represents rational management skills, EQ (Emotional Quotient) represents people skills, and PQ (Political Quotient) represents an organization's structural approach skills. Each management skill needs to be balanced and emphasized depending on the situation in the workplace.

Quality management is not a separate territory from standard business management. However, quality management needs to emphasize specific aspects and skills for *CS (customer satisfaction)*. In this perspective, MQ (Management Quotient) for CS needs to be emphasized as in Figure 3.9, which shows the quality management skills in a relationship with other management quotient elements.

In other words, each management quotient can be redirected into the aspects of customer satisfaction as follows:

$$\text{QM (Quality Management) Skills} = (\text{IQ} \cap \text{EQ} \cap \text{PQ}) \text{ for CS} \tag{3.3}$$

3.2.3 Critical Actions of Quality Management

Management is an action, focusing on deploying a company's strategies and systems throughout all functional teams based on the quality deployment methodology discussed in Chapter 2. The previously mentioned CS-minded quality management skills need be demonstrated through critical actions for effective quality management. The right-side diagram of Figure 3.9 shows how each management skill can lead to make critical quality management actions.

The critical actions for QM (Quality Management) can be formulated as follows:

$$\begin{aligned}\text{QM Actions} = {} & \text{Quality Communication} + \textit{Genba} \text{ Leadership} \\ & + \text{Quality Action Deployment}\end{aligned} \tag{3.4}$$

3.2.3.1 Quality Communication

Communication is a key part of intelligent management, which starts by sharing information with relevant individuals and parties including employees,

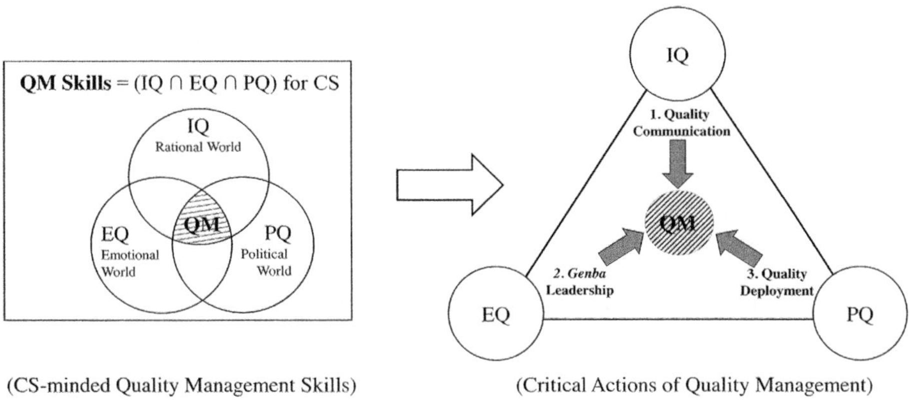

FIGURE 3.9
Quality management skills and critical actions.

suppliers, and customers. Quality communication begins with listening to the customers.

> *S-Company requires all their managers to regularly visit the customer call center to listen to the customers' voices directly. They share the headphones with the call center employees to understand what customers are demanding in the market. Some customers are angry and yelling for their quality problems to be resolved, and some make very excellent suggestions on how product design and quality can be improved. Hearing the customers' voices is an invaluable process that every company should adopt to increase the efficiency of quality management.*

Quality information needs to spread throughout the organization. Quality strategy and targets to satisfy the customers in the market need to be shared, discussed, and deployed.

There are several practical methods for formal quality communication: daily or weekly quality meetings, quality newsletters, the company's intranet, and mobile communication. However, informal communication with employees, colleagues, and team members are more effective. *Intelligent* quality leaders always inspire people with the live knowledge of the company and the competitors concerning products and quality data. The purpose of quality communication is not just to share the information, but also to drive forward the necessary actions to improve quality both in-house and in the market.

> *T-Company use a tray liner and computer screen saver for an informal way of quality communication, which contains company quality strategy, targets, weekly quality indices, and new model's launch situation in the market. The information also contains questions and surveys to understand employees' overall knowledge and interest in the company's quality activities. They also regularly give out awards for the best answers or ideas.*

New product development requires focused communication and coordinated actions between each function: design studio, R&D engineering, accounting, suppliers, manufacturing, and sales and service. During development, difficulties in meeting the project targets may occur, or unexpected quality issues may arise, potentially jeopardizing the product's development. In such a situation, some project members would like to wait and see until they are proven guilty for the problem. As a result, problem-solving may take an unusually long time, and the project launch could be delayed.

Good communication means not only sharing the quality information but also good coordination and motivation of people *in an intelligent way* in solving the issues as they arise.

3.2.3.2 **Genba** ***Leadership***

Some managers are very good at presentation, documentation, theory, and logic in the operation of a company. However, sometimes smart managers fail to understand the real situation on the ground floor. They receive nice reports from their team members and focus on making excellent documents. However, the reality is that the plant is overloaded with many repairs, too much WIP (work in process) stock, and high defect ratios in the incoming parts from their suppliers. Why do they have such a huge mismatch between the paper report and the *Genba*? Because such leaders rarely visit *"Genba," resulting in a lack of understanding of the reality on the ground floor.*

Genba is the actual workplace where the company values are created: the plant floor, design center, field test, the market, etc. *Genba-oriented quality management* requires emotional interactions with people in the actual workplace. Management should listen to the voices from the workplace and work with the people in the real situations.

> *The R&D project team of N-Company was proud of launching a new model with three-point engine mounts instead of four, saving costs. The role of the engine mounts not only holds the engine in place to the vehicle's frame or subframe but also absorb engine noise and vibration.*
>
> *Customers' complaints regarding the high-idle vibration on the steering wheel were abnormally high in the market as a result of this change. The R&D engineers and managers did not understand because the design and testing passed all their internal specifications. A quality manager arranged a service network visit for their R&D people to check the customers who complained about their vehicle in Genba directly. After having checked several customers' complaints and their vehicles, they were able to understand that their internal development standards did not meet customers' expectation levels in the cold start conditions that happen in the mornings.*

Knowledge has power, but Genba has bigger power. Quality management leadership should be demonstrated in *Genba* with subsequent actions following suit, *which brings emotional attachment to the working level of people.* When the leaders know little about *Genba*, what they plan and direct are not well-accepted by the people in *Genba*.

People truly listen when their leader speaks with the knowledge of *Genba*. People are motivated when their boss gives them practical guidance.

3.2.3.3 Quality Action Deployment

Knowledge is power, and so is information. However, information that does not flow has no power. 'Constructal Theory' developed by Adrian Bejan in 1996 states that "for a flow system to persist in time (to live) it must evolve freely such that it provides greater access to its currents."[18] When things flow, they create energy. Information regarding the company strategy, quality policy, and quality target of projects should flow to the bottom of the organization systematically. Quality action deployment should *convince* people in the organization to create values throughout the value stream.

An excellent example of quality information and action flow is demonstrated in Figure 3.10. Once a company's corporate quality policy and goals are established, specific quality pillars should be set up to support the quality goal (Step-1), where the quality pillars should represent primary quality activities to support a company's quality goal. The quality pillars need to be broken down to each division or function groups (Step-2). In this step, each function group should take ownership of its responsible areas and its portion of quality targets broken down from the quality pillars' target. Then

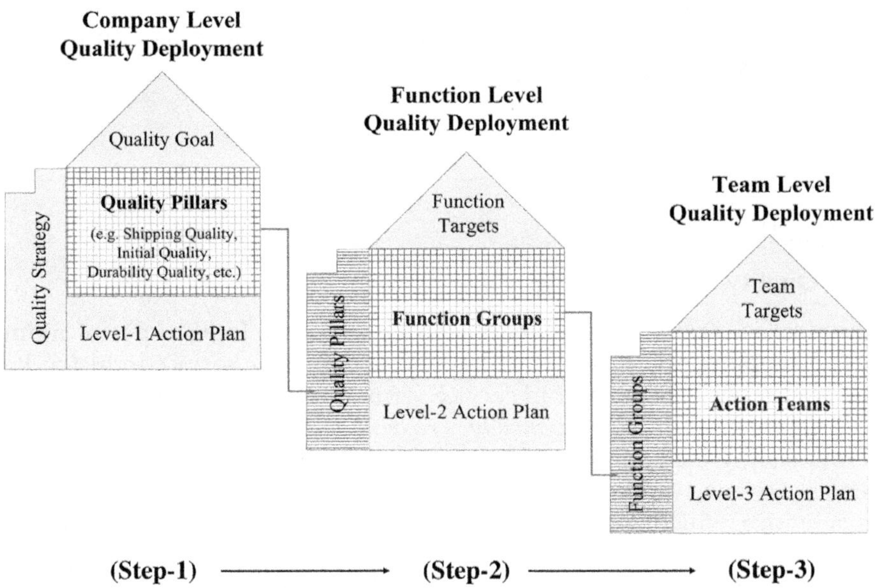

FIGURE 3.10
Quality action deployment flow.

each division's or department's quality target is cascaded into relevant action teams (Step-3). Such a flow is the target breakdown process, and the process should be done *through consensus meetings and discussions* with the relevant teams. The outcome of this process is that the targets are clearly understood and allocated to each action group in the organization. After this step, specific action plans to achieve the target need to be made by each relevant team.

R-Company was struggling to meet quality targets. The quality plan was made thoroughly and well, but results were not achieved; moreover, there was a significant gap between the target and the result. They decided to benchmark N-Company and found a crucial difference between N-Company and their company. In the case of N-Company, the corporate targets always reach the bottom of the organization, to the team levels. However, that was not the case for R-Company, where the bottom of the target actions did not align with the top targets. They decided to adopt N-Company's systematic method as described in Figure 3.10. Moreover, they improved upon the evaluation methods for their managers such that they no longer looked only at the final quality results but also evaluated the processes of how they systemically deploy their targets to the bottom of their team level.

During the quality action deployment, the role of a quality manager is to coordinate the actions *vertically and horizontally*; to make sure the targets are allocated to each action group and each action group *makes a specific action plan* to achieve the targets. Moreover, follow-up actions should be taken in a timely manner through Plan-Do-Check-Action *iteratively* to achieve the corporate or project target.

Later in Chapter 8, "Structured Quality Deployment," we will have an in-depth discussion on this topic.

3.3 Quality Engineering

3.3.1 Quality by Engineering

A product is created through engineering activities. In all types of industries, a company, whether in the manufacturing or the service industry, has products in physical form or in service form. *A company is known by its product. If the product is good, the company is recognized to be good as well.*

Previously, we emphasized the importance of a quality system and management for good quality outcomes. Products are created through engineering activities, which must be linked systematically with the quality system and management to enhance customer satisfaction. Effective quality systems and management help to build good design and engineering processes. However, the quality outcomes of engineering activities are not necessarily guaranteed only by the system and management.

TABLE 3.2

Activity Scope of Engineering vs. Quality Engineering

Engineering	Quality Engineering
Boundary Conditions	**Viewpoint**
• Science, technology, cost, functions, features, and delivery schedule	• Customer satisfaction
• Engineering optimization	
Requirements	**Requirements**
• Fundamental design and engineering skills	• Converting customers' viewpoints into engineering requirements
• Technical capabilities	
Focus Activities	**Focus Activities**
• Conformance to the regulations	• Prevention of past problems
• Features	• VOC (voice of the customer)
• Functions	• Initial quality
• Performance	• Appeal quality
• Attributions	• Warranty
• Specifications	• In-house manufacturing quality
• Design and product verification	- Straight shipping ratio
• Product validation	- Defects/unit
• Manufacturability	• Failure mode avoidance activity
• Salability	- Design review
• Reliability	- Failure mode and effects analysis
	- Fault tree analysis
	- Marketability testing

Engineering requires a wide range of activities from product planning and design to production. Engineering is not only a creation process of a new product but also requires finding solutions to the quality problems of the created products. *Engineering is a highly creative discipline that draws on the knowledge of product development, technical design, prototyping, and manufacturing techniques to convert ideas into a tangible value form in the market.*

Science and technology are the fundamental bases for continual improvement of engineering activities. Quality engineering helps engineering activities by using scientific and systematic engineering methods to build quality into product design and manufacturing.

Table 3.2 shows engineering vs. quality engineering activities. Engineering activities are deployed within limited boundary conditions like science, technology, cost, functions, features, and delivery schedule. On the other hand, the viewpoint of quality engineering goes beyond the boundary conditions. Its mission is to guarantee customer satisfaction with the products. Quality engineering is focused on how to achieve the quality objective through engineering activity.

3.3.2 Organization of Quality Engineering

The activity scope of engineering is extensive, ranging from product design to the manufacturing process. The structure of quality engineering should mirror activities with an engineering organization structure. There are three main categories in quality engineering activities: DQE (Design Quality Engineering), SQE (Supplier Quality Engineering), and MQE (Manufacturing Quality Engineering)–AS Quality Engineering is optional depending on the company's strategy. In addition to these functions, PQM (Project Quality Management) is an essential function that a new project goes through for a robust quality development process. These quality engineering activities can be located under the quality engineering department structure as represented in Figure 3.11.

3.3.2.1 DQE (Design Quality Engineering)

Product development starts with design. Design engineering defines a product's fundamental characteristics. During this process, it demands and fulfills all technical requirements to realize a product by design concept and specifications through drawings, prototype making, and verification and validation tests. The purpose of DQE is to make sure the engineering design is completed robustly to realize customers' expectations and the needs of the market.

DQE activity is mainly performed through the "Design Review" process. During the process, DQE members check design concerns based on past problem history, design guidelines of the IQS (Initial Quality Survey), Perceived Quality Manual, FMEA, FTA and so on. Once they find concerns they provide the feedback to designers before drawings are released at each stage. At the same time, DQE members participate in virtual build events with design engineering and manufacturing engineering to avoid any design concerns that might happen during the build of a product.

3.3.2.2 SQE (Supplier Quality Engineering)

Most industrial engineering activities require suppliers to design and manufacture parts. The quality aspects of the suppliers' product development and manufacturing process need to be confirmed by SQE. The viewpoint of SQE is to confirm that a supplier's engineering design and manufacturing processes are robust enough to develop the parts and then produce them to meet proper quality standards as defined by design.

There are specific tools used by the industry like APQP (Advanced Product Quality Planning) and PPAP (Production Parts Approval Process) defined by AIAG (Automotive Industry Action Group).[19] We will discuss these details later in Chapter 5.

> *SQE has a communication function to review a part's design concerns or any difficulties suppliers might encounter in realizing a product in the manufacturing process. This feedback activity from suppliers is called a "red-pen drawing meeting" in the case of H-Company—the origin of the word "red-pen" is from actual red-pen markings on paper drawings that used to be practiced a long time ago. SQE members meet with each supplier to listen to their concerns or suggestions on how to improve the quality, productivity, or cost of a relevant part. Once the concerns and ideas are confirmed, SQE members provide recommended actions, including design changes, to their respective R&D engineering teams.*

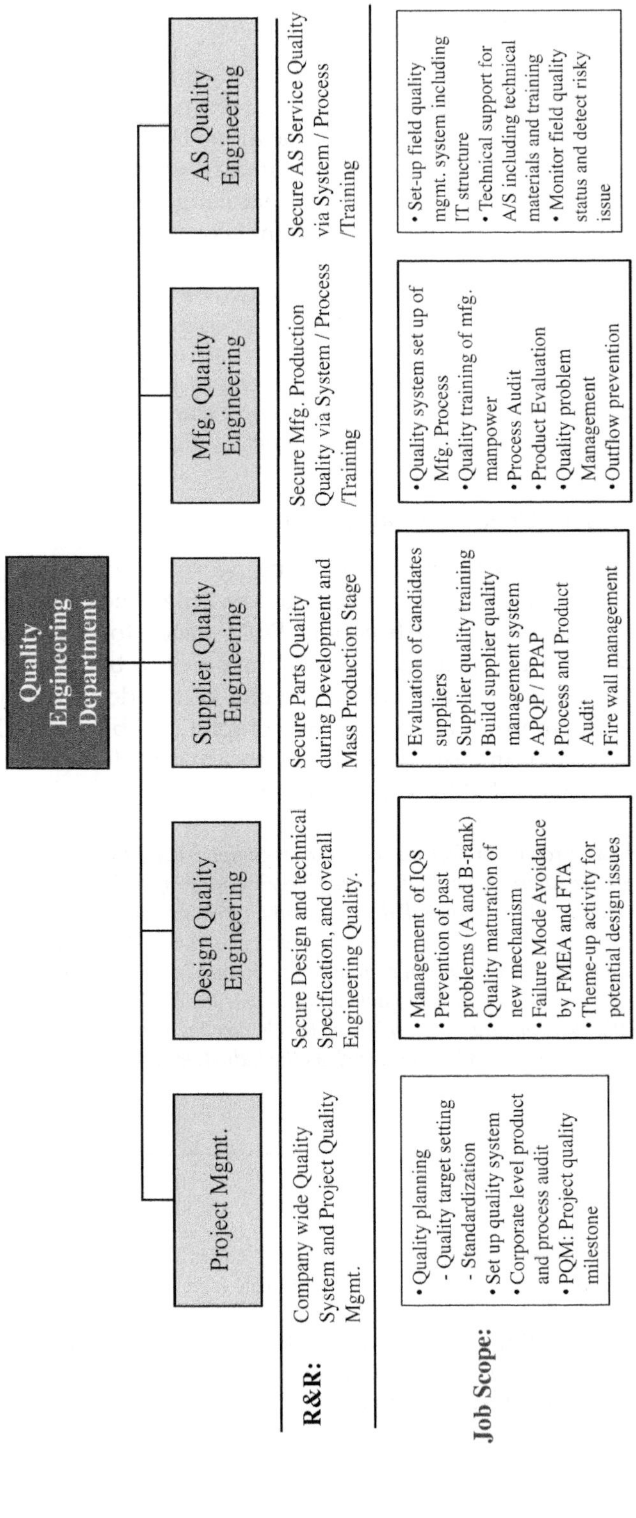

FIGURE 3.11
Organizational structure of quality engineering.

3.3.2.3 MQE (Manufacturing Quality Engineering)

In-house product manufacturing processes have as many substantial challenges as design engineering processes. Manufacturing's goal is not only to make parts aligned to drawings and specifications but also with the lowest variations through a production process. Standard work procedures, training for the workforce, and necessary manufacturing equipment need to be clearly defined for quality assurance of the final completed product.

> *Z-Company has many small volume plants in different cities in China. The MQE works with the manufacturing department to make sure the right manufacturing process and quality assurance procedures are in place. For example, PFMEA (Process Failure Mode and Effects Analysis) and SPC (Statistical Process Control) are the necessary tools to design a robust process and control manufacturing processes as part of quality assurance activities.*

3.3.2.4 PQM (Project Quality Management)

Each new product is made through a new engineering project. Each project has a timeline and milestones. Typically, quality milestones in each critical stage are managed by the PQM function. PQM needs to make sure project teams follow the right procedures and outcomes to be able to achieve the quality targets. For example, before the end of the development stage, PQM needs to confirm product designs are completed robustly so that suppliers and in-house manufacturing departments are confident going into the industrialization. This is PQM's responsibility.

> *R-Company has a robust project milstone process emphasizing PQM's role. The PQM manager regularly attends major engineering meetings. The PQM manager needs to understand the overall development as well as quality maturity of a new product. As a project's quality milestones approach, the PQM manager organizes quality meetings with relevant engineering function teams to check the quality progress of a project. At official quality milestones the PQM manager makes a decision with the company's top leaders as to whether the project can move forward to the next stage of development, such as the industrialization process.*

Notes

1. "Toyota and Its Component Suppliers—Case Study," https://elsmar.com/pdf_files/Toyota and Component Suppliers.pdf (accessed January 17, 2019).
2. SCDigest Editorial Staff, Global Supply Chain News, "Toyota Taking Massive Effort to Reduce Its Supply Chain Risk in Japan," Says it will Reduce Its Time to Recovery from Major Disruption from Six Months to Two Weeks," March 7, 2012 www.scdigest.com/ontarget/12-03-07-2.php?cid=5576 (accessed January 17, 2019).

3. Ohno, Taiichi. *Work Place Management*, McGraw Hill, 2013.
4. Design the Future, Gain the Edge, http://blog.mobis.co.kr/220648823494 (accessed January 17, 2019).
5. Sinek, Simon. *Start with Why*, Portfolio, 2009.
6. Verasai, Anna. *The HR Digest*, September 29, 2018, "The Southwest Airlines HR Success Story You'll Want To Read Today," www.thehrdigest.com/the-southwest-airlines-hr-success-story-youll-want-to-read-today/ (accessed January 17, 2019).
7. Aguayo, Rafael. *Dr. Deming: The American Who Taught the Japanese About Quality*, First Carol Publishing Edition, 1990.
8. Nicholls, Henry. The truth about bees, June 15, 2015, www.bbc.com/earth/story/20150615-the-truth-about-bees (accessed January 11, 2019).
9. George, Stacy. September 1, 2017, "Why Are Honeycomb Cells Hexagonal?," www.sciencefriday.com/educational-resources/why-do-bees-build-hexagonal-honeycomb-cells/ (accessed January 11, 2019).
10. Chandler, Alfred. *The Visible Hand*, The Belknap Press of Harvard University Press, 1977. Chandler studied the business cases of Du Pont, General Motors, Standard Oil of New Jersey, and Sears Roebuck to investigate how these company grew through diversification strategy by adopting the revolutionary multi-division form.
11. Red Bead Experiment, https://deming.org/explore/red-bead-experiment (accessed January 17, 2019).
12. www.iso.org/iso-9001-quality-management.html (accessed February 26, 2019).
13. Quality Management Principles, www.iso.org/files/live/sites/isoorg/files/archive/pdf/en/pub100080.pdf (accessed February 26, 2019)
14. ISO 9001 Quality management, www.iso.org/iso-9001-quality-management.html (accessed January 12, 2019).
15. ISO 9000, https://en.wikipedia.org/wiki/ISO_9000 (last modified June 21, 2019).
16. Surveyed in 2012 by a Japanese consulting company, APPLIWAR.
17. Owen, Jo. *How to Manage*, Pearson Education Limited, 3rd Edition, 2012.
18. Bejan, Adrian. The Physics of Life, St. Martin's Press, New York, 2016.
19. Automotive Industry Action Group, https://en.wikipedia.org/wiki/Automotive_Industry_Action_Group. The Automotive Industry Action Group (AIAG) is a not-for-profit association founded in 1982 and based in Southfield, Michigan. It was originally created to develop recommendations and a framework for the improvement of quality in the North American automotive industry (last modified May 30, 2019).

4

Principle-Based Quality Management

In the previous chapter, we discussed that there are three major quality pillars: Quality System, Quality Management, and Quality Engineering. In this chapter, we would like to delve into the pillar of Quality Management.

In every society, there are specific forms of customs and rules that people respect and observe. Charles Duhigg emphasized the importance of habit in his book, *The Power of Habit*;[1] according to the Duke University research, more than 40% of the actions people performed each day were not decisions, but habits. In the military, the most important part of training is to build the habits of soldiers to react properly under fire.[2]

This habit principle also applies to the business world. People are inclined to follow specific principles and procedures rather than making case-by-case decisions through elaborate thinking. This principle-based decision-making process becomes a *business habit*. Good business habits help to solve our daily issues effectively.

Quality managers confront issues and problems during daily quality management. For example, they may have to make a tough call to stop shipping products due to potential quality concerns. When we are under high pressure, we tend to make rushed judgments by relying on our intuition, which is based on experience and knowledge. However, this kind of judgment does not necessarily lead to the right decision.

Daniel Kahneman, a Nobel laureate, mentioned two different ways of thinking and decision making in his book, *Thinking, Fast and Slow*. Thinking fast leads to a more intuitive decision based on the *automatic operation* called "System 1," and thinking slow leads to a more rational decision based on the *controlled operation* called "System 2." System 1 is an *automatic system* informed by knowledge and experience, and System 2 is an *effortful system* used to consciously think through an issue in a systematic way.[3]

In a typical situation of quality management, only a minimal amount of time and information are available for decision making. A right decision at the right time is critical for customer protection while minimizing losses to the company. Judgment by System 1 is fast, but it can lead to a biased decision. System 2 is accurate, but it is slow. To prepare for demanding situations in quality management, we need to train ourselves for effective decision making by *using both System 1 and System 2 together*. The issue is how we can connect these two systems.

Theories and principles in science are essentially universal laws supported by long-term observations of phenomena. In law, principles are like the constitution, and the detailed processes are like the rules and regulations. Management should have certain principles and guidelines for efficient business and quality management—such principles should be the ones *proved by long-term successful practices. Such principles* can then effectively bridge the two different thinking systems, System 1 and System 2, *to be used in the actual workplaces.*

When an organization has a specific policy, principles, and rules, employees can prioritize issues promptly and make decisions effectively. Most of all, *principle-based management helps the organization members cooperate better because they think in a similar fashion.* Such a judgment and decision-making process is rational and still prompt and reliable. As a result, we become less likely to be biased by emotions under stress, pressure, or anger.

On the other hand, case-by-case management without preset principles is not efficient because paths for decision making can be varied and are difficult to predict by the organization members so that *they cannot prepare the proper steps in a cooperative way* necessary for the right decision.

The organization cannot merely react based on a decision from the top bosses. In quality management, many types of problems require the right decision at the right time based on a situation on the work floor rather than an elaborate discussion and judgment in the office or meeting rooms.

> *R-Company have a special customer care team under their CRM (Customer Relationship Management) team. The team's mission is to monitor a cyber and hotline for 24 hours concerning any critical quality problems or customers' complaints in emergency, such as PL (product liability) issue.*
>
> *They need to react quickly to such emergency situations without approval from management. They have a separate budget and relevant resources to treat customers in trouble. Such speedy action can protect customers and the interests of a company. The members are highly trained with preset principles, scientific problem-solving skills, extensive technical knowledge, and many years of customer treatment experience in field.*

In this chapter, we would like to review two exemplary management principles. One is set by W. Edwards Deming at the beginning of the 1980s when American manufacturing was experiencing a quality crisis. His principles were based on his observations of the American and Western styles of management. The other principle is Joseph Juran's 10 Steps to Quality Improvement. Juran emphasized the role of management and workforce through a holistic view of quality (moving from "Little q" to "Big Q")—a larger scope of quality than conformance to standards only.[4]

In this chapter, the author would like to share his quality management experience, summarized as *Lim's 5 Principles for Quality Management.* These five principles are based on the observations and practices of engineering and production activities at global manufacturing companies. They reflect the lessons from situations and problems *commonly rooted* in multinational companies, including American, Japanese, European, Korean, and Chinese companies. These principles also address the common issues of the global quality crises that have occurred in the past decades.

1. Deming's 14 Points for Management
2. Juran's 10 Steps to Quality Improvement
3. Lim's 5 Principles for Quality Management

4.1 Deming's 14 Points for Management

After the Second World War, American industry was booming. America was the only country whose industry remained intact during the war. There was no competition against products "Made in U.S.A." However, as a result, complacency began to grow among the American industry leaders, who did not have much motivation to improve their products and quality competitiveness. During the oil shock in the 1970s, Japanese products rushed onto American soil, and products such as cars imported from Japan impressed the American consumers.

There was an American behind such impressive quality improvement of the Japanese products. This man was W. Edwards Deming, who taught quality to Japanese scholars and industry leaders. Deming was an engineer and statistician who went to Japan in 1947 with the Supreme Command for the Allied Powers (SCAP) to help prepare for the 1951 Japanese Census. Deming taught statistical quality control techniques to the Japanese scientists and engineers through formal and informal meetings. Japanese industry began to transform through these new quality management methods and techniques, such that the image of "Made in Japan" began to change from a synonym for junk to the last word for quality.[5,6]

In his old age, upon his return to America, Deming began to work on helping American industries. He firmly believed that the failure of American products was due to the failure of the American and Western styles of management. In 1982, he published a book, *Out of Crisis*, to inspire the American industry leaders to transform their management philosophy. During many meetings and seminars in America, Deming made acute comments on the Western corporate culture and management style. Deming developed his "14 Points" as the fundamental management principles to provide guidelines for a shift in management thinking in the Western world to ensure quality, productivity, and success.[7] He consulted and helped the American industries by teaching 14 Points for management transformation and statistical quality control.[8] Through Deming's efforts, American manufacturing began to be globally competitive again.

Unfortunately, some bad practices of the Western management style spread globally, including to the East. Nowadays, many companies worldwide are utilizing a short-term, profit-focused management—many of the recent recalls at a global scale evidence this.

The time is now ripe to revisit Deming's intentions and relearn his 14 management principles.[9] Through the detailed discussion that follows, the author would like to convince industry leaders that Deming's principles remain relevant and valuable even today for companies aiming to be competitive in the long term. *For each of the 14 Principles, the author will share both good and poor examples practiced in today's global companies of the West and the East.*

1. Create constancy of purpose for improving products and services
2. Adopt the new philosophy
3. Cease dependence on inspection to achieve quality
4. End the practice of awarding business on price tag alone
5. Constantly improve the system of production and service (forever)
6. Institute training on the job
7. Institute leadership
8. Drive out fear
9. Break down barriers between departments
10. Eliminate slogans, exhortations, and targets for the workforce
11. Eliminate numerical quotas. Substitute leadership
12. Remove barriers to pride of workmanship
13. Institute a vigorous program of education and self-improvement
14. Take action to accomplish the transformation

4.1.1 Create Constancy of Purpose for Improving Products and Services

A company's long-term goal should be to stay in business and create jobs rather than to merely make money. Short-term, profit-driven management is not sustainable. Long-term growth should be the objective and can be achieved by being in a competitive position, which comes from constant improvement of current products, innovative products, new materials, new production technology, and new ways of marketing.

Innovation requires research and education on a foundation of quality and productivity, as Deming states: "Innovation, the foundation of the future, cannot thrive unless the top management has declared unshakable commitment to quality and productivity."[10]

4.1.2 Adopt the New Philosophy

Only a new philosophy can change minds to achieve transformation of management. Quality must become a new religion. Without the change of a company's philosophy, the culture does not change. A short-term, profit-driven philosophy needs to change for higher competitiveness based on the quality of workmanship and services in the market. According to Deming: "Short-term profits are not a reliable indicator of the performance of management. Anybody can pay dividends by deferring maintenance, cutting out research, or acquiring another company."[11]

4.1.3 Cease Dependence on Inspection to Achieve Quality

Quality should be right in the first place. Any subsequent activities cannot guarantee quality assurance, and they are a waste of time and resources. To assure quality in production, quality design and reliable processes are the fundamentals that must be implemented in the first place to remove any inspection in later stages.

Any critical safety process needs quality assurance methods by using smart manufacturing methods rather than inspections. For example, QA (Quality Assurance) machines or methods can be used: a foolproof method (*Poka-yoke* in Japanese terminology) or an automatic check method by a machine can be built into the manufacturing processes.

For a process that still requires human action, like torque checking, we can do sampling and auditing checks with the collection of numerical data instead of "Good" or "NG" judgments only.

Figure 4.1 illustrates the evolution of quality inspection procedures in the manufacturing process. Model A and Model B are primitive methods using

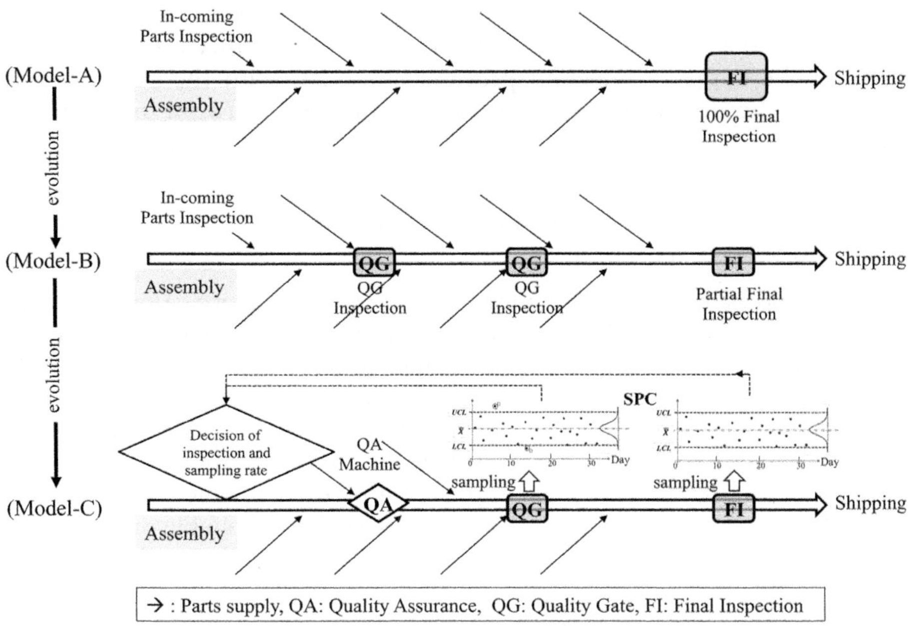

FIGURE 4.1
Quality inspection vs. quality assurance process.

100% or partial inspection. Model C is the SPC (statistical process control) QA method using sampling data, from which we can make a control chart to observe and control the process. From the SPC we can study the sources of the variances and control the causes. When the process becomes continually stable with less variance, we can decrease the sampling rate or eliminate the inspection process itself.

4.1.4 End the Practice of Awarding Business on Price Tag Alone

Part price is a critical parameter for the construction cost of a product. For a typical industrial product, the cost of parts or materials makes up 60% to 80% of the manufacturing cost of a product. A number on a price tag is not ambiguous; a purchasing department can judge the competitiveness between each contender without dispute.

However, a lower price tag does not guarantee a lower total cost when we *include* the cost of product development, manufacturing, and after-sales service. For instance, slow reaction or less capable engineering supports from a supplier during new product development can jeopardize a good quality product's launch time. Moreover, a high defects ratio once in the market increases warranty expenses for a manufacturer and decreases customers' loyalty to the brand. Such hidden risks can *increase* the total cost, offsetting the initial cost savings of purchasing a cheaper part.

Figure 4.2 describes a supplier selection process and decision model. The old model (Model A) represents a process focused on cost. Here, the quality department's opinion on the supplier's quality capability is only a minor factor in the supplier selection process. However, with the new process (Model B), the quality department's opinion becomes crucial in "sourcing for quality."

4.1.5 Constantly Improve the System of Production and Service (Forever)

There should be no end in the improvement cycle. Japanese call it *"Kaizen,"* or *continual improvement*. The purpose of quality is not to make things within specifications only. Quality is to reduce the variances of processes and products, which can lead to higher customer satisfaction.

Management must lead the way for improvement towards perfection. Improvement does not mean hard work. Principle-based management, quality engineering, and a statistical approach are critical elements in the continual improvement of a system, a robust design, and a reliable process.

4.1.6 Institute Training on the Job

People can change and improve through education and training. Without a qualified workforce, a company cannot prosper. Nobody knows his or

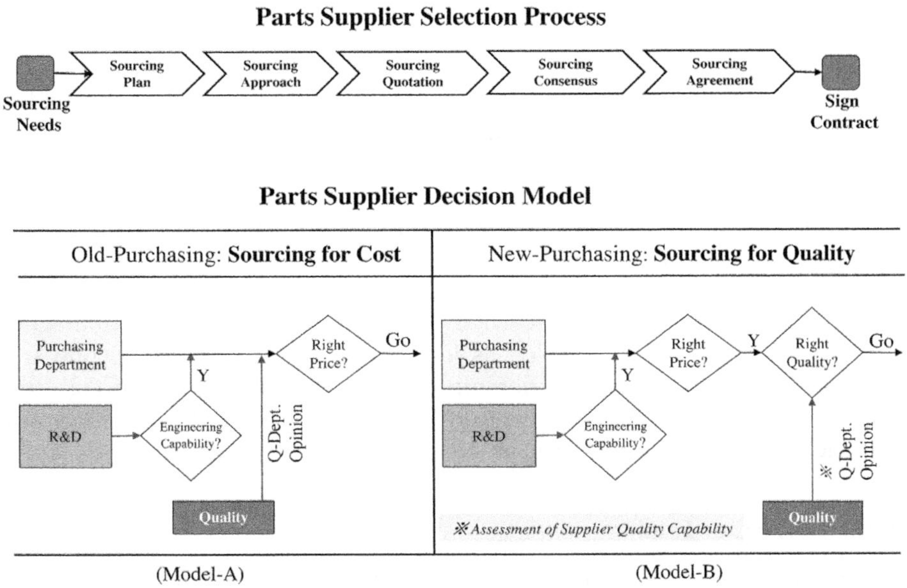

FIGURE 4.2
Supplier selection process. (Example: G-Company.)

her job perfectly from the beginning. Even highly educated or experienced people need to learn a new system and working procedures of a new company when they are hired. They need to learn from someone *qualified*.

First, a company should have a system manual and working procedures in place. A company should make sure *only qualified persons* train people. In 1911, Taylor, in his book, *The Principles of Scientific Management*, emphasized that a company should train employees only by a master who has the best performance record of an actual job.[12]

A proper training system requires an appropriate workforce evaluation system that can effectively judge their skill levels.

> *N-Company uses an I, L, and U grading system: "I" represents a person who cannot do the job by oneself without a coach; "L" represents a person who can do the job by oneself; and "U" represents a person who can do the job by oneself and teach other people.*

Despite advanced computer and IT technology, coaching and mentoring people on the job is the most effective method to motivate people to learn and self-improve. A proper teaching approach should be combined with the principle, *"Learning by Doing."* Like learning to drive a car, people can only truly learn something while they are doing the actual job.

4.1.7 Institute Leadership

Deming stressed, "The aim of management should be to help people, machines, and gadgets to do a better job."[13] He emphasized that good leadership comes from good knowledge and experience in a job. However, in the Western style of management, there is frequent job rotation of managers, putting them in a weak position since they never know the reality on the floor.

True leadership requires understanding a real situation and barriers on the floor such as inefficient process layout, poor product design, frequent breakdowns of machines, and problems with incoming materials.

A company should be concerned with the job rotation of managers. Managers should be recognized as long-term investments for a company. Let them demonstrate real leadership on a working floor long enough that they can earn respect from the ground-level employees. Such experience and knowledge can make them true assets to a company in the future.

4.1.8 Drive Out Fear

In the Western management style, there are fears that result from short-term, profit-driven, erratic decisions, such as company takeovers and laying off people by merging departments. For example, we frequently hear about certain news events such as "Ford Motor Co. has announced a planned reorganization that will cut jobs, … to squeeze costs and improve efficiencies."[14]

In many cases, people on the floor are reluctant to reveal issues mainly because they know that such matters will not be resolved no matter what they say. Instead, most likely, the finder of a problem is doomed to take over responsibility to resolve it without management support. There is a fear of telling the truth.

In the case of excellent quality-leading companies, there is an award program that encourages employees to voluntarily find issues, solve problems, and improve systems and processes.

H-Company has a Hawk-eye and a Kaizen-Circle program, through which employees are encouraged to find any abnormal issues, voluntarily improve their workplaces with respect to cost, productivity, technology, and quality. In this program, relevant managers also participate in coaching and supporting the employees' activities.

4.1.9 Break Down Barriers between Departments

Deming said, "People in the research, design, sales, and production must work as a team, to solve problems in the production line and market."[15]

Today, many global companies have KPI (key performance indicator) systems to evaluate divisional or departmental performance. The KPI method aims to make sure the organization understands its responsibilities and strives to achieve its goals.

However, it can also be a blind management tool if management believes that simply providing targets can bring in good results for the company. Unfortunately, the reality is, in most cases, there are conflicts in KPIs between departments. For instance, with respect to warranty expenses, the quality department has a responsibility to control the warranty budget. Therefore, they need to prevent any over usage of the warranty spending by service networks. However, the service department has KPIs concerning increasing customer service satisfaction so that they want to be generous in treating customers with good warranty schemes including goodwill.

There could be many such cases in the KPI way of management. Therefore, we should systematically improve this kind of concern: at the beginning of a department's goal setting, potential conflicting targets between departments need to be resolved to a common goal.

4.1.10 Eliminate Slogans, Exhortations, and Targets for the Workforce

Deming's intention in making this statement is clear—simply creating slogan or targets without providing the means to achieve them has no meaning. They are just vague words—no message of substance is delivered to the employees. It is the management's responsibility to improve a system and process to make the slogan or the target achievable.

For example, let us say there are some defects of an electronic part, where the cover is frequently assembled in the wrong direction. Instead of pushing the employees to achieve "Defect Zero" via such a slogan, management should initiate a design or process change by *Poka-yoke,* i.e., a foolproof design. That is the way the plant can *actually achieve zero defects* rather than relying on a "Defect Zero" slogan.

4.1.11 Eliminate Numerical Quotas. Substitute Leadership

A strict management method by numerical quotas per manager or employee causes side effects. Many organizations have a habit of evaluating the performance of its members by measurable numbers. Numbers seems to be a very convenient tool, and the numerical outcome is easy to judge. However, the numbers alone do not tell the whole story. Without a qualitative performance indicator, pushing for a numerical outcome alone could result in problems. For example, employees may blindly pursue the number targets without respecting teamwork or the right procedure.

> *In 2018, a shocking banking cheating scandal broke out. Wells Fargo, one of the big-four banks of the United States, imposed sales quotas on its employees. An investigation by the Los Angeles Times revealed that the company put intense pressures on its bank managers and individual bankers to produce sales to hit extremely aggressive and even mathematically impossible quotas. Beginning in 2011, the branches of Wells Fargo created millions of fraudulent savings and*

> *checking accounts on behalf of Wells Fargo clients without their consent.* "The Federal Reserve on Friday imposed unusually harsh penalties on Wells Fargo, punishing it for years of misconduct and barring it from future growth until the bank fixes its problems."[16] "Various regulatory bodies, ... fined the company a combined $185 million as a result of the illegal activity, and the company faces additional civil and criminal suits."[17]

Such practices and their potential side effects can occur in many types of business. For example, typically in the automobile industry, many manufacturing plants are given a target number of parts or vehicles to produce daily or weekly. In that case, the given number is the hard target that a plant manager must meet. In such a circumstance, any delay or stopping of production to improve quality is recognized as an enemy to the productivity and production quota. Such a kind of culture destroys the quality culture.

Any numerical quotas beyond the organization's capabilities have no meaning. A company's leadership should focus on improving its capabilities rather than pushing a quantitative number for the organization members to achieve. We should also respect qualitative approaches like teamwork, process improvement, robust design changes, and training, which are difficult to measure quantitatively.

4.1.12 Remove Barriers to Pride of Workmanship

When managers push for unreasonable targets, the employees become frustrated because they know there are many barriers hindering them from doing a proper quality job. When they focus on the target numbers, they give up their pride in workmanship to meet management's requirements.

Without respecting the workmanship of the employees, it is difficult to expect excellent quality outcomes. The employees will design and produce mediocre products or services. This relates to the eleventh point. Employees must feel their skills and expertise are respected and fully utilized, and then they will execute their works professionally. As a result, their performance can excel and outperform the competitors.

4.1.13 Institute a Vigorous Program of Education and Self-Improvement

Deming emphasized the importance of education, saying "What an organization requires is not just good people; it needs people that are improving with education."[18] This principle directly connects to the sixth point. In the age of the knowledge economy, knowledge is the power that advances a company's competitive position for the future.

Improving knowledge is everyone's responsibility: a company's responsibility to its employees and employees' responsibility to themselves for self-improvement. Employees are motivated to enhance their expertise and knowledge when a company systematically supports their career development or education program. Furthermore, they are committed to going the distance when a company values them with investments in their professional careers.

> *Many global competitive companies like Samsung and Honda have an excellent education reimbursement program. When their employees take outside courses or programs, whether they are job-related programs or not, the company pays most of the cost. For example, in the case of Samsung, the company subsidizes expenses for outside computer or language courses, or even sports lessons. They do not treat their employees as a cost but as a value for the future. The capability of the employees becomes the capability of the company.*

4.1.14 Take Action to Accomplish the Transformation

Transformation is the work of everyone. Any knowledge without action has no value. However, action does not necessarily mean a significant revolutionary change. A small correction of existing problems can ignite a big transformation of an organization.

For example, short communication sessions in the morning can make a friendly and efficient work environment for a company. Cleaning workplaces and completing a progressed job before taking a break can build good habits for the cultural transformation of a company. Regularly following up daily or weekly activities by comparison to the original plan can initiate evolutionary changes for an organization. These are small steps for change; however, starting little things today brings about significant changes in the future.

Any organization must build new habits and new cultures to take a big step towards quality transformation in the future.

4.2 Juran's 10 Steps to Quality Improvement

In 1924, Juran worked as an electrical engineer in the Western Electric Company in the same place where Deming was working. In the early 1950s Juran consulted for Japanese scientists, engineers, and company leaders. He cited the loss of market share, failure of products, and waste as the results of poor quality planning and management. Juran advocated for important quality tools like the *vital few and trivial many,* called *Pareto principles,* and the *quality trilogy* to improve quality from *quality planning* to *quality control* and *quality improvement,* called "Juran's trilogy."[19]

Juran proposed the concept of the "Big Q," where multifunctional teams should participate in quality improvement activities. This approach later became the core concept of TQC (Total Quality Control) in Japan. Juran emphasized people, teamwork, and responsibility of management to achieve quality. He said the biggest roadblock to process improvement is not tools, but the creation of a cultural shift among the people involved. Unlike Deming, he focused on top-down management and technical methods rather than worker pride and satisfaction.[20]

Juran's 10 Steps to Quality Improvement originate from a different viewpoint than Deming's points. Deming's 14 points address fundamental philosophies and policies for a transformation of the management into quality. Juran's 10 steps[21] are rooted in "Juran's trilogy" and address the practical approaches and necessary steps that management should take for quality improvement.

These 10 steps are self-evident and clear, and today many successful companies adopt them into their new product development and problem-solving systems. The 10 steps are introduced here with practical methods to deploy them in an organization.

4.2.1 Build Awareness of Need and Opportunity for Improvement

This is the situation analysis step of the current business operation. We collect data and information from internal customers and external customers in the market. After the analysis, we will be able to see opportunities for improvement.

4.2.2 Set Goals for Improvement

Goal setting is the starting point of quality planning. Goals should be in line with the company's policies and strategies. They should be challenging, achievable, and measurable.

4.2.3 Organize to Reach Your Goals

Detailed methods and means are necessary to achieve goals. An organization should be set up and appropriate resources should be allocated with the exact scheduling of required activities.

4.2.4 Provide Training

This step is essentially equivalent to Deming's sixth point. Without education or training, improvement cannot be achieved. Specific and systematic programs should be designed, and only qualified people should train the workforce.

4.2.5 Carry Out Projects to Solve Problems

Before we endeavor to improve quality on a large scale, we should solve known problems. We need to prioritize problems to solve by taking Juran's approach, focusing on a vital few problems.

4.2.6 Report Progress

Analysis of the progress made is an essential process to achieve a goal. During the process, gaps and risks should be identified and reported to management. With management support, corrective actions may need to be taken to bring the project to the planned path.

4.2.7 Give Recognition

Recognition is a means of management. Without recognition by management, quality improvement activities lose momentum, making it difficult to go forward against difficulties. Top management should recognize a team's activities and support them.

4.2.8 Communicate Result

We learn from good results or mistakes—both are lessons. Lessons are assets to an organization. There should be formal communication meetings and databases set up to share lessons learned within an organization conveniently. Mr. Honda said, "Even if you make mistakes, the lessons will be utilized when you make a leap in the future."[22]

4.2.9 Keep Score

This is relevant to points 2 and 6—targeted goals should be measurable so that we can check them and score the progress. An objective judgment and scoring provide leverage to control and improve planned activities.

4.2.10 Maintain Momentum by Making Annual Improvement Part of the Regular Process of the Company

Companies have yearly business planning, which allocates resources to achieve the next year's goal. Quality improvement should be a big part of an annual business plan—this step of action leads back to point 1 for continual improvement.

4.3 Lim's 5 Principles for Quality Management

Deming's 14 points and Juran's 10 steps caused a fresh impact on American industry in the 1980s and are still valid in today's business environment.[23] Despite the positive transformations made in the Western industry sector, some bad practices of the West also spread globally.

During the past decade, quality lost its attraction among manufacturing industry leaders. Recently, we have witnessed many severe quality issues around the globe in the automobile industry—for example, in 2009, Toyota's unintended acceleration recall; in 2014, GM's ignition switch recall; in 2014, Takata's airbag recall; in 2015, Volkswagen's emission scandal; in 2016, Mitsubishi's fuel mileage cheating scandal; and in 2018, BMW's engine fire recalls.

These pandemic quality issues were not just coincidental; there were fundamental reasons behind them. First, the entire industry became complacent with its current level of quality; the focus was no longer on continual improvement of quality management, but on the continual increase of profit management. Second, most industrial manufacturing became global with the same product designs, the same supply chains, and the same production processes. For example, GM has more than 150 manufacturing plants in 30 countries[24], and Toyota has global operations in 28 countries outside of Japan.

Third, the product development cycle was becoming shorter, with sophisticated technologies. However, quality management methods did not keep up with the engineering speed, complex technologies, and business challenges.

In addition to these historical factors, we expect intensive challenges in the future: innovative technology designs, software-focused products, and IT and machine-integrated manufacturing industry.

It is time for us to revisit our quality management principles and methods to prepare for the future. For this process, we need to use our knowledge, intelligence, technology, and education. Our quality management methods must be put on an advanced scientific foundation to meet the needs of the future.

Even though there are some differences in the system and cultures of global companies, there are common issues behind the problems we have witnessed in the past decade. *The value of quality for future society will be much more demanding.* Unless the industries addresses common issues now, we will experience larger-scale problems in the future.

Corporate business cultures and habits need to be transformed, which is possible only by changing management habits. In this section, the author would like to introduce the fundamental quality management principles to correct common mistakes and misconducts of industry in the past. The lessons and remedies need to be systematically reflected in a business constitution, preparing for the upcoming manufacturing and software industries in the future. Summarized here are *Lim's 5 Principles for Quality Management,* which is pictographically represented in Figure 4.3.

1. Start from the customer
2. Cultivate a quality culture
3. A closed-loop management
4. Focused and detailed management
5. Transparent management

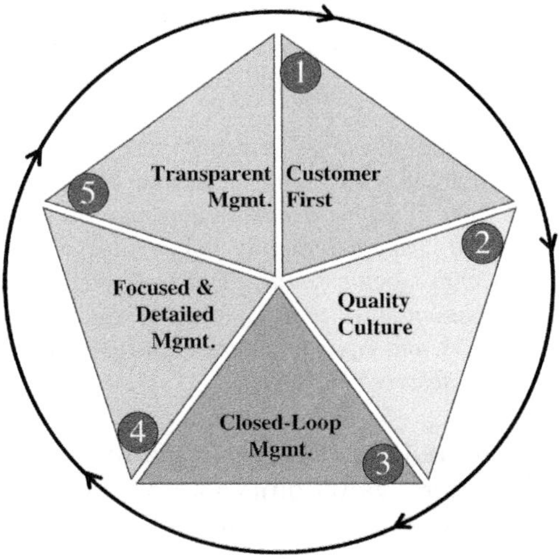

FIGURE 4.3
Lim's 5 management principles.

4.3.1 Start from the Customer

Without customers, there is no business. Customers' complaints about the company's quality issues can lead to an opportunity to solve potential problems. Any system, management, or engineering activities separated from interaction with customers are destined to create a product made in isolation that cannot survive in the market. Management that is attentive to the customer's needs should be the ultimate starting point for any businesses.

4.3.1.1 Is the Customer Number One?

All business leaders say that customer satisfaction is their number one priority. Unfortunately, history tells us that what they say and what they do are different. Many historical product failures in the market were mainly attributable to the company's poor management. The management in such cases did not start from the customer; instead, they started from a mere hope or short-term interest. Here is a historical example of the consequences that arise when a product does not meet what customers want to purchase:

> *Nissan was founded in 1934 and has the longest history of an automobile manufacturing company in Japan (before Toyota in 1937). Nissan made substantial industrial supports for the Japanese army during the Second World War by supplying military trucks.[25] They prospered in sales volume and technology. Apparently Nissan's manufacturing and engineering prowess were world-class. Many elites from the top Japanese universities wanted to enter Nissan*

rather than Toyota. The company was driven by the Japanese demonic need for excellence. The executive ranks were dominated by graduates of the prestigious University of Tokyo, which typically churns out Japan's top government bureaucrats.[26]

Nissan's top decision makers stayed in a cozy office building environment in Tokyo's expensive district. They were removed from their development center or manufacturing plants. They became more bureaucratic in decision making. Despite Nissan Maxima's superb durability and quality, sales became sluggish. The old-style design with lagging features and technologies did not appeal to potential customers. Nissan failed to make what the market wanted. The company's sales in the U.S.A. and Japan plummeted. In 1999 the company (with its $22 billion in debt) was absorbed by Renault.

The lesson is clear—technology or superb quality does not matter unless it is rooted in the customers' viewpoints. Good quality unperceived by the customer has no meaning. Only the quality that customers or the market appreciate counts as meaningful to the company and society.

4.3.1.2 What Is Real Globalization and Localization?

In the 21st century, global marketing, operations, and manufacturing became the focus of many international companies. Building manufacturing plants or R&D centers in local markets is an excellent strategy to get close to the market and the customers. Unfortunately, companies' real driving force for localization is not for customers, but for cost and business benefits, which does not bring about real benefits to the customer and company.

M-Company had an automotive manufacturing plant in Brazil. The design specification of most parts was carried over from Europe. Typically in Europe and America, a horn button on the steering wheel is used as a panic button in an emergency. However, in some countries, it is more than a panic button. For example, Brazilians also use the horn to greet people in the street. However, the design specification of the horn spring force was too stiff to use the horn in a way to greet people in the street. The Brazilian market and plant operation of M-Company had to fight for three years to get a unique design specification for the Brazilian market.

A physical presence close to the market does not mean a business operation automatically becomes customer-oriented. *A real change should occur in a soft manner: system, process, and culture should listen to the voice of the customer (VOC).* It also needs to train employees the principles and procedures to access customers actively. Real globalization and localization should foresee customers' minds quickly and accurately so that a company can make decisions about better product designs, production, and marketing to increase the customer's satisfaction by reflecting local customers' needs.

4.3.2 Cultivate a Quality Culture

Farmers do not sow seeds in any soil. First, they cultivate soil with fertilizer to make it fertile and then they sow the seeds so that the valuable seeds can sprout, grow, and bear good fruit. In the same way, quality activities and systems without a supportive culture and mindset in the organization cannot survive or thrive. Quality starts from more attitude, principle, policy, philosophy, and methodology rather than technology. To enrich the quality culture and mindset of the organization, top management's example, leadership, and management methods are the first steps followed by continual education and training. Management practices and education inspire people and change the organization. Successful companies always have a rich quality culture and mindset starting from management.

4.3.2.1 Culture Cannot Be Copied

The foundation of quality improvement activity is to *develop* a "quality culture" or mindset within the organization and integrate it throughout the company. A quality-focused culture creates a healthy work environment and leads to satisfied customers.[27] *Culture makes quality. Culture cannot be copied but must be developed.*

Many Japanese quality systems have been copied by American companies. However, the output with respect to the quality of the products was not the same.

> Ford bought Mazda; Ford first took a 7% stake in Mazda during 1979 to learn its small car magic. By 1996, it owned 33.3%, enough for managerial control.[28] Ford had contracted a Japanese supplier, Mazda, to make a certain portion of their front-wheel-drive automatic transmissions, with the balance of production made at a U.S. plant in Batavia, Ohio. Both sites were making transmissions to the same set of blueprints, and the transmissions were being installed in American cars. Mazda's version, as warranty records showed, had substantially lower claims than the Batavia version. Ford investigated the phenomenon and found that Mazda's transmissions were made much more consistently than their own. Mazda was building the major components to the center specifications, i.e., using only 27% of the allowed design tolerance range, while Batavia was using 70%.[29]

Why did this happen? Was it due to a difference in manufacturing technology between Ford and Mazda? One may claim that Mazda's manufacturing engineering was more advanced than Ford's. However, a fundamental issue here is not about technology or engineering capability but a quality culture difference between Mazda and Ford. Ford's mentality was that as long as they were making things within the drawing specifications, it was sufficient. Mazda's thinking and attitude sought continual improvement for perfection to the nominal specifications, which reflected their craftsmanship culture.

4.3.2.2 *The Power of 5S Culture*

> *The American suppliers that the author visited in North America had substantial differences in their working environment compared to that of Japanese suppliers. It was quite common for American suppliers to have snacks and drinks in the assembly lines, while this was impossible to imagine for Japanese suppliers.*

5S that started in Japan in the 1970s is still the fundamental pillars for Japanese industries today. Company owners and managers clean their own desks, pick up garbage on the plant floors, and clean the machines.

5S is a workplace organization method that uses a list of five Japanese words: *seiri, seiton, seiso, seiketsu, and shitsuke.* These have been translated as *"Sort, Set in order, Shine, Standardize, and Sustain."* As for the details of 5S activity in the workplace, we will have an in-depth discussion in Chapter 8.

Hiroyuki Hirano, the developer of 5S, said "A company that cannot successfully implement the 5S's cannot expect to effectively integrate JIT (Just in Time), re-engineering, or any other large-scale change. Good workplaces start with the 5S culture."[30]

Quality output comes first from the minds and attitudes of the leaders, managers, and employees of the company, even before technical skills. *Quality starts with culture. A culture cannot be copied but must be built.* Building a quality culture starts from top management by *"doing leadership"* rather than *"talking leadership."* Furthermore, continual education and training to put the customer first and perfection is also necessary to cultivate a quality culture from the bottom up.

4.3.3 Closed-Loop Management

How many activities we initiate does not matter if we cannot close the loop. Once something has been undertaken, we should close it. Such closed-loop management will eventually deliver proper value to customers. Some managers request many activities from their team members, but they rarely check their outputs in order to coach them. This kind of poor management practice demotivates employees. *Minimize the number of requests to the employees, and maximize the outputs by utilizing closed-loop management.*

4.3.3.1 *P-D-C-A Cycle*

In every matter, we always make a *plan* and *do* to implement the plan. In the real world, problems and gaps between the target and the result always arises. Humans cannot make a perfect plan to forecast all possible events in the future. Unexpected situations may occur systematically or randomly. When we do not *check* the reality in the middle of a project, we cannot notice any symptom of potential problems until it becomes too late to solve them.

If we can notice some concerns in an early stage, we can take corrective *actions* to minimize the impact. This type of corrective action is possible by

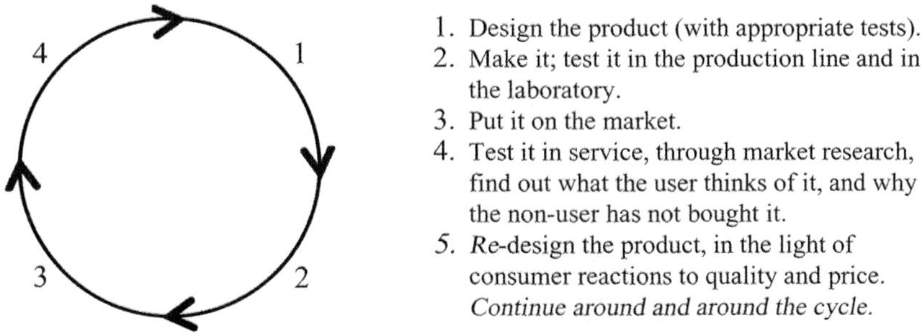

FIGURE 4.4
Original Deming Cycle (new product development).

closing the loop of our activities by using a concept like the P-D-C-A cycle, i.e., Plan-Do-Check-Act.

The concept of the P-D-C-A cycle was originally developed by Shewhart and Deming in the 1950s.

"Deming stressed the importance of constant interaction among design, production, sales, and research and that the four steps should be rotated constantly, with the quality of product and service as the aim." Deming's Shewhart cycle was modified slightly in 1951 as shown in Figure 4.4.[31]

The Deming Cycle has been evolved as P-D-C-A, a four-step problem-solving cycle, while he was interactively working with Japanese scholars. Figure 4.5 shows the continual cycle of the P-D-C-A and actions required at each stage: Plan (making an action plan), Do (implementing), Check (evaluating the results), and Act (improving or standardizing).

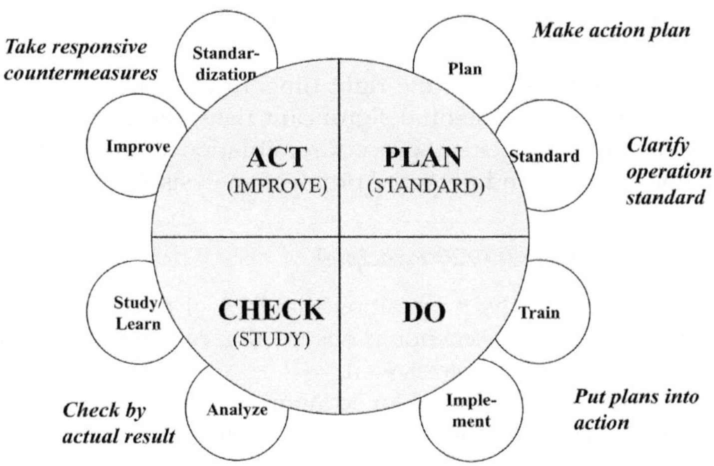

FIGURE 4.5
P-D-C-A diagram (four-step problem-solving).

4.3.3.2 Visual Management

How does management recognize the progress of resolving quality issues and problems occurring in one's workplace? Employees in workplaces often rely on computers, saying, "the problems are managed well, and the relevant files and data are on the computer." However, problem management *only by reliance on a computer* is not efficient or effective enough; people should turn on the computer, search for the files, display the files, and confirm the contents of the data or information. Sometimes there could be many poor management cases hidden in the computer until someone opens and checks them.

Closed-loop management should be supported by simple and clear confirmation methods like visual display. Visual management is the most convenient and efficient method to confirm the status of problems and issues. *Our visual cognition is the fastest and most potent method to receive and process information.*

> *Critical data, files, graphs, and parts should be visually displayed or located in a workplace where actual jobs are being done. There is no need to open the computer to search for records or go to somewhere to retrieve information. Critical information or parts are immediately available on the spot.*

4.3.4 Focused and Detailed Management

Only a few core causes exist for many sorts of problems. Some managers treat every problem equally and try to resolve it without prioritization. Prioritization is an essential management principle and skill. This process requires critical thinking, expertise, extensive knowledge, and experience. Good management focuses on core issues rather than many trivial matters. Through focused management, we can take a small hammer approach rather than a sledgehammer approach. Focused management provides the ability to make the right decisions at the right time. In utilizing focused management, we can better see potential significant risks and prepare in advance through a rational management procedure: situation analysis, problem analysis, decision analysis, and potential problem analysis.[32]

4.3.4.1 Pareto Principle (80/20 Principle)

The Pareto Principle is the proposition that 80% of problems occur due to 20% of causes. Such a phenomenon is observed in real life, and this has been proved in statistics and science as well.

In 1897, Vilfredo Pareto, an Italian economist, noticed that a small minority earned a substantial majority of the total income of a society. The income distribution showed almost the same pattern and ratio whatever the time period and also whichever country he examined: 20% of the people owns 80% of the wealth of society. This observation became the "Pareto Principle,"

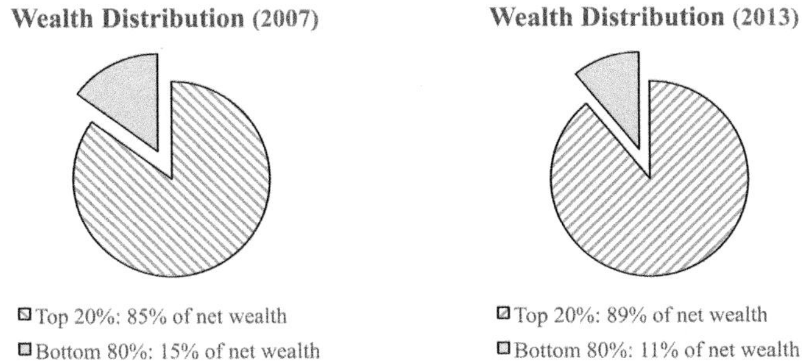

FIGURE 4.6
Pareto analysis of the U.S. wealth distribution.[34,35]

also known as the 80/20 Principle[33], which is a universal law in statistics, economics, sociology, quality, and so on.

The 80/20 Principle is also conspicuously observed in today's wealth distribution. Figure 4.6 illustrates the U.S. wealth distribution in 2007 and in 2013, which shows that the top 20% of the population owns more than 80% of the total wealth of society.

Some other examples that demonstrate the Pareto Principle are:

- 80% of a product's usage involves 20% of its features
- 80% of a town's traffic is on 20% of its roads
- 80% of company revenue comes from 20% of its products
- 80% of innovation comes from 20% of the people
- 80% of a company's sales are made by 20% of its sales staff
- 80% of progress comes from 20% of the effort
- 80% of errors are caused by 20% of the component
- 80% of a company's complaints come from 20% of its customers

Juran called this principle the *"Vital Few"* rule in his book *Quality Control Handbook*, published in 1951, and it was also summarized in his later, 1992 book, *Juran on Quality by Design*.[36] Juran emphasizes a management principle focusing on the vital few.

For example, during new product development and its launch, it is normal that many quality problems occur. It is virtually impossible to resolve all of the issues within the timeline of the project. Figure 4.7 explains a real case of how the Pareto Principle can be used for quality actions with priority. The figure shows the kinds of defects found at the final inspection line at a new model startup. Utilizing the Pareto concept, we can group similar characteristic quality problems with a priority focusing on the top 20% of the most frequently occurring issues.

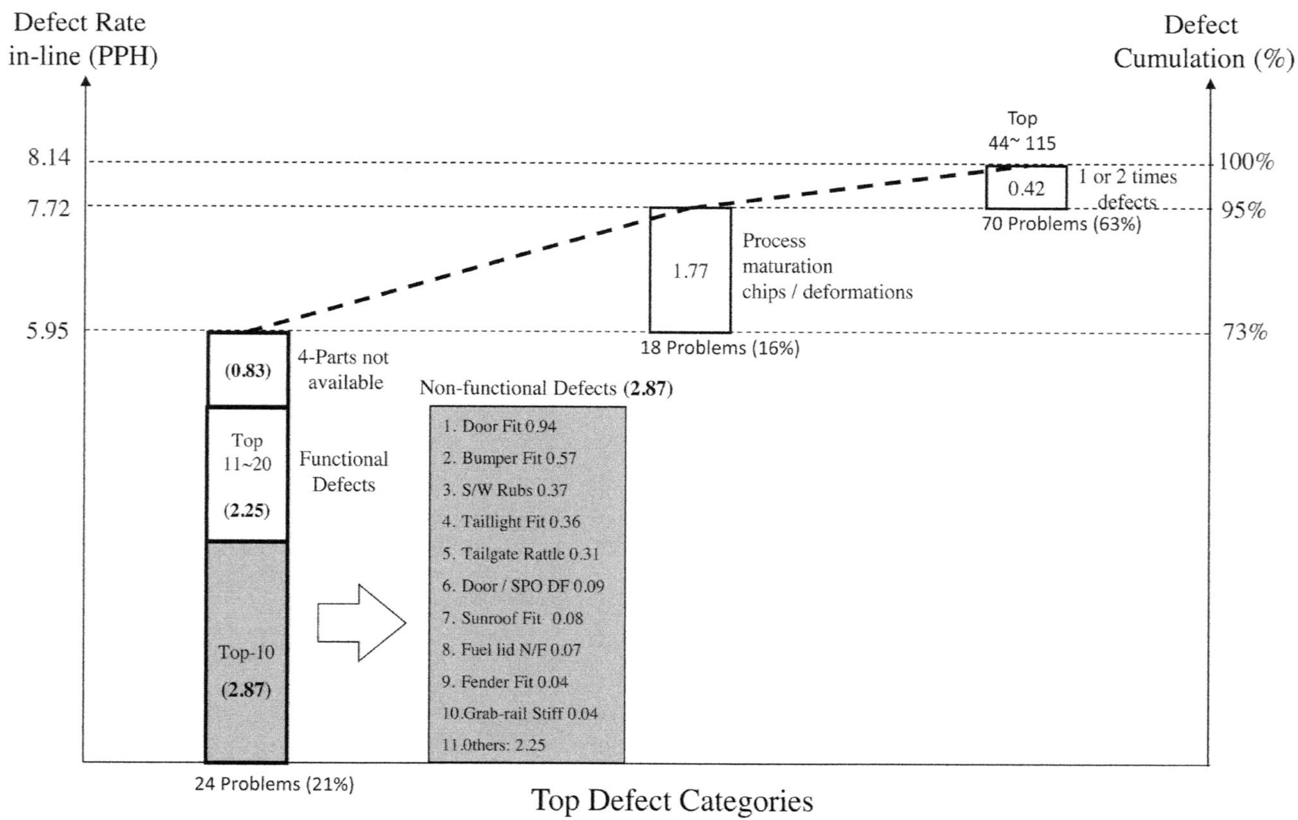

FIGURE 4.7
Analysis of startup problems by Pareto principle. (Example: SUV Model launch at H-Company.)

During a new model launch, an automotive company experiences many varieties of problems. These problems are difficult to resolve all at the same time. First, the quality department needs to analyze the situation with production teams. Figure 4.7 shows typical defects occurring at a production line and the breakdown of the data from the largest to the smallest in PPH (problems per hundred units). Then the situation becomes clearer as to which defects should be focused on. Figure 4.7 shows that the top problems are "fit and finish issues," "functional problems (e.g., radio not working or airbag warning light on)," and "parts availability issue due to suppliers' delivery problem." These three categories make up 73% of the total defects. Then the quality management team can work with production teams to determine how these top three categories can be resolved—these items require focused and detailed management.

4.3.4.2 Prioritization Ranking by Business Needs

One of the most useful methods of prioritization is to use a ranking system. The ranking of issues or problems can be done in several different ways based on types of business or industry. The simplest model of ranking system can be done by business needs as in Figure 4.8. This method can be used in most of business operations; from daily decision makings to milestone event management.

In this ranking method, the importance represents the level of potential severity in losses to customers or the company. The urgency represents the possible consequence of the timing effect of resolving issues to meet a project timeline, a particular event timing, testing schedule, or mass production timing.

First, we can concisely list up issues or problems in a spreadsheet or database like (b) of Figure 4.8. Then, the leader of the team review the issues *with the team members*, and assign priority ranking *through a consensus* by using the criteria (a). Each of A and B-rank items need focused management with the assignment of PIC (Person In Charge) and specific time line to complete.

After PIC has been assigned to each issue, *detailed management steps are followed*, requiring management's supports on higher ranking issues, such as a control plan and more allocation of resources, including more budget, and higher skilled workforces.

To solve the issues and problems effectively, we also need thorough scientific processes as much as good skills. Specific approaches and techniques for problem solving will be discussed in Chapter 6, Scientific Problem Management.

4.3.4.3 Prioritization Ranking by Risk Level

When we deal with technical and engineering problems, we can prioritize the problems or issues by potential risk levels to *customers*. When prioritizing the potential risks, we can use the RPN (Risk Priority Number) concept. The RPN concept is from FMEA (Failure Mode and Effects Analysis) developed in the 1960s—we will discuss details relating to FMEA in Chapter 5. FMEA

• Importance Analysis:
- What are the impact to the business?

• Urgency Analysis:
- When is the critical time line to resolve the issues?

Importance / Urgency	Important	Less Important
Urgent	**A-rank**	**B-rank**
Less Urgent	**B-rank**	**C-rank**

※ (Ranking Criteria)

- A-rank for a severe impact on business
- B-rank for a substantial impact on business
- C-rank for a minor impact on business

(a) Ranking method by business needs

Control No.	List of Issues	Rank	PIC	Target Timing	Progress
Xx-x1	No parts available	A	John	1-day	------
Xx-x2	Door testing robot broken	A	Chris	1st week	------
Xx-x3	Navi supplier not nominated	B	Mary	2nd week	------
Xx-x4	Parts data delayed	C			
Xx-x5	Training delayed	C			

Focus items (Xx-x1 to Xx-x3)

(b) Examples of control method by ranking

FIGURE 4.8
Prioritization by the ranking of business needs.

is a scientific method used in designing a product or manufacturing process to prevent potential failures.

In FMEA, the prioritization of potential failure mode is based on RPN:

$$RPN = Severity\ (S) \times Occurrence\ (O) \times Detection\ Capability\ (D) \tag{4.1}$$

where × is multiplication.

With the above formula the potential risk levels of each failure mode can be quantified.

To prioritize problems for management purposes, we can decide potential risk levels by using only two of the parameters: the level of *Severity* and *Occurrence* as illustrated in Figure 4.9. In this example, for a simple and clear management purpose, we can use a 1 to 3 scale instead of the 1 to 10 scale of the *AIAG (Automotive Industry Action Group)* standard.[37]

The *Severity* represents the potential impact of the risk on the customer when a problem occurs: 3 can be assigned to the most severe potential cases like a safety or a regulation issue, 2 to critical functional defects, and 1 to general defects. The *Occurrence* represents the probability that a problem will occur: 3 is assigned to a problem of new technology, new design, and new process (since we do not have history, we have a higher risk); 2 to a medium-frequency item that is mainly rooted in design or historical problem parts; and 1 to a low incident item based on the historical data experienced by the company. Actual assignment values can vary based on the company's standards and product characteristics.

Actual control and management method of each problem will be similar to (b) of Figure 4.8. We also need to differentiate our management focus and supports depending on the potential risk level determined by Figure 4.9.

- **Severity Analysis**
  - *What are the effects to the customers?*

- **Occurrence Analysis**
  - *How frequently can the problem occur?*

Severity / Occurrence	Safety / Regulation (3)	Critical Defects (2)	General Defects (1)
High (3)	S-rank	A-rank (RPN = 2×3 = 6)	B-rank (RPN = 1×3 = 3)
Medium (2)	S-rank	A-rank (RPN = 2×2 = 4)	B-rank (RPN = 1×2 = 2)
Low (1)	S-rank	B-rank (RPN = 2×1 = 2)	C-rank (RPN = 1×1 =1)

※ (Ranking Criteria)

- S-rank for a safety or a regulation issue
- A-rank for a critical risk to customers
- B-rank for a significant risk to customers
- C-rank for minor annoyance to customers

FIGURE 4.9
Prioritization by ranking of risk levels.

4.3.5 Transparent Management

Looking at historic catastrophic quality issues, we see that the most significant commonality of the problems was "*moral hazard*." This terminology originates from economic and financial terms: moral hazard occurs when one party proceeds to undertake a certain type of risky transaction, knowing the other party will bear the cost of the consequences.[38] Moral hazard also applies to quality issues, when a company proceeds without addressing or by hiding potential risk issues, knowing that those problems or risks may cause losses to customers in the form of nuisance, trouble, financial damages, injury, and even fatality.

Generally, risks and problems are intentionally hidden from customers by a company that is aware of the problems. Hiding the problems or cheating may work temporarily, but eventually, they are exposed to the public, making the situation even worse.

Once the risks of problems have been understood, sharing and communicating the problems internally and externally is the first step of transparent management. The principle of transparent management boosts a company's ethical image, giving reasons for customers to believe what a company says. Such a way of management also increases the culture of transparency inside a company, motivating ethical practices among its employees.

Every company faces daily challenge on this topic from its top management to the ground-level employees. *When senior management does not practice transparency with customers, bad practices can spread to the level of each department and team.* Employees can cheat test data to their managers or not report problems promptly, while still creating an excellent paper report that is highly discrepant from the real situation.

To promote a culture of transparency and transparent practices in an organization, we need to address two types of transparency issues: one is *internal transparency* inside the organization, and the other is *external transparency* with society and the customer.

4.3.5.1 Internal Transparency

During the development of a new product, product verification is one of the key operational processes. Many issues inevitably occur, and relevant departments need to participate in resolving them. Sometimes function managers and engineers are reluctant to disclose all the problems that surface and hope those problems will not become a severe issue or will be resolved internally before their leaders discover them. Since they hide the issues, they are not registered in a problem tracking system for closed-loop management, i.e., P-D-C-A. As a result, such problems may not have a chance to get fixed in time. They linger in the product design, seep into the mass-produced products, and are eventually sold to the customers.

A primary concern of the internal transparency issue is that it can evolve into a bad organizational habit like cancer, which will tend to develop and spread throughout the organization. *Without internal transparency, no external transparency is possible for the customer.*

Therefore, several systematic practices need to be implemented in the organization. First, there should be a system and culture not to punish individual employees when they share bad news like design mistakes or problems detected during testing. Management should focus on resolving issues as well as finding systematic weaknesses behind them. Secondly, there should be a system and process for transparent communications between managers and employees. For example, having a daily short communication meeting in the morning or a wrap-up meeting at the end of a day can help to build a sound communication culture.

4.3.5.2 External Transparency

The project cycle time for new product development is becoming shorter due to fierce competition in the market. Despite advanced technologies like CAE and the accelerated testing methods, not all problems are technically possible to detect by engineers before mass production and sales.

> *A new industrial product is launched with weaknesses in the design or validation tests. Surprises are found after the new product has been launched into the market. Then, a firefighting process starts when the quality team receives complaints in the form of a warranty or PL claim. The relevant team analyzes the problems and reports the seriousness to management. When senior management gets the report, they realize that the scale of the corrective action and the accompanying costs are enormous. Then they become hesitant or reluctant to take necessary measures. They hope that the problem will remain undetected for a long enough time to be forgotten while their customers continue to suffer in the market.*

This is a typical scenario as evidenced by the catastrophic quality issues happening in the market. In the past decades, consumers have seen, read, and experienced pandemic quality issues of industrial products all over the world. *The commonality behind this problem is a transparency issue.* Such habits are rooted in *moral hazard*, which is dangerous to society as well as resulting in massive losses to the company.

In 2009, Toyota finally made recall announcements regarding unintended acceleration. However, in spite of many deaths from accidents, for a long time prior to the recall, the company dismissed any suggestions of technical issues with their vehicles until hard evidence was presented by the customers and the authorities: "Car manufacturer Toyota has agreed to pay a staggering $1.2 billion to avoid prosecution for covering up severe safety problems with 'unintended acceleration',... FBI said, 'The disregard Toyota had for the safety of the public is outrageous, ... Not only did Toyota fail to recall cars with problem parts, they continued to manufacture new cars with the same parts they already knew were deadly.'..."[39]

In 2014, there was a massive recall announced by GM. The spring force of the ignition key cylinder was weak, unable to hold the key position to "run" and prevent switching to the "accessory" position. GM denied any knowledge of the problem in spite of earlier design changes issued before the recall in 2014.[40] Their internal document revealed, "General Motors allegedly knew about faulty ignition switches for more than a decade."[41]

In 2014, the Takata Air Bag company had to make a safety recall. By that time many people had already been killed and injured due to the faulty airbag. In spite of Takata's initial denial of pre-knowledge of the problem, "February 28, 2017, the company has finally pleaded guilty to deceiving automakers about the safety of its airbags."[42]

In 2015, Volkswagen was forced to recall more than 550,000 vehicles in the United States alone because of high nitrogen oxides during real-life driving conditions. The top management of the company *approved* such a cheating emission program. Volkswagen pleaded guilty at the United States.[43]

In 2016, Samsung, the world's biggest mobile phone manufacturer, recalled 2.6 million Galaxy Note 7.0 phones, and shortly after pulled the product from the market entirely. Samsung's hasty treatment of the problem to underplay the issue resulted in a bigger problem later.[44]

History repeats. On March 10, 2019, Ethiopian airline of Boeing 737 Max crashed, killing all 157 people aboard.

Boeing knew that there was a problem with one of the safety features on its 737 Max planes back in 2017—Boeing engineer discovered a problem on the attack angle senor system well before the Lion Air crash in October 2018 and the Ethiopian Airlines crash in March.[45]

The commonality of the historical issues is apparent—the problems were not transparently handled internally or externally by the companies. Then the problems became bigger and more significant, bringing catastrophic losses to society, while the companies were still hiding the information about the problems from their customers.

We need to pause and think about why the company leaders made such risky decisions, not taking prompt corrective actions to protect customers. The core reasons behind such absurd decisions are mainly related to *the company leaders' mindsets and the company's culture*. Such company leaders are focused on short-term profit; they are more concerned about the significant financial impact on the company. Also, the company's culture and its systems are more profit-oriented than customer-oriented.

When looking at the disastrous outcomes of such historical cases, the lesson is very obvious. Those companies that cheated and lied to customers lost money and sales opportunities, not to mention the damages to the company's image. The company's top management should look at the issue from a long-term perspective rather than short-term profit. *Dodging a problem now is not a solution. The real answer is to confront it, acknowledge it, and solve it.*

There should be specific principles that the company management should respect and follow. First, the company should have *a crisis management protocol or manual* to systematically handle safety or potential brand damage issues. Second, a prompt and thorough investigation should be initiated according to the crisis manual by the *senior management's transparent leadership*. Third, the company should *communicate with customers* about what happened and how the company intends to fix it. Fourth, the company should deploy *risk management* to minimize the scope and severity of potential damages on the customers. Fifth, the company should provide *generous compensation packages* for the already affected customers.

The company should not wait until someone comes forward with complaints or evidence of a problem. The company needs to know that, once they have found a problem, it is more than likely that someone else has also found the problem or will in the future. *The truth always comes out.*[46]

> *Honda is famous for an open office culture. Nobody in Honda owns a separate office—not even the CEO. This open space concept creates Honda's unique culture for open communication between the top to the bottom and department to department. The CEO and the top executives also share a pool of secretaries. For example, two secretaries serve four or five executives. In such an environment, communication between executives is fluid, and it is more difficult to create or hide secret information. The result is that senior management's decisions and relevant information can be shared with other decision makers transparently.*

In this chapter we discussed several management principles. Management by principles is powerful and effective in *bridging the fast and slow thinking systems*. Application of particular principles can depend on the situation of a company and its business environment. However, the intention of principle-based management is universally applicable to any type of business. Management principles should be aligned with the company's culture and policies, and they should be embedded in the company's quality management system (see Section 8.8).

Each company should work on adaptation or creation of its management principles. As Deming said, *"Each company must work out its own adaptation, suitable to corporate culture."*[47]

Notes

1. Duhigg, Charles. *The Power of Habit: Why We Do What We Do in Life and Business*, Random House, 2014.
2. Verplanken, Bas and Wendy Wood. "Interventions to Break and Create Consumer Habits," *Journal of Public Policy and Marketing* 25, no. 1 (2006): 90–103; David T. Neal, Wendy Wood, and Jeffrey M. Quinn, "Habits—A Repeat Performance," *Current Directions in Psychological Science* 15, no. 4 (2006): 198–202.
3. Kahneman, Daniel. *Thinking, Fast and Slow*, Straus and Giroux, 2011
4. Juran, J. M. *Juran on Quality by Design*. The Free Press, A Division of Simon & Schuster, Inc. 1992

5. Reif, Rita. Oct. 9, 1994, "Made in Japan (Without the Inferiority Complex)," www.nytimes.com/1994/10/09/arts/made-in-japan-without-the-inferiority-complex.html (accessed January 15, 2019).
6. W. Edwards Deming, https://en.wikipedia.org/wiki/W._Edwards_Deming (last modified March 24, 2019).
7. Deming, W.E., *Quality, Productivity and Competition Position*, Massachusetts Institute of Technology, Centre of Advanced Engineering Study, 1982.
8. Deming, Edward, *Out of Crisis*, The MIT Press, 1982.
9. Walton, Mary, *The Deming Management Method*, The Putnam Publishing Group, 1986.
10. Deming, Edward, *Out of Crisis*, The MIT Press, 1982.
11. Deming, Edward, *Out of Crisis*, The MIT Press, 1982.
12. Taylor, Frederick. *The Principles of Scientifics Management*, 1911.
13. Deming, Edward, *Out of Crisis*, The MIT Press, 1982.
14. "Ford to Cut Jobs as It Reorganizes Salaries Workforce," *The Wall Street Journal*, October 6, 2018.
15. Deming, Edward, *Out of Crisis*, The MIT Press, 1982.
16. Flitter, Emily. February 2, 2018. "Federal Reserve Shackles Wells Fargo After Fraud Scandal," www.nytimes.com/2018/02/02/business/wells-fargo-federal-reserve.html (accessed January 15, 2019).
17. Wells Fargo account fraud scandal, https://en.wikipedia.org/wiki/Wells_Fargo_account_fraud_scandal (last modified May 9, 2019).
18. Deming, Edward, *Out of Crisis*, The MIT Press, 1982.
19. Juran, J. M. *Juran on Quality by Design*. The Free Press, A Division of Simon & Schuster, Inc. 1992.
20. Introduction to Total Quality Management, June 7, 2009, https://totalqualitymanagement.wordpress.com/2009/06/07/dr-joseph-juran/ (accessed January 15, 2019).
21. "Juran's 10 Points Theory for Quality Improvement," August 20, 2011, http://goodmorningishan.blogspot.com/2011/08/jurans-10-points-theory-for-quality.html (accessed January 16, 2019).
22. Fukui, Takeo, President and CEO of Honda. "By going back to the origin, we start seeing the present and the future," Top Talks—Honda's Origin, p. 152., November 17, 2006.
23. Walton, Mary. *The Deming Management Method*, The Putnam Publishing Group, 1986.
24. General Report, Supplier Chain, www.gmsustainability.com/manage/supply.html
25. Nissan Motor Company founded, www.history.com/this-day-in-history/nissan-motor-company-founded (accessed February 14, 2019).
26. Gibney Jr., Frank, "Nissan Stalls Out," *Time*, March 8, 1999.
27. Staffy, Shady E.I. "Five essential ingredients for a quality culture," www.processexcellencenetwork.com/lean-six-sigma-business-transformation/articles/key-ingredients-for-quality-culture-development (accessed January 16, 2019)
28. Williams, Casey. September 23, 2015, "Life after Ford: Mazda, Land Rover, Volvo poised to grow," Chicago Tribune, www.chicagotribune.com/classified/automotive/sc-fords-former-brands-autos-0924-20150921-story.html (accessed January 16, 2019).

29. Ross, Phillip J. *Taguchi Techniques for Quality Engineers,* attributed to Ford Motor Company, Dearborn, Michigan, 1987.
30. Hirano, Hiroyuki. *5 Pillars of the Visual Workplace,* English translation, Productivity Inc., 1995
31. Moen, Ronald. *Foundation and History of the PDSA Cycle,* Associates in Process Improvement-Detroit, https://deming.org/uploads/paper/PDSA_History_Ron_Moen.pdf (accessed January 16, 2019).
32. Kepner and Tregoe, *The New Rational Manager,* Princeton Research Press, 1981.
33. Koch, Richard. *The 80/20 Principle and 92 Other Powerful Laws of Nature,* Bookwell, Finland, 2014.
34. Wolff, E. N. "Recent trends in household wealth in the United States: Rising debt and the middle-class squeeze" Working Paper No. 502. Annandale-on-Hudson, NY: The Levy Economics Institute of Bard College, 2017.
35. Wolff, E. N. *A Century of Wealth in America,* Harvard University Press, 2017.
36. Juran, J. M. *Juran on Quality by Design,* Juran Institute, 1992.
37. FMEA, 3rd Edition, AIAG, 2001.
38. Moral hazard, www.investopedia.com/terms/m/moralhazard.asp (accessed January 16, 2019).
39. Ross, Brian., Joseph Rhee, and Angela M. Hill. "Toyota to Pay $1.2B for Hiding Deadly 'Unintended Acceleration'," March 19, 2014, https://abcnews.go.com/Blotter/toyota-pay-12b-hiding-deadly-unintended-acceleration/story?id=22972214 (accessed January 16, 2019).
40. Reynolds, Talesha and Rich Gardella. "Document Shows GM Engineer Approved Ignition Switch Change," April 1, 2014, www.nbcnews.com/storyline/gm-recall/document-shows-gm-engineer-approved-ignition-switch-change-n68371, (accessed January 16, 2019).
41. Business.com Editorial Staff. "6 Lessons in Corporate Ethics from the GM Recall," February 22, 2017, www.business.com/articles/6-lessons-in-corporate-ethics-from-the-gm-recall/ (accessed January 16, 2019).
42. "Takata Airbag Recall: Everything You Need to Know," *Consumer Reports,* April 05, 2018.
43. The Associated Press, "Former Volkswagen CEO charged in U.S. over diesel emissions cheating—Martin Winterkorn is the 9th person charged by U.S. authorities in the case," www.cbc.ca/news/business/volkswagen-martin-winterkorn-charged-1.4647233 (accessed January 16, 2019).
44. Mozur, Paul. "Galaxy Note 7 Fires Caused by Battery and Design Flaws, Samsung Says," *The New York Times,* January 22, 2017.
45. NPR, May 6, 2019, "Boeing Knew About 737 Max Sensor Problem Before Plane Crash In Indonesia," www.npr.org/2019/05/06/720553748/boeing-knew-about-737-max-sensor-problem-before-plane-crash-in-indonesia (accessed May 13, 2019)
46. Business.com Editorial Staff. "6 Lessons in Corporate Ethics from the GM Recall," February 22, 2017, www.business.com/articles/6-lessons-in-corporate-ethics-from-the-gm-recall/ (accessed January 16, 2019).
47. Walton, Mary, *The Deming Management Method,* The Putnam Publishing Group, 1986.

5

Quality Engineering and Methods

Quality is built by engineering activity. Engineering has real power to design and build to create tangible products. Without design and manufacturing engineering, robust quality systems and good quality management are not as meaningful.

Quality engineering began to prosper after the era of quality control in the 1950s. We were able to change the fundamental structure of product design and production methods by the power of engineering to improve the level of product quality significantly. *Before engineering parameters are fixed either in design or in process, we have much more freedom and flexibility to change, update, and improve the characteristics of products and processes whether they are industrial products, software, or services.*

As defined in Eq. (3.1), the mission of quality management is to make sure we effectively and efficiently manage quality activities through quality planning, quality control, quality assurance, and quality improvement. To achieve these steps effectively, practical usage of quality engineering methods and tools is essential. Further, it is necessary for quality engineering, engineering functions, and quality management teams to *work together through each product development and process design,* leveraging the power of quality engineering methods and tools.

1. Quality Engineering Scope and Process
2. Quality Engineering Methods and Tools
3. Quality Management in Engineering

5.1 Quality Engineering Scope and Process

As discussed in Chapter 3, quality engineering activity is comprised of three major categories: design quality engineering, supplier quality engineering, and manufacturing quality engineering. There are two scopes of quality engineering activities: one is "Product scope," and the other is "Process scope." Product scope is related to product concept and design, and process scope is how we realize the product design through production processes in a stable and reliable way with *minimum variance.*

TABLE 5.1

Product and Process Quality Engineering for "q" and "Q"

	Product Quality Engineering	Process Quality Engineering
Little q	• Design Conformance to: - Defect-free - Regulations - Specifications	• Process Conformance to: - Meet the standards - Quality control - Quality assurance
Big Q	• Customer Satisfaction: - Features - Performance - Robust design	• Good Manufacturability: - Stable process - Robust process - Process improvement

Juran explained two engineering approaches to quality by "Little q" and "Big Q." The goal of Little q is to make defect-free products, and the goal of Big Q is to create salable products by enhancing customer satisfaction.[1] Table 5.1 explains how these two "qualities" are related to quality engineering activities.

The deployment flow of quality engineering is closely linked with new product development. Quality engineering methods and tools need to be used appropriately at each stage of the quality engineering process in respect of Little q and Big Q. Figure 5.1 explains how each process and method is linked to each other through the timeline of the new product development. It starts with customer requirements and customer voices coming from the market.

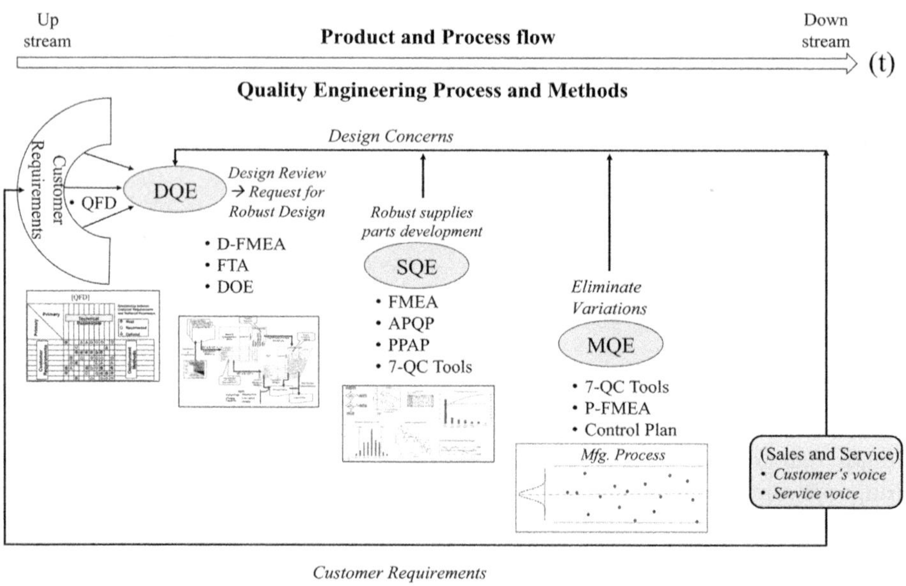

FIGURE 5.1
Quality engineering methods and processes.

Any product design or process concerns need to be addressed by the upper stream activities as soon as they are identified by quality functions, suppliers, manufacturing, sales, or service. Product and process designs need to be correlated; therefore, any process concerns must be addressed by product design changes or process improvements. The success of quality engineering activities is measured by *how many design and engineering process changes are undertaken to improve the design of products and processes at the planning and design stage before fixing the design.* Any subsequent activities to alter the fixed designs are reactive activities, i.e., recovery activities.

5.2 Quality Engineering Methods and Tools

Quality engineering activity is the backbone of scientific quality management in engineering. Quality engineering requires engineering methods and tools to realize product quality and production process design in tangible form. Quality managers and engineers should be able to understand and use those tools effectively to manage quality in engineering.

5.2.1 Statistics and Quality Control

Statistics is mathematics and science. Any function whose value can be determined from sample observation is called *statistics*. Reliable sampling, data analysis, and scientific forecasting are possible only through a statistical approach.

There are two branches of statistics:[2]

- Descriptive Statistics: This is the part of statistics concerned with the description and summarization of data. *The focus is on understanding the current situation (data).*
- Inferential Statistics: This is the part of statistics concerned with the drawing of conclusions from data. *The focus is on forecasting based on the analysis of the data.*

Statistics began to be applied to the quality control area in the manufacturing industry in the 1940s. This branch of statistical application to quality, "SQC (Statistical Quality Control)," was developed and taught by Shewhart and Deming. SQC is data-driven scientific quality control.

5.2.2 Seven QC Tools

Seven QC tools are fundamental tools for the SQC approach and method using numerical sampling data. These tools are graphical, easy to understand, and simple to implement in the field without the need for a high level of statistical competence. These basic tools have been used for more than 70 years, and they have proven effective for data and problem analysis, process control, and quality improvement activity in the engineering and manufacturing industry.

5.2.2.1 Flow Chart or Stratification Chart

- *Flow Chart:* Each of the engineering, manufacturing, and service industries has its own unique processes. The overall process flow can be represented graphically by a flow chart.

  Figure 5.2 illustrates an example of a flow chart of "Engine Block Die Casting Process" with CCPs (critical control points) attached. In this example, the quality of the engine block is made through each process—the manufacturing and quality control processes work together to build quality.

  Once the flow chart has been completed with CCP, we can delve into the detailed contents of the CCPs in PQCT (process quality

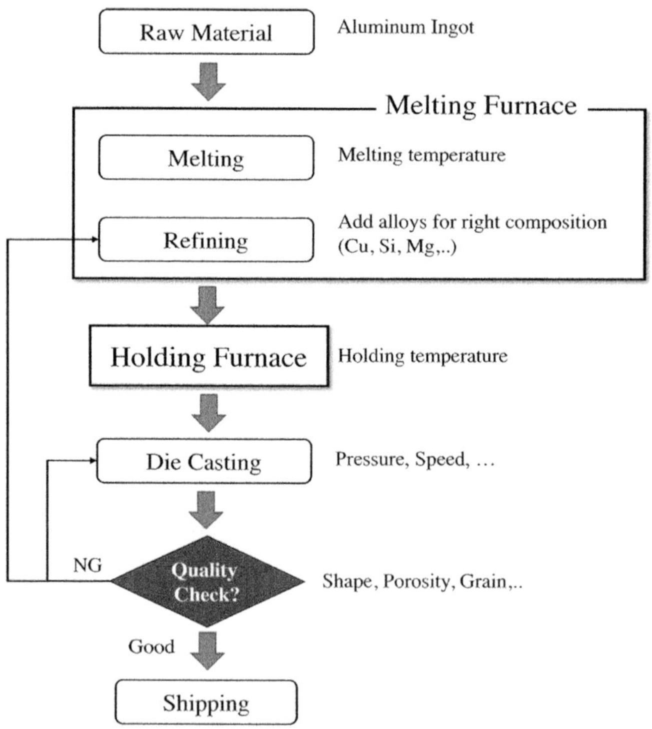

FIGURE 5.2
Flow chart (Engine Block Die Casting Process).

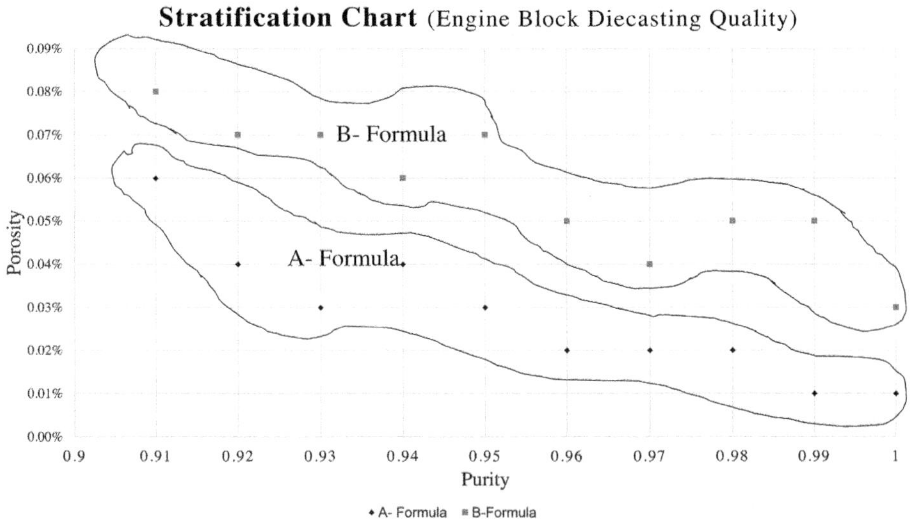

FIGURE 5.3
Stratification chart (Engine Block Die Casting Quality).

control table), such as the holding temperature of the holding furnace, die casting pressure, etc.

- *Stratification Chart:* Frequently, data on different characteristics are mixed in a sampling data group and therefore it may be difficult to understand the phenomena. Specific characteristics of the situation can be understood better when the data are grouped based on particular attributes or traits through a stratification process.

  Figure 5.3 shows a stratification chart of "porosity ratio of engine block" vs. "purity of aluminum material" between two different alloy composition formulas. When the data of the two formulas are grouped separately, the phenomenon becomes clear—the "A-Formula" of alloy composition shows lower porosity, i.e., better quality.

5.2.2.2 Check Sheet

Collecting data is a fundamental step in statistical analysis. *The first important step is to make a plan about how to collect data in a systematic way.* Then a check sheet is created to collect data in the simplest and easiest way.

Figure 5.4 shows how a very simple check sheet can be created and filled with data related to the focus areas. The check sheet shows different types of defects each day. Even by glancing at the data, we can quickly capture substantial information about the defect types and the significance of each defect type.

Check Sheet : Door Defects

Data Collector: James Cooper
Department: Line Quality, Assembly Dept.
Date: Jan. 7 – Jan. 11, 2019

Defect Types	Jan. 7 (Monday)	Jan. 8 (Tuesday)	Jan. 9 (Wednesday)	Jan. 10 (Thursday)	Jan. 11 (Friday)	Total
Fit and Finish	*IIIII II*	*IIIII*	*II*	*III*	*III*	20
Switch not working	*I*		*II*		*I*	4
Dent	*III*	*IIII*	*II*	*III*	*III*	15
Abnormal Noise	*I*	*II*	*I*	*II*	*I*	7
Paint Defect	*III*	*III*	*II*	*I*	*IIII*	13
Others	*IIII*	*II*	*III*	*II*	*III*	14
Total	*19*	*16*	*12*	*11*	*15*	73

FIGURE 5.4
Check sheet (door defects).

5.2.2.3 Pareto Chart

As we discussed in Chapter 4, 80% of phenomena are attributable to 20% of the causes. This 80/20 principle is a universal law called the Pareto Principle and it is observed in statistics, economics, sociology, quality, and many other areas.[3] In quality management, we need to remember that 80% of problems are attributable to 20% of the issues or causes.

Figure 5.5 shows defects types and the number of defects in each process location. The table is ordered from the most common defect type to the least common and also shows the accumulated percentage.

The data can be graphically displayed as in Figure 5.6, i.e., in Pareto chart format. The Pareto chart is simple; however, it delivers critical information in a precise manner. For instance, by crisscrossing the line of 80% accumulated defect with the defect types, we can identify the top issues that need to be resolved as a priority. In this case, the Pareto chart reveals that "screen scratch," "slow operation," and "switch button" are the major quality issues that cause more than 80% of the total problems.

	Defect Source							
Defects Type	In-coming Parts Defects	Operation Mistake	Transport	Software Problem	Design Problem	Others	Total	Accum. (%)
Screen Scratch	15	24	7	0	0	4	50	58%
Slow Operation	0	0	0	3	9	0	12	72%
Button Sticking	5	2	0	0	0	1	8	81%
Garnish Defects	2	4	0	0	0	1	7	90%
Mapping Problem	0	0	0	4	0	0	4	94%
Others	2	0	1	1	0	0	4	99%
No Power	0	1	0	0	0	0	1	100%
Total	24	31	8	8	9	6	86	

FIGURE 5.5
Summary of defects by type and source (navigation unit).

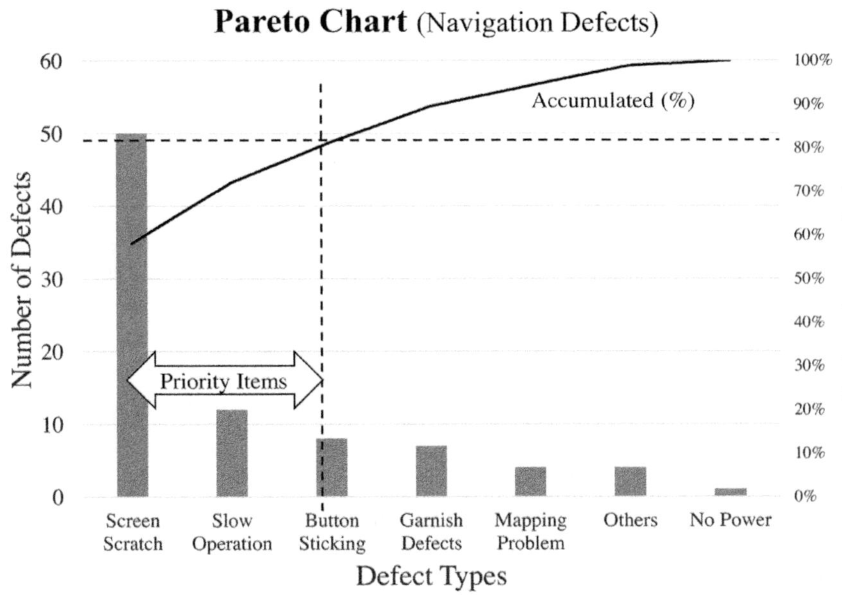

FIGURE 5.6
Summary of defects by types and sources (navigation unit).

5.2.2.4 Histogram

One of the main reasons for sampling data analysis in statistics is to see how the population is distributed. To understand the distribution, we need to know the mean and the variance. Without using complex statistical analysis and calculations, we can easily see the distribution of the samples and forecast the population characteristics by using a graphical method, i.e., a histogram.

First, the raw measurement data can be rearranged by frequency by using a spreadsheet like MS Excel as shown in Figure 5.7.

Length of the Ignition Switch Spring (Spec.: 5.2mm +/0.1mm)

Sample (n = 50)				
5.19	5.24	5.11	5.30	5.05
5.24	5.17	5.17	5.13	5.25
5.12	5.31	5.23	5.30	5.11
5.29	5.07	5.01	5.06	5.18
5.18	5.23	5.19	5.15	5.24
5.23	5.09	5.12	5.28	4.97
5.03	5.29	5.35	5.18	5.22
5.19	5.12	5.07	5.22	5.09
5.09	5.37	5.28	5.13	5.15
5.17	5.17	5.15	5.19	5.06

Frequency

	Bin	Bin Label	Freq.
Min	4.90	<= 4.90	0
	4.95	4.93	0
	5.00	4.98	1
	5.05	5.03	3
	5.10	5.08	7
	5.15	5.13	7
	5.20	5.18	14
	5.25	5.23	8
	5.30	5.28	7
	5.35	5.33	2
Max	5.40	5.38 <	1

FIGURE 5.7
Measurement data and rearrangement (length of ignition switch spring).

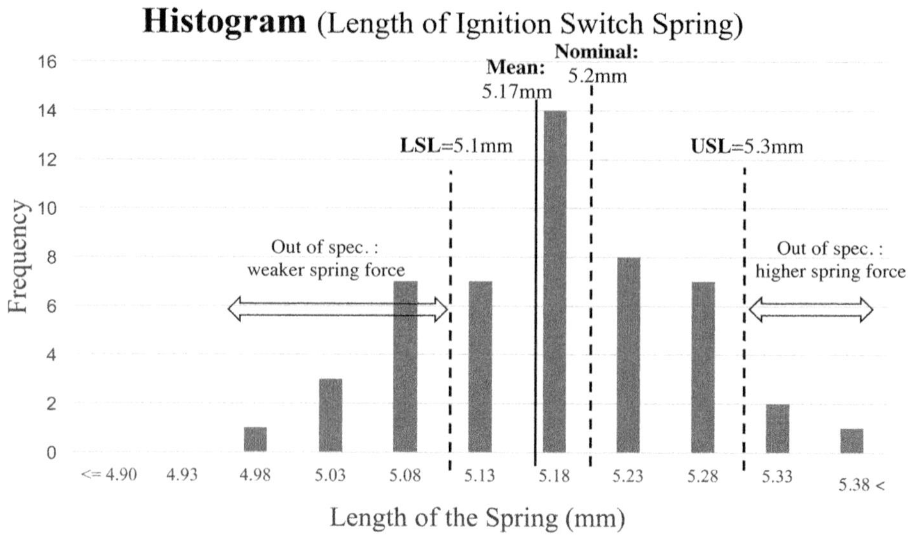

FIGURE 5.8
Histogram (length of ignition switch spring).

Using graphic software, we can display the data in a bar chart as shown in Figure 5.8. The horizontal axis shows the intervals of the data, and the vertical axis represents the frequency, i.e., the number of occurrences at each interval of the spring length. By inserting the specification limits of the part into the bar graph, such as upper specification limit (USL) and lower specification limit (LSL), the distribution of the data becomes more explicit. The overall parts distribution is slightly skewed to the lower specification side. In other words, a more significant portion of the parts is shorter in length leading to a weaker holding force of the ignition switch.

5.2.2.5 Scatter Diagram

Sampling method of data is supposed to be random, and measurement data values always contain some amount of error due to the inherent variation in the measurement system and noise (outer, inner, and product noses).[4] Furthermore, a small sample of data may not necessarily follow theoretical trends or forecasts precisely.

However, merely obtaining a bigger sample size can be expensive without proper statistical analysis. Figure 5.9 shows the actual temperature measurements inside a paint oven. The paint oven temperature is critical for the paint quality of automobiles. Since this company has a one-shift operation, the oven is turned on and off every morning and evening on weekdays.

At first glance, the measurement temperatures in the data table seem random, without any particular trend. However, when the measurement data is

Painting Oven Temperature: Standard: 250±10°C

Time (hour)	Jan. 7	Jan. 8	Jan. 9	Jan. 10	Jan. 11	Jan. 14	Jan. 15	Jan. 16	Jan. 17	Jan. 18
9:00	235	245	246	240	237	228	229	235	240	237
10:00	240	240	246	243	238	229	233	240	243	238
11:00	245	251	247	247	240	245	235	245	247	240
12:00	246	253	250	249	243	246	243	250	249	243
13:00	250	259	255	250	254	248	247	255	250	254
14:00	252	258	257	252	258	252	251	260	252	258
15:00	250	261	259	255	259	255	256	261	255	259
16:00	255	262	260	257	261	260	256	260	257	255
17:00	260	258	257	259	257	262	261	258	259	257

FIGURE 5.9
Random data distribution (temperature of paint oven).

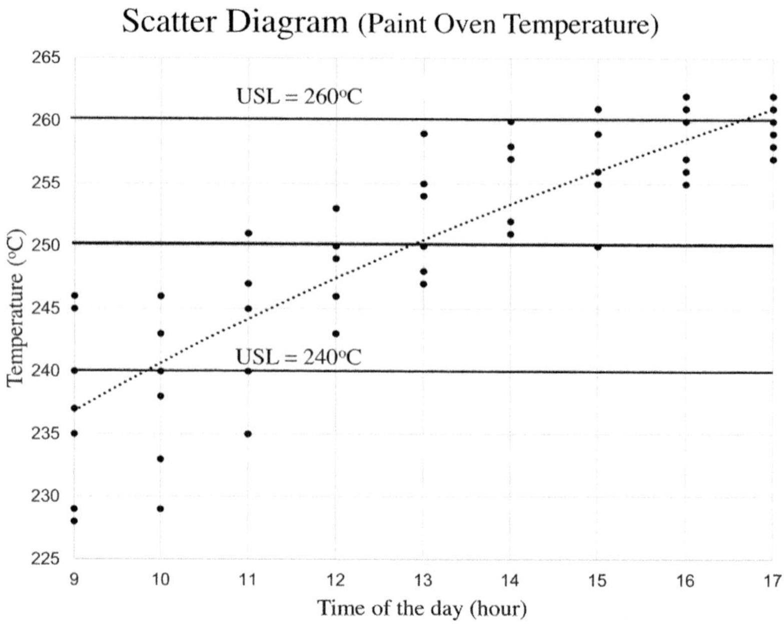

FIGURE 5.10
Scatter diagram (trend of paint oven temperature).

plotted, there seems to be a trend, as shown in Figure 5.10. This type of diagram is called a scatter diagram. The scatter diagram helps to identify the relationship between two variables and helps analyze the situation so as to prepare the necessary countermeasures as necessary. In this particular case, the oven temperature seems to be low in the morning and higher in the late afternoon.

The scatter diagram shows a trend—the oven temperature increases as time passes. At the beginning of the day, before 13:00, the overall temperature was below specification (250°C). Therefore, a possible remedy is to keep the oven warm by minimizing heat loss during the night or to turn on the oven 2.5 hours before the work begins.

5.2.2.6 Cause & Effect Diagram

With the help of basic QC tools, we can effectively understand the overall phenomenon of our interest areas. Every phenomenon has its causes. Most of the daily quality problems on the manufacturing floor need prompt countermeasures. First and foremost, it is essential that a systematic approach be taken to identify the root causes of the problems and subsequently apply the appropriate countermeasures.

The cause and effect diagram developed by Kaoru Ishikawa in 1968 has been widely used across various industries to analyze the causes of a specific event[5]– this is also called a "fishbone diagram" due to its shape.

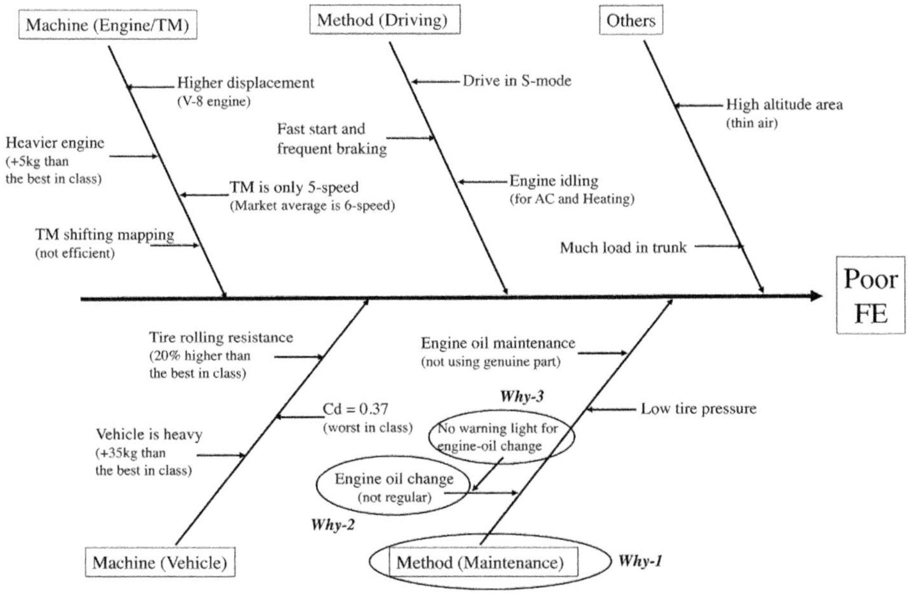

FIGURE 5.11
Cause and effect diagram (poor fuel economy).

A fishbone diagram is straightforward and effective. Figure 5.11 explains the cause and effect diagram method, through which possible causes of poor fuel economy are identified. Since the structure of the diagram is like a fishbone, we start from a big bone first, from *why-1*, then *why-2*, *why-3*, and so on.

5.2.2.7 Control Chart

Inherent variations exist in any process in nature or human-made systems. Today's process does not guarantee the same quality results tomorrow due to the variations existing in its systems and processes. There are two kinds of variations: *special cause* and *common cause*.[6] The special causes are specific to some workers, or to a specific machine, or to a specific local condition. Therefore, the variation is unpredictable in both time and magnitude. In contrast, the common causes of variation are inherent in the performance of a product and the process of the manufacturing, where it remains until corrected or reduced. Therefore, the variation levels can be measured and predicted using statistical techniques.

The control chart is the core tool of SQC (Statistical Quality Control) or SPC (Statistical Process Control). *The control chart starts with collecting data from a* stable process *where a special cause of variation is not present.*

Figure 5.12 shows the data collected from the toner filling process of a laser printer cartridge. Once the stability of the process has been confirmed, we

Toner Filling Process of Laser Printer Cartridge

Weight of Laser Printer Cartridge (unit: kg)

Date	07-Jan	08-Jan	09-Jan	10-Jan	11-Jan	14-Jan	15-Jan	16-Jan	17-Jan	18-Jan	21-Jan	22-Jan	23-Jan	24-Jan	25-Jan	28-Jan	29-Jan	30-Jan	31-Jan	01-Feb	04-Feb	05-Feb	06-Feb	07-Feb	08-Feb
Group	1	2	3	4	5	6	7	8	9	10	11	12	13	14	15	16	17	18	19	20	21	22	23	24	25
x1	2.4	1.6	2.0	2.1	2.1	1.9	2.0	2.2	2.2	1.8	1.9	1.9	2.1	2.2	2.1	2.4	2.3	1.6	2.1	2.3	2.2	2.1	2.0	1.9	1.7
x2	2.0	2.3	2.1	2.0	2.2	2.0	2.1	2.4	2.2	2.0	2.1	2.5	1.6	1.8	1.8	2.3	2.4	1.8	2.0	2.0	2.1	2.3	2.0	2.0	2.0
x3	2	2.0	2.0	1.9	1.8	1.9	2.1	1.7	2.4	2.4	2.2	1.7	1.8	1.9	1.9	2.3	2.0	2.1	2.0	2.1	1.8	2.1	1.6	1.9	1.9
x4	2.4	2.3	1.8	2.2	1.7	1.8	2.0	2.1	1.6	1.9	2.1	2.1	2.0	1.8	2.2	2.4	1.9	1.7	1.8	1.9	1.9	2.2	1.8	2.2	1.8
x5	1.8	1.9	1.8	2.3	2.0	2.2	1.9	2.1	2.0	2.2	1.8	2.0	2.3	1.6	1.9	2.3	2.2	2.0	2.1	2.2	2.2	2.0	1.6	2.1	2.2
Sum	10.6	10.1	9.7	10.5	9.8	9.8	10.1	10.5	10.4	10.3	10.1	10.2	9.8	9.3	9.9	11.7	10.8	9.2	10.0	10.5	10.2	10.7	9.0	10.1	9.6
$\bar{x}$ (average)	2.1	2.0	1.9	2.1	2.0	2.0	2.0	2.1	2.1	2.1	2.0	2.0	2.0	1.9	2.0	2.3	2.2	1.8	2.0	2.1	2.0	2.1	1.8	2.0	1.9
R (range)	0.6	0.7	0.3	0.4	0.5	0.4	0.2	0.7	0.8	0.6	0.4	0.8	0.7	0.6	0.4	0.1	0.5	0.5	0.3	0.4	0.4	0.3	0.4	0.3	0.5

$\bar{X}$ (average of $\bar{x}$) = 2.0
$\bar{R}$ (average of R) = 0.5

$UCL = \bar{X} + A_2\bar{R} = \bar{X} + 0.577\bar{R} = 2.27$
$LCL = \bar{X} - A_2\bar{R} = \bar{X} - 0.577\bar{R} = 1.72$

$UCL_R = D_4\bar{R} = 2.114\bar{R} = 1.06$

Sample Size (n)	A_2	D_4
2	1.880	3.268
3	1.023	2.574
4	0.729	2.282
5	0.577	2.114

FIGURE 5.12
Data from toner filling process of the laser printer cartridge.

can calculate the average and the range of each sample group, i.e., ($\bar{x}$) and R. Then the expected *UCL* (upper control limit), *LCL* (lower control limit), and UCL_R (upper control limit range) can be calculated from the statistical formula as shown in the bottom of Figure 5.12. In the calculation, proper values for the constants (A_2 and D_4) are used depending on the sample size of each subgroup; in this case, $n = 5$ samples.

From the data shown in Figure 5.12 we calculate each average ($\bar{x}$) and R, $\bar{X}$ (the average of $\bar{x}$) and $\bar{R}$ (the average of R). Then we can get $UCL = 2.27$, $LCU = 1.72$, and $UCL_R = 1.06$ as shown in the bottom of Figure 5.12.

Now we can plot x-bar and R-Chart with each data point of $\bar{x}$ and R with *UCL*, *LCL*, and UCL_R as in Figure 5.13. Here the purpose of the control chart becomes clear. By plotting the data in the chart, we can identify any abnormal changes in the toner filling process of the laser printer cartridge over time. When we see an "out of control point" in the control chart, we recognize it as a warning sign and need to focus on identifying any special causes of the variation. In this way, we can use the control chart for quality improvement activity by removing the special causes of variation as well as reducing common causes of variation.

5.2.3 QFD (Quality Function Deployment)

The purpose of quality engineering activity is to increase the customer's perception of the quality of products or services through customer-focused engineering activities—the focus is on Big Q as discussed in Table 5.1. In order to achieve this purpose, we need to use systematic tools to translate customer requirements and VOC (voice of the customer) into engineering specifications.

In this perspective, QFD (Quality Function Deployment) is the first step of the quality engineering tool for scientific quality planning and structured quality deployment. This was first conceptualized under the umbrella of Total Quality Control by Shigeru Mizuno and Yoji Akao in 1966.[7] Then the method was applied at a Japanese shipbuilding firm, the Kobe Shipyards of Mitsubishi Heavy Industry, in the 1970s for new product development. It provided a systematic way of dealing with the many complexities and trade-offs between the ship design parameters faced by product developers. Later, the QFD approaches and methods migrated to the Japanese auto industry and then to the U.S. auto industry by the mid-1980s.[8]

When QFD is applied to new product development, the project will have fewer mistakes and errors during the product and process design. As a result, the project will require fewer engineering changes, considerably reducing start-up costs and shortening the product development cycle.

Figure 5.14 shows the format of QFD: QFD is a matrix type format, composed of horizontal rows and vertical columns. The intersection areas display the relationships between the rows and columns. Through this structured approach, we can prioritize our activities by discerning technical parameters

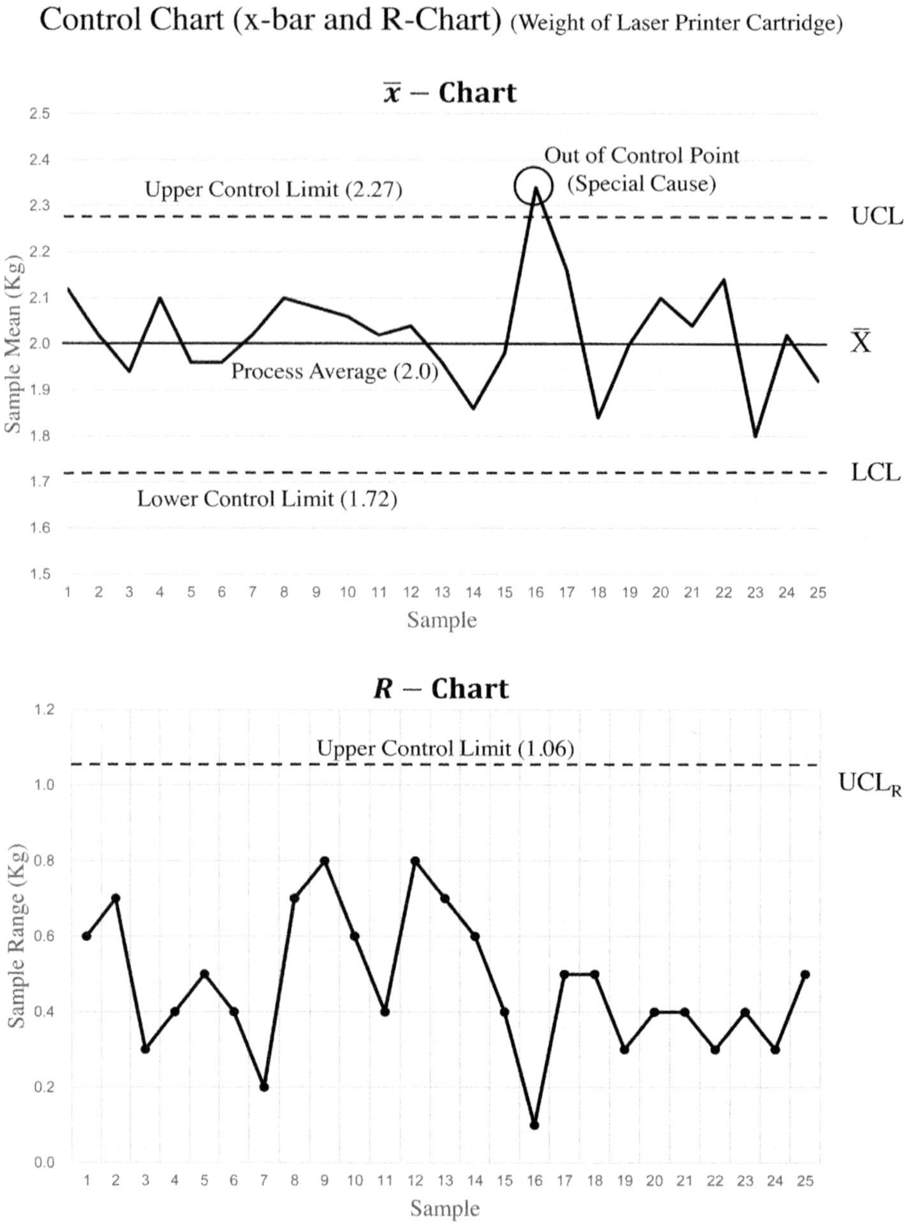

FIGURE 5.13
Control chart of laser toner filling process.

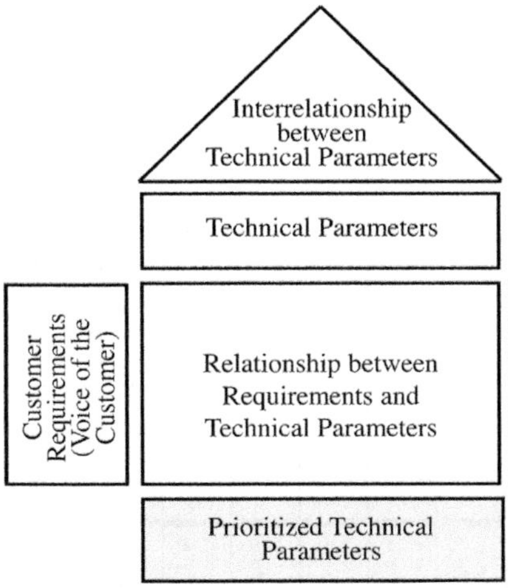

FIGURE 5.14
QFD (Quality Function Deployment) structure.

and customers' requirements. With this advantage, we can reduce potential engineering changes downstream as well as considerably reduce start-up costs due to fewer tool and equipment changes during production.

Figure 5.15 shows an application of QFD for a car cup holder design. The cup holder is becoming a critical part of car design because globally more number of customers demand it due to lifestyle changes. The left column shows the customers' viewpoints and the top row shows the engineering design parameters. The inside matrix shows the *relevance* by symbols between these two parameters; for instance, "double circle" means a strong relationship between customers' requirements and engineering parameters. The *degree of relevance* needs to be decided through logical analysis and discussion between team members.

Then, a number for each matrix section is calculated by multiplying the weight ratio (①) and the degree of relevance (②), and the total number is summarized in the bottom row as the "Total (③ = ① × ②) ." Through this process, we can understand which engineering parameters are the most sensitive to the customers' overall satisfaction with the cup holder.

From the summarized number in the row, "Total," *we can prioritize key engineering parameters* to maximize the quality values to customers in the following order:

1. Size and depth of the cup holder
2. Two cup holders with two different sizes
3. Adjustability of the size
4. Shutter cover and operation mechanism
5. Other

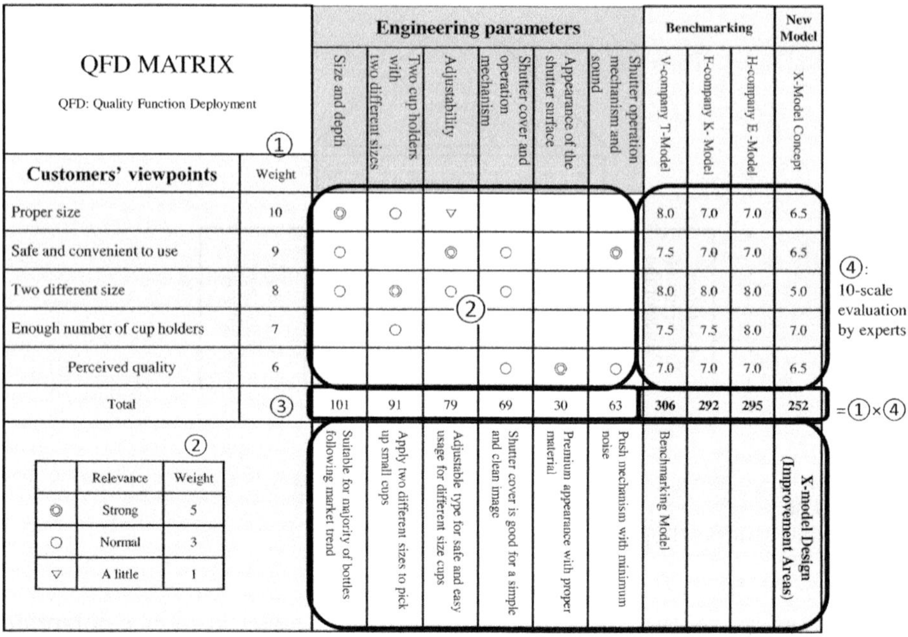

FIGURE 5.15
QFD. (Example: car cup holder.)

From this QFD process, we can also perform a benchmarking activity. First, through experts' evaluation of the benchmarking models with customers' viewpoints, we judge the achieved ratio of each viewpoint on a 1 to 10 scale. Then we multiply the number and the weight value from the left-side column. Then the final summation number is written at the bottom of the benchmarking section as "Total." From this activity, we can see that *our future X-model concept design needs improvement compared with the other competitive models in the market.*

This demonstration shows a straightforward application of a simple QFD method for quality engineering activities of new product development. A more complex form of QFD can be used for the development of more complex products.

5.2.4 FMEA (Failure Mode and Effects Analysis)

FMEA is a scientific method used to prevent potential failure modes in a product or process design. The FMEA technique can be used in a variety of industries from manufacturing to software and service development. The purpose of FMEA is to identify the potential failures of a product or process and implement appropriate countermeasures to mitigate the risks.

FMEA was developed by NASA in the 1960s, and since then it has been widely used in various industries, including the automotive industry in particular. There are two types of FMEAs: design FMEA, *D-FMEA, which is for a robust product design*; and process FMEA, *P-FMEA, which is for robust process design.*

FMEA is a bottom-up inductive procedure starting from a system, a part, or the function of a product or process. It should consider potential failures of a system, a part, or process and identify the range of their effects on the system or the product. The general process of deploying D-FMEA is represented as an example in Figure 5.16. The focus of this particular example is about the O-ring failure of the Challenger spacecraft as discussed in Figure 2.7 in Chapter 2.

The process flow is as follows: ① select a system, part, or function; ② list potential failure modes; ③ describe the impact of the failure; ④ determine severity level of the failure (low to high; 1 to 10 scale); ⑤ determine the class based on the company standard; ⑥ describe potential causes of the failure; ⑦ determine the expected occurrence of the failure (low to high; 1 to 10 scale); ⑧ describe the current design control of failure prevention; ⑨ describe the current design control of failure detection; ⑩ determine the detection ability of the failure (*high to low*; 1 to 10 scale); ⑪ prioritize the potential failure risk by RPN (Risk Priority Number): Severity (S) × Occurrence (O) × Detection (D); ⑫ take action to reduce the risk; ⑬ assign responsibility; ⑭ final action to reduce the risk; ⑮ recalculate RPN; and ⑯ repeat the process if RPN is still high.

Here, the RPN criteria for a countermeasure requirement depends on the category of industry and each company's policy.

In the FMEA, the prevention of the potential failure mode is based on RPN:

$$RPN = Severity\ (S) \times Occurrence\ (O) \times Detection\ Capability\ (D) \tag{5.1}$$

With the formula above, Eq. (5.1), the potential risk levels of each failure mode can be quantified by AIAG criteria.[9] *"Severity"* represents the level of the potential failure's impact on the customer or regulations if the problem occurs; "10" is for the most severe cases representing failures such as a safety issue without warning, and "1" for no discernible effect. *"Occurrence"* represents the probability that the failure will occur; "10" is used for new technologies or new designs, and "1" for no-expected-failure cases. *"Detection"* represents the detectability of a problem before it fails; "10" is used for a case where no detection is possible, and "1" for a case where failure detection is virtually certain.

Figure 5.16 shows a particular example of D-FMEA of the Challenger spacecraft. Using a thorough deployment of FMEA, the cause of the potential failure, the O-ring leaks, could have been identified and counter measures taken. The same design for the spacecraft had been used for 29 missions[10] before the disaster. However, this short history cannot justify using lower numbers for potential failure occurrence and detection rates. That is why a

Failure Mode and Effects Analysis
(Design-FMEA)

Item / Function	Potential Failure Mode	Potential Effect(s) of Failure	SEV	CLASS	Potential Cause(s)/ Mechanism(s) Of Failure	OCCUR	Current Design Controls Prevention	Current Design Controls Detection	DETEC	RPN	Recommended Action(s)	Responsibility & Target Completion Date	Action Results: Actions Taken	Action Results: SEV	Action Results: OCC	Action Results: DET	Action Results: RPN
O-ring: Sealing of hot-gas during ignition	O-ring leaks	Rocket blowup	10		Loss of sealing effect	9	Rubber sealing performance confirmed at normal temperature 10 < T < 40°C	Leak test at laboratory at 10 < T < 40°C	9	**810**	Leak test at laboratory at -10 < T < 40°C	NASA Safety Test Engineer Jan. 1985	Based on the test results, design change of the sealing structure and material	10	2	5	**100**
①	②	③	④	⑤	⑥	⑦	⑧	⑨	⑩	⑪	⑫	⑬	⑭				⑮

⑯ ← repeat ← ⑮ (⑯ → ⑫)

FIGURE 5.16
FMEA. (Example: Challenger Spacecraft.)

high number, "9," should be used for the occurrence rate and detection risk in this case study. Through the FMEA of the spacecraft we can identify the potential risks; therefore, we need the countermeasure actions (12, 13, and 14), which should have taken place before the launch.

FMEA activity is an essential engineering process that can improve the robustness of a product design or a process from the very beginning. However, the FMEA method has some weaknesses, which stem from the subjective determination of the risks that may lead to improper priority-setting based on RPN significance.[11] Therefore, for an entirely new design or a new technology product, FMEA should be complemented by a deductive procedure, such as the FTA (Fault Tree Analysis) method.

5.2.5 FTA (Fault Tree Analysis)

FTA is the reverse method of FMEA and can be a complementary tool to FMEA. FTA is a deductive procedure, a top-down approach. It starts with a top failure event rather than starting from a system or a part level like FMEA. This top-down failure analysis is used for discovering the root causes of failures or potential failures of a system or a component.

Once the top event is defined, the fault tree is constructed by working downwards through the system, one level at a time, to determine which failures could lead to the failure event. The top event is logically analyzed by looking at the potential contributing factors of design, systems, processes, and/or parts to the top failure event. The failure of each triggering factor is logically traced by using Boolean logic to combine a series of lower-level events. The main logic gates used are similar to the symbols used by electronic circuit designers: "AND" or "OR" gates. Through the fault trees, the probability of each failure mode can also be calculated based on the failure frequency or probability of each contributing factor.

FTA was first used by Bell Telephone Laboratories in connection with the safety analysis of the Minuteman missile launch control system in 1962.[12] Currently, the method is widely used for new technology designs by part suppliers in the automotive industry. This approach can also be used for safety-critical industries, such as the chemical equipment industry, nuclear engineering, space engineering, medical engineering, etc.

Figure 5.17 shows the FTA deployment on the Challenger spacecraft.[13] This FTA approach has a contrast logic in the deployment (top-down approach), unlike the FMEA deployment logic (bottom-up approach) discussed in Figure 5.16. This time, we start with the top failure event instead of starting from the part, the O-ring. This top-down deductive procedure can lead to broader scopes of potential failures through intermediate events leading to final basic events.

Using FTA, we can also calculate the probability of the top failure event by logic gates (such as "AND" and "OR") from the probability of the basic failure events. For example, to calculate an "AND" event we use $A \cap B = A \times B$ and for an "OR" event, $A \cup B = A + B - (A \cap B)$.

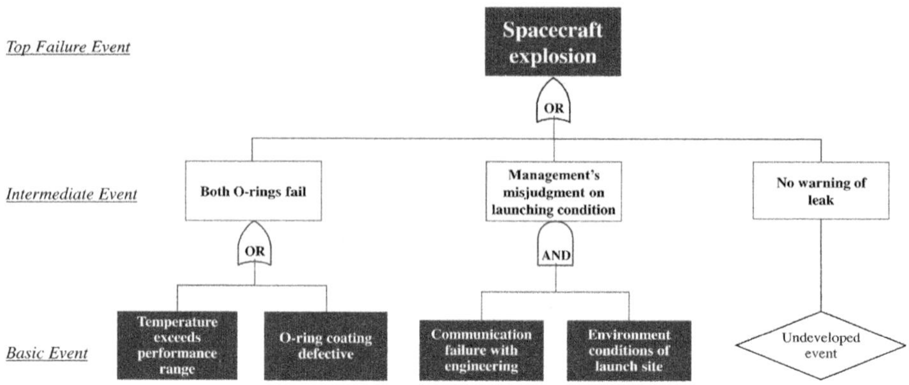

FIGURE 5.17
FTA. (Example: Spacecraft.)

There are two advantages of FTA: it can logically capture basic failure events and avoid subjective calculations of RPN like in FMEA. Both FTA and FMEA are very potent quality engineering tools for a product and process design.

5.2.6 DOE (Design of Experiment)

Once we have identified the priority of the high-level engineering parameters by QFD, we go into a detailed engineering design process. For example, when we design the cup holder cover mechanism as shown in Figure 5.15, a smooth, quiet operation is necessary. As a next step, we need to determine key design parameters that most sensitively influence this operation performance within engineering boundary conditions such as weight, cost, and durability.

A traditional way of finding critical design parameters starts with our experience and knowledge of a specific product, which we often call "know-how." However, our expertise is useful only within a limited category of products that we have already designed and produced.

In the modern high technology world, we need to design and launch innovative products in the market to be competitive; therefore, it is inevitable that we enter into uncharted technical and engineering territories. However, navigation through unchartered territories requires a scientific approach rather than a trial-and-error approach. Such a scientific process needs to be undertaken via systematic experiments with the power of statistics. We call this engineering technique "Design of Experiment (DOE)."

DOE was invented by Ronald A. Fisher in the 1920s and 1930s for agricultural research. Fisher's approach was to draw valid conclusions efficiently from experiments with natural fluctuations such as temperature, soil conditions, and rainfall, in the presence of nuisance variables.[14]

Any type of experiment or test requires a substantial amount of time, money, and personnel. Therefore, an effective way of performing experiments and proper analysis of data are critical to determining the right values for the design parameters and the tolerance values of a system or a component within the relevant economic constraints.

There are several methods of DOE: OFaat (One factor at a time) Experimental Design, Full Factorial Design, Fractional Factorial Design, and the Taguchi Method.

- *OFaat (One factor at a time) Experimental Design:* This experimental method tests all levels of each factor while freezing the others at a nominal level. Such a simple experimental method has its economic advantages; however, this type of unbalanced experiment design is not able to see any average or interaction effects between design parameters.
- *Full Factorial Experimental Design:* This type of experimental design tests all combinations of the design parameters. The test changes more than one parameter at a time in the test runs. Generally, this type of test requires too many experiments; in the case of k design factors with n levels each, the total number of experiments becomes n^k. The primary benefit of this type of design is that interactions between design parameters can be detected and measured.
- *Fractional Factorial Experimental Design:* The weak point of the full factorial design is that the number of experiments increases exponentially, n^k, as we increase the design parameters (k) with more levels (n). To maximize the effectiveness of the experiment in terms of time and cost, one idea is to take a less useful piece of information and confound it with a more useful piece of information. In other words, we fractionate the larger design to a smaller size that we can afford to run. Through this process, we can significantly reduce the number of experiments, saving both cost and time.
- *Taguchi Method:* During a product or process design, the robustness of the design is critical to ensure the function or performance of a product despite noises that result in quality variances. The inner noises come from sources within our control; however, the outer noises are not controllable—they are from uncontrollable manufacturing process variability or the usage conditions of the product by the customer.

  The Taguchi Method uses the experimental design to study responses associated with both the means and variances of a product's characteristics. The goal is to choose appropriate factor settings so that both the variance and bias become smaller simultaneously.

The appropriate design factor setting comes from the minimization of the expected "loss" of each different experiment by using the loss function as we discussed in Eq. (2.1).

Case Study: Photo Image Quality

A built-in digital camera is now an essential part of a smartphone. The photo image quality of such a camera relies on many technical factors. The connection of those technical factors to the image quality judged by the human is a complex task. The overall judgment of image quality is dependent on a subjective measure called psychometric response.

The DOE method can be used as a scientific tool to make an engineering correlation of the technical factors with the photo image quality. For demonstration purposes, we will focus on only primary design parameters influencing the overall photo image quality: *sharpness, dynamic range,* and *color accuracy.*

Sharpness is the single most important factor influencing the overall photo quality. It determines the amount of detail an image can convey. The most sensitive parts influencing sharpness are the lens quality, the capability of the image sensor (number of pixels), and signal processing (sharpening and noise reduction).

Dynamic range is the range of light levels a camera can capture, which is defined as the ratio between maximum light intensity measurable (at pixel saturation) to minimum light intensity measurable.[15] The dynamic range of a digital camera depends on hardware performance and processing effects, e.g., the capability of the image sensor.

Color accuracy refers to the exact color tones or shades of gray demanded by signals sent by the source object. Color accuracy of a photo image is affected by a camera's color response: its color shifts, saturation, and white balance effectiveness. Color accuracy is affected by the color filter array and the signal processing of the camera.

Factorial Experimental Design:

For DOE demonstration purposes, we are going to use the most typical experimental method, i.e., factorial experimental design. This method is used to investigate the effects of many different factors by varying them simultaneously instead of changing only one factor at a time. Factorial designs allow estimation of the sensitivity to each factor and also to the combined effect of two or more factors.[16,17]

As mentioned above, we choose the three most significant factors influencing photo image quality: *sharpness, dynamic range,* and *color accuracy.* Photo

Test \ Parameters	Sharpness (A)	Dynamic Range (B)	Color Accuracy (C)	Photo IQ
1	Medium	Medium	Medium	7.5
2	High	Medium	Medium	9.4
3	Medium	High	Medium	8.4
4	High	High	Medium	9.7
5	Medium	Medium	High	8.3
6	High	Medium	High	9.5
7	Medium	High	High	9.0
8	High	High	High	9.9

FIGURE 5.18
Factorial design and photo IQ (full factorial design).

image quality is the psychometric response judged by customers subjectively and summarized in Figure 5.18. This summary represents each combination of the test and the results of the photo image quality (IQ). "Medium" refers to base performance and "High" refers to enhanced performance.

Figure 5.19 is the summary of the *average photo IQ effect with a different combination of design parameters*. From this result, we can make a separate conclusion on the effect of each design parameter on the photo IQ. The information can be graphically represented by *f(A, B, C)* in Figure 5.20 to show the effect of each separate design parameter concerning the photo IQ.

The power of the factorial design is more than simply being able to see the effect of each separate parameter. More importantly, through this full factorial experiment, we can check any possible interactions between each parameter.

Figure 5.20 shows how the combination of each design parameter can influence the IQ. In other words, *IQ can be represented as a function of f(A, B, C); A (sharpness), B (dynamic range), and C (color accuracy)*. For instance, as represented in the graph *f(B, C)*, there is no interaction observed between the dynamic range (B) and the color accuracy (C), because both the high and medium of color accuracy is linearly proportional to the dynamic range. However, as represented in graph *f(A, C)*, we observe a strong interaction between the sharpness (A) and color accuracy (C), i.e., a strong non-proportionality exists between high and medium of color accuracy (C) with respect to the sharpness (A).

Design Parameters \ Photo IQ	Average IQ			Conclusion
	Medium	High	Difference	
Sharpness (A)	8.3	9.6	1.3	Higher sharpness has the biggest effect
Dynamic Range (B)	8.7	9.3	0.6	Dynamic Range has a medium effect
Color Accuracy (C)	8.8	9.2	0.4	Color Accuracy has the smallest effect

FIGURE 5.19
The average IQ effect of key design parameters.

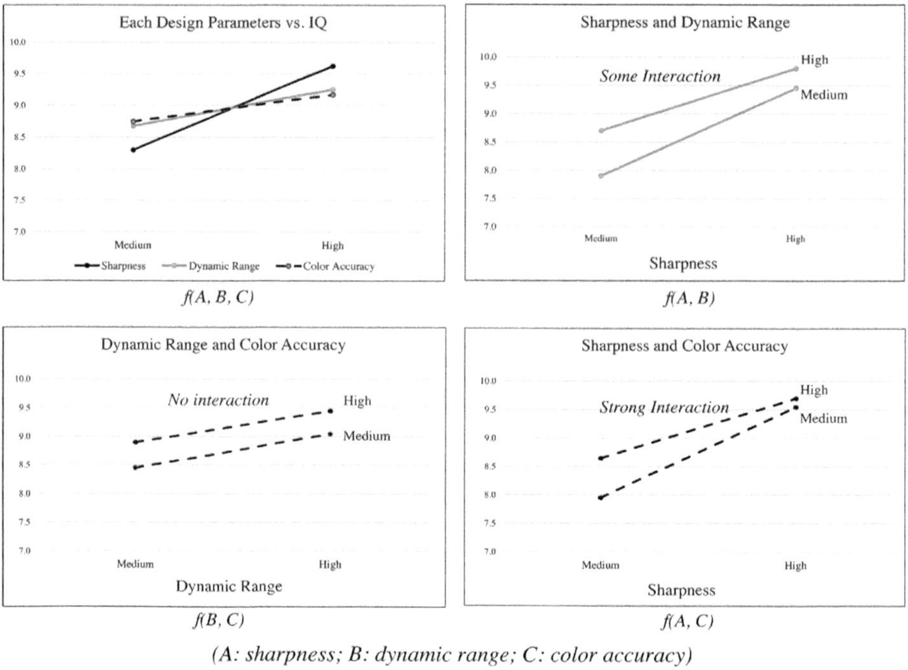

FIGURE 5.20
The effect of key design parameters on photo IQ.

The conclusion from this analysis is:

- Sharpness is the most critical factor affecting the overall photo image quality (see Figure 5.19 and *f(A, B, C)* in Figure 5.20).
- When sharpness is "Medium," using "High" value of the dynamic range and color accuracy becomes more effective for IQ.
- When the sharpness is "High," the effect of the dynamic range and color accuracy is weak.

Based on the conclusion of such an experiment, we can focus on more critical design parameters rather than every parameter, balancing value and cost.

5.2.7 APQP (Advanced Product Quality Planning)

The SQE (supplier quality engineering) activity makes sure the parts from suppliers meet the right level of quality with respect to the client's requirements, technical specifications, quality, delivery timing, quantity, etc.

The SQE activity is mainly composed of three stages of activities: the assessment of suppliers, the product development quality, and the production quality.

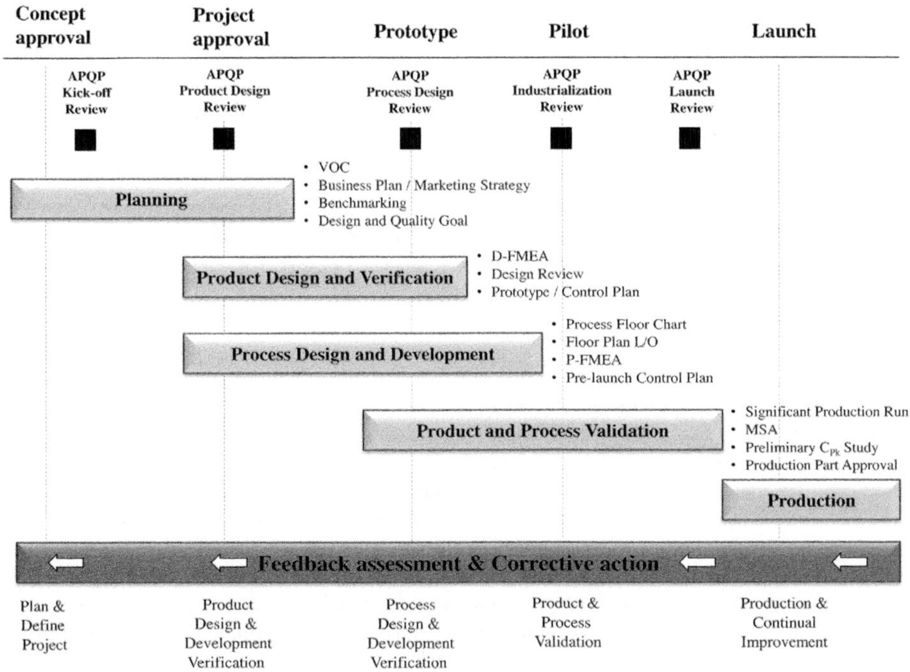

FIGURE 5.21
APQP deployment process.

APQP (Advanced Product Quality Planning) and PPAP (Production Parts Approval Process) are quality part development processes required by suppliers. *APQP* and *PPAP* as standard quality planning methods were developed in the late 1980s and 1990s by a commission of experts gathered around the "Big Three" in the U.S. automobile industry: Ford, GM, and Chrysler.[18]

According to the procedure defined by APQP, the supplier should follow the procedure and submit the relevant engineering planning document and deploy quality activity for quality part development.

Figure 5.21 explains the overall APQP process[19] that suppliers should follow, and the SQE team should evaluate the process and the activities of suppliers through APQP milestones. During the APQP activity, relevant quality engineering tools that we discussed previously should be appropriately used: QFD, D-FMEA, FTA, P-FMEA, and DOE.

5.2.8 PPAP (Production Part Approval Process)

PPAP is the output of APQP activity, and the process is triggered in Phase 4, Product & Process Validation of APQP. Actual measurements are taken of the parts produced and are used to complete the various test sheets of PPAP.

The purpose of PPAP is to ensure that products made from production tools and processes meet engineering requirements during an actual production run. The supplier should understand all customer engineering design and specification requirements, and the process should have the potential to produce products that consistently meet these requirements during an actual production run at the quoted production rate.[20]

We need to use data on both the subject product or part and the corresponding process captured from a significant production run to complete the 18 Elements of PPAP document summarized in Table 5.2.

TABLE 5.2

PPAP Package Elements[21,22,23]

No.	Document	Contents
1	Design Record	Technical evidence of the saleable product; CAD/CAM data, drawing, specification, performance data, etc. The "Design Records" could be a drawing of the part. This is normally provided by the customer because they are the ones who would normally design the part and understand its end use.
2	Authorized Engineering Change Documents	When an engineering change occurs in the product or tooling, the change request and approval must be included in the PPAP package.
3	Customer Engineering Approval	This usually refers to a positive outcome of a trial at the customer's facility conducted with the production parts from the supplier.
4	D-FMEA	Identified design risks should have corresponding corrective actions with design changes in the drawing, specification, performance data, etc.
5	Process Flow Diagram(s)	Process Flow Diagrams (or flowcharts) document the details of the production process, sequence, quality assurance, etc.
6	P-FMEA	Identified process risks should have corresponding corrective actions with the evidence of the improvement through trials.
7	Control Plan	The control plan defines all methods for process control and check methods of special characteristics and specifications with a sampling method.
8	MSA (Measurement System Analysis) Studies	The supplier should confirm the reliability of the measurement system or test equipment used for the development and production of the part.
9	Dimensional Results	Dimensional verifications required by the design record and the Control Plan should be completed and documented.
10	Records of Material / Performance Test Results	The supplier should have records of material and performance test results for tests specified in the Design Record or Control Plan.
11	Initial Process Studies	These studies evaluate the process stability and capability to meet special characteristics by the customer; C_{pk}, P_{pk} from 300 consecutive parts.

(continued)

TABLE 5.2 (*Continued*)

PPAP Package Elements[21,22,23]

No.	Document	Contents
12	Qualified Laboratory	Inspection and testing are to be performed by a lab showing evidence that the laboratory is qualified or accredited.
13	Appearance Approval Report (AAR)	An AAR is required for each part that has appearance requirements noted on the Design Record—it requires the customer's approval.
14	Sample Production Parts	The supplier must provide the customer with a sample part for approval—the part must come from a significant production run.
15	Master Sample	A Master Sample required by the Design Record and Control Plan is to be approved by both the customer and the supplier.
16	Checking Aids	Checking aids are special tools used to inspect, test, or measure parts during the production process—the reliability shall be confirmed by MSA studies.
17	Customer-Specific Requirements	The supplier should have records of compliance with customer-specific requirements, which should be included in PPAP documentation.
18	Part Submission Warrant (PSW)	The PSW is a summary of the PPAP submission complete with all data for the 18 elements for each customer part number; for example, information such as part number, name, materials, weight, dimensional measurement data, quality checking equipment, etc.

5.3 Quality Management in Engineering

A big part of the knowledge economy comes from the power of innovative technologies. Innovative technologies are not possible without science, mathematics, technology, and engineering. In the tech industry, engineering knowledge is a necessary resource for the effective operation and management of a company. In quality management in the tech industry, quality engineering methods and tools are indispensable.

> *In the Eastern style of management of tech companies, engineering knowledge is indispensably required for most management positions. For example, in Japan and Korea, CEOs or presidents of most tech companies have engineering backgrounds. Historically, most CEOs of H-Company had engine development or manufacturing experience. In the same way, almost all CEOs of S-Company are from engineering or manufacturing. Such a practice is in stark contrast to the Western style of management positions, where many top managers have business or finance backgrounds. This difference drives different approaches for company operations: Western management focuses on short-term financial outputs, while Eastern management focuses on long-term growth focusing on products.*

As for product design, QFD, D-FMEA, FTA, and DOE are must-use tools to manage the quality of new product development. The statistical approach can make engineering design processes scientific and effective. For instance, the statistical tolerance design is more potent than a conventional tolerance design method.

As for manufacturing process control, 7-QC Tools, P-FMEA, Control Plan, and statistics are essential tools. For instance, when measuring and adjusting production equipment statistical analysis is essential. In the 1930s, Shewhart developed a process-focused quality control method when he was in Bell Labs: instead of 100% inspection or sorting, Shewhart's approach was to use samples. When the process variability is less than the design requirement, we can trust that we can accept the products made from the process, instead of measuring or inspecting every part. Today, Shewhart's approach is known as SPC (statistical process control).[24]

> *SPC in Quality Management* requires the process to be controlled statistically based on sampling data analysis from a stable process. The purpose of manufacturing process control is to produce the right quality parts *consistently with minimum variance*—the variance cannot be eliminated. How much we can minimize the variance depends on the process capability. Regular industrial products require *minimum* 3-sigma capability; in other words, 99.73% of manufactured parts are within three standard deviations of their specifications. However, some safety parts require 6-sigma process capability: 3.4 defects per million parts.

The control chart is an indispensable tool for a SPC approach. The control chart can track changes in the process over time and determine if the process is stable and *capable* within current process conditions.

The control chart uses the average and the standard deviation of sampling data from an established stable process. The most popular chart type is *x*-bar and *R*-Chart as discussed in the previous section. For example, the previously discussed PPAP process requires a minimum of 25 subgroups containing at least 100 readings from consecutive parts of the significant production run.[25]

Figure 5.22 shows how we can use the control chart for quality management: *quality plan, quality control, quality assurance, and quality improvement* (see Eq. 3.1). A traditional method of process control is designated only by the design requirements of *USL* (upper specification limit) and *LSL* (lower specification limit); however, this cannot guarantee a stable manufacturing process to ensure that only good parts are made in the first place (note that USL and LSL are different from *UCL* (upper control limit) and *LCL* (lower control limit)). The process control chart explained in the previous section can identify potential process instability issues in advance before they go beyond the design specification of *USL* and *LSL* by using *UCL* and *LCL* and UCL_R.

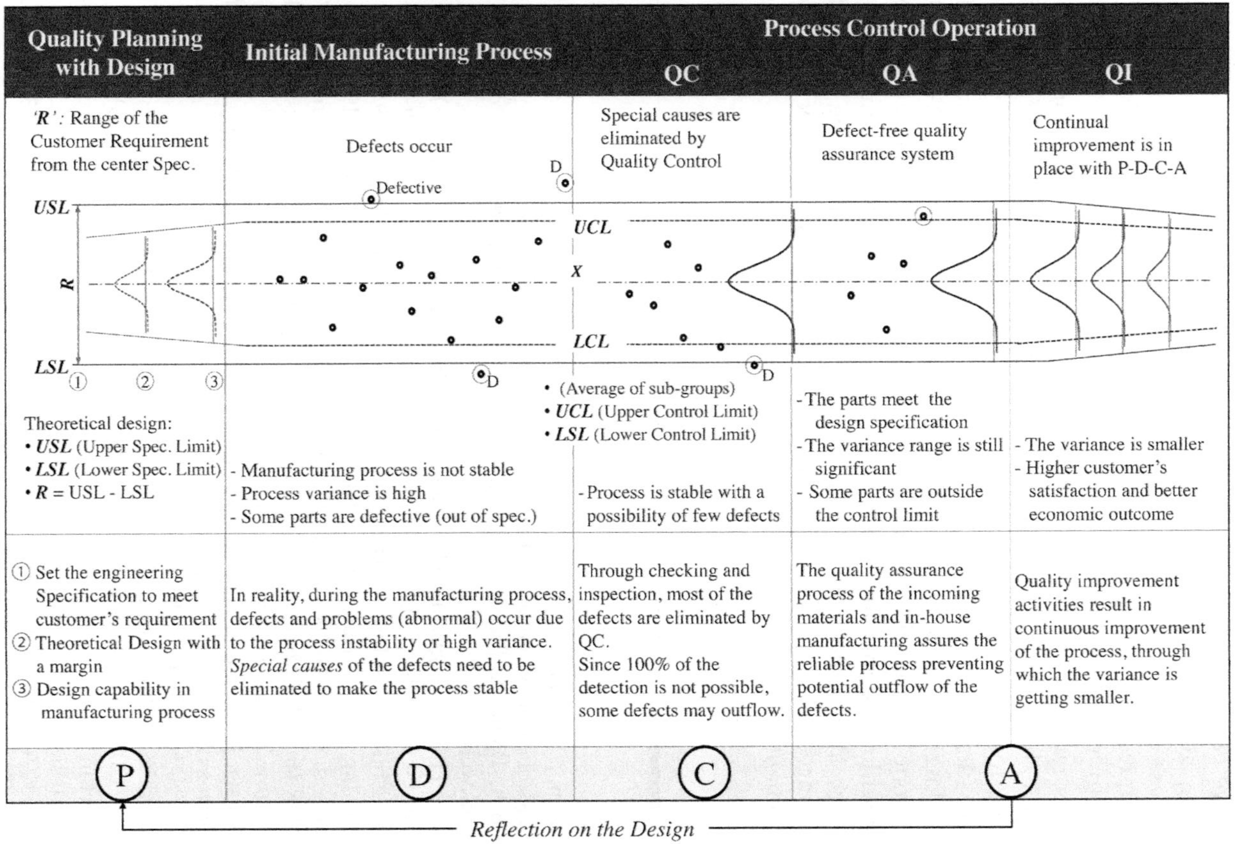

FIGURE 5.22
Application of SPC for quality management.

The general procedure of quality management by SPC is shown in Figure 5.22. First, any special causes for the unstable process should be eliminated by *QC (Quality Control) activity*. Once the process becomes stable, we can create x-bar and R-Chart with UCL, LCL, and UCL_R. By inputting UCL, LCL, and UCL_R in *QA (Quality Assurance) process activity*, the process becomes increasingly reliable in producing only good parts. However, we do not stop there; we then initiate *QI (Quality Improvement) activity* to minimize the variance in the process. Such QI activity increases customer satisfaction and brings about better economic outcomes for the company, resulting in cost savings with fewer defects and better-performing products that satisfy customers.

As we discussed earlier, Quality System, Quality Management, and Quality Engineering activities are linked together intrinsically. We cannot address one without the others. As for quality engineering activities, the author will delve into an in-depth discussion in his future sequel book, *Technology Development and Quality Engineering*,[26] which will be focused on advanced engineering approaches and methods with technology development.

Notes

1. Juran, J. M. *Juran on Quality by Design*. The Free Press, 1992.
2. Ross, Sheldon M. *Introductory Statistics*, Elsevier, 2005.
3. Koch, Richard. *The 80/20 Principle and 92 Other Powerful Laws of Nature*, Bookwell, 2014.
4. Ross, Phillip J. Taguchi Techniques for Quality Engineering, McGraw-Hill, 1988.
5. Cause-effect Diagram, https://sixsigmastudyguide.com/cause-effect-diagram-aka-ishikawa-fishbone-herringbone-fishikawa/ (accessed January 19, 2019).
6. Deming, W. Edwards, *Quality, Productivity, and Competitive Position*, 1982.
7. Jaiswal, Eshan S. "A Case Study on Quality Function Deployment (QFD)," *IOSR Journal of Mechanical and Civil Engineering (IOSR-JMCE)*, ISSN: 2278-1684 Volume 3, Issue 6 (Nov-Dec. 2012), pp. 27–35.
8. Hauser, John R. and Don P. Clausing (1988), "The House of Quality," *Harvard Business Review*, 66, 3, (May-June), 63–73.
9. *FMEA*, 3rd Edition, AIAG, 2001.
10. SPACE.com Staff, June 27, 2011, "Space Shuttle Mission Chronology," www.space.com/12025-space-shuttle-missions-1981-2011.html (accessed January 20, 2019).
11. Egerton Consulting, "Choosing between Failure Modes and Effects Analysis (FMEA) and Fault Tree Analysis (FTA)," February 20, 2015, https://egertonconsulting.com/fmea-v-fta/?doing_wp_cron=1537265428.9716870784759521484375 (accessed January 20, 2019).
12. Fault tree analysis, https://en.wikipedia.org/wiki/Fault_tree_analysis#History (last modified February 14, 2019).

13. Jenab, Kouroush and Saeid Moslehpour. "Failure Analysis: Case Study Challenger SRB Field Joint," *International Journal of Engineering and Technology*, Vol. 8, No. 6, December 2016.
14. Telford, Jacqueline K. "A Brief Introduction to Design of Experiments," *Johns Hopkins APL Technical Digest*, Volume 27, Number 3 (2007).
15. Dynamic Range in Digital Photography, www.cambridgeincolour.com/tutorials/dynamic-range.htm (accessed January 20, 2019).
16. Telford, Jacqueline K. "A Brief Introduction to Design of Experiments," *Johns Hopkins APL Technical Digest*, Volume 27, Number 3 (2007).
17. Barker, Thomas B. *Engineering Quality by Design*, Marcel Dekker, INC, 1990.
18. AIAG, "APQP (Advanced Product Quality Planning), www.aiag.org/quality/automotive-core-tools/apqp (accessed January 20, 2019).
19. "Advanced Product Quality Planning and Control Plan," APQP 2nd Edition, AIAG, July, 2008.
20. "Production Part Approval Process," *PPAP*, 4th Edition, AIAG, June, 2006.
21. "Advanced Product Quality Planning and Control Plan," *APQP 2nd Edition*, AIAG, July, 2008.
22. Quality Training Portal, https://qualitytrainingportal.com/resources/apqp-resource-center/what-is-ppap/ (accessed January 20, 2019).
23. PPAP Documents (Production Part Approval Process), http://ppapdocuments.com (accessed January 20, 2019).
24. Pyzdek, Thomas. *Quality Engineering Handbook*, Marker Dekker, Inc., 2003.
25. Production Part Approval Process, *PPAP*, 4th Edition, AIAG, June, 2006.
26. Lim, Jong S. *Technology Development and Quality Engineering* (forthcoming).

6

Scientific Problem Management

Every day we experience issues we have to resolve, whether they be big or small—so too in business. From new product development to mass production, as well as in the market after the launch of a new product, a variety of problems can occur. The typical quality problems can be categorized by *timeline*: design problems, manufacturing problems, initial quality problems, and durability problems. Further, the types of problems can be categorized by their *characteristics,* such as appearance problems, noise problems, functional problems, performance problems, conformance problems, safety problems, and so on.

Major quality problems occurring in the market can be grouped by the source types as shown in Table 6.1. We should look at the problems not only from a technical perspective (Type-1) but also from a management perspective (Type-2). The technical issues, Type-1, are comprised of a product design problem (Type 1-1) and a process problem (Type 1-2). Shipping defective products is a management issue (Type-2 problem). They are shipped because the problem is not detected, which is a Type 2-1 problem. Even where the problem is detected, if no robust countermeasure is applied, this is another management issue, which is a Type 2-2 problem. *Type 1 and Type 2 issues are always linked together; therefore, the problems should be analyzed in both respects: technical and management perspectives.*

Quality problems can occur at any time under any circumstance. It is essential to react to such issues *promptly to mitigate the scale and severity* of the fallout from the problem. At the same time, we need to create a mechanism to ensure that the same type of problem *does not recur* in the future. In this respect, a systematic problem management method, which the author refers to as "Scientific Problem Management," should be prepared and instilled throughout the organization.

In this chapter, we will focus on introducing *"scientific thinking and methods"* for reactive and proactive quality problem management.

1. Scientific Thinking and Method
2. Problem Solving Methods and Tools
3. Preventive Quality Management

TABLE 6.1

Quality Problem by Source Types

Category	Source (Reason)
Type-1: Technical Issue – Why is the problem created?	(Type 1-1): Product design problem (Type 1-2): Process problem
Type-2: Management Issue – Why is the problem shipped?	(Type 2-1): Detection problem; the problem is not detected (Type 2-2): No Robust CMR applied; the problem is detected; however, no robust countermeasure is applied.

6.1 Scientific Thinking and Method

Bertrand Russell emphasized the importance of science in his book, *The Impact of Science on Society*. Science has had substantial effects on our way of life, thinking, and technological development throughout history. Science has two functions: (1) to enable us to know things; and (2) to enable us to do things.[1] We are living in an age of uncertainty. Forecasting the future based on scientific thinking is a necessary ability for survival.

Scientific thinking is a rational way of thinking through the mathematical, statistical, logical, and psychological processes supported by evidence. Scientific thinking leads to engineering inventions. Scientific thinking is a universal language in the development of new technology through engineering activities. At the same time, *scientific thinking produces scientific skills and methods, which can be applied to solving issues and problems that arise in the daily operation of a business.*

Science, technology, and engineering are strongly correlated; the more scientific power we have, the more technology and engineering capability we can have.

6.1.1 The Scientific Process and Methods

To solve problems in science, technology, and engineering, scientific methods are indispensable. Scientists develop generalized theories or solutions for problems by using the scientific method as illustrated in Figure 6.1.

The scientific method has five fundamental steps and repeated processes to generate a precise model or theory:

1. Make an observation.
2. Analysis through a question.
3. Form a hypothesis or testable explanation.
4. Make a prediction based on the hypothesis.
5. Test the prediction.

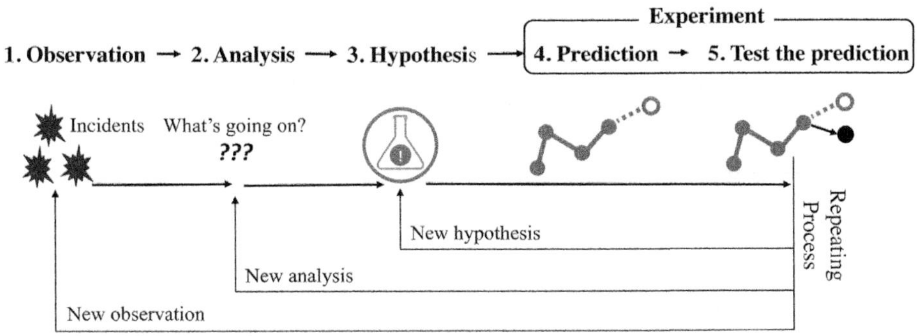

FIGURE 6.1
Scientific process and methods.

These five steps can be summarized as three core steps: observation, hypothesis, and experiment. *We are all scientists consciously or unconsciously; when we come across interesting areas or problems, we make an observation, form a hypothesis, and perform an experiment.*

In spite of thorough analysis and logical thinking during the process, we cannot be perfect the first time in the observation, hypothesis, or experiment. Therefore, feedback and corrective processes are necessary, as shown in Figure 6.1. For instance, if the predicted results of the experiment are consistent with the hypothesis, there is evidence to support the hypothesis. If the two do not match, we adjust methods, seek an alternative explanation, and redesign the experiment.

When we use scientific processes and methods in our quality management, *the process can be repeatable, correctable, and improvable in a systematic way.* These scientific skills and methods to solve problems can be termed *"scientific problem management,"* which needs to be taught to managers and engineers.

6.1.2 Fast and Slow Thinking

In 2011, Daniel Kahneman, a Nobel laureate, described two modes of problem-solving methods in his book *Thinking Fast and Slow.*[2] Our brain has two fundamental thinking modes: System 1 is the "Automatic System," informed by knowledge and experience, and System 2 is the "Effortful System," used to consciously think through an issue in a systematic way.

- System 1 is best to use when *the issue is simple,* the cost of being wrong is low, and the potential consequences are acceptable. For instance, when we confront an issue that we have seen many times before, we can decide on an action quickly. We can judge many things based on our expertise and experience. That is the power of knowledge and experience.

- System 2 is best to use when *the issue is complex*, the cost of being wrong is high, and the consequences are unacceptable. For instance, when we confront an issue that we have never seen before, we need time to judge through Effortful System thinking.

In the real world, problems are oftentimes a combination of both the simple and the complex. When many factors are mixed together, it is difficult to see the whole picture of the problem. Figure 6.2 demonstrates a practical process to solve the problems by using System 1 and System 2, which produces fast thinking and slow thinking, respectively. The flow chart represents the following procedures:

- remove non-important issues through prioritization and ranking, which was discussed in Chapter 4
- define the complexity of the issues
- solve the simple issues by using System 1
- work on the complex issues by using System 2

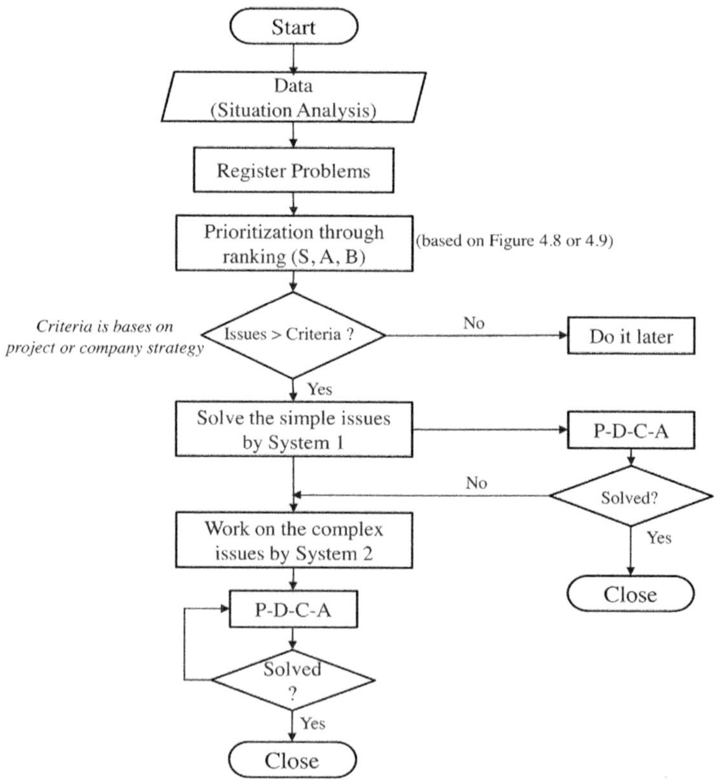

FIGURE 6.2
Problem solving process by fast and slow thinking.

In many cases, the real problems remain in a complex form where more than one parameter is interacting with each other independently.[3] These issues need to be addressed through rational and scientific methods.

6.1.3 Kepner-Tregoe Method

In the book *The New Rational Manager*,[4] Charles Kepner and Benjamin Tregoe explain a rational way of problem solving and decision making. The Kepner-Tregoe (KT) method was developed in 1958 and has been widely used for management in industry.[5] This *rational management model* has been mainly used in business management; however, the methodology can also be applied to quality management.

As we discussed earlier, the problem of our fast way of thinking is that it can bias our judgment, which comes from the wrong interpretation of the information, emotions, or cognitive bias, etc. Thus, the System 1 thinking is not a proper approach for complex issues or when the stakes are high.

As for complex issues, we should take a systematic approach by using the slow thinking process like the KT method, which *fundamentally follows the scientific process and method* as explained in Figure 6.1.

Figure 6.3 graphically represents the four patterns of thinking and procedures of the KT method.

1. *Situation appraisal* is clarifying the situation, prioritizing concerns, and choosing a direction—similar to "Observation" in Figure 6.1.
2. *Problem analysis* is defining the problem and determining its root cause—similar to "Analysis" in Figure 6.1.

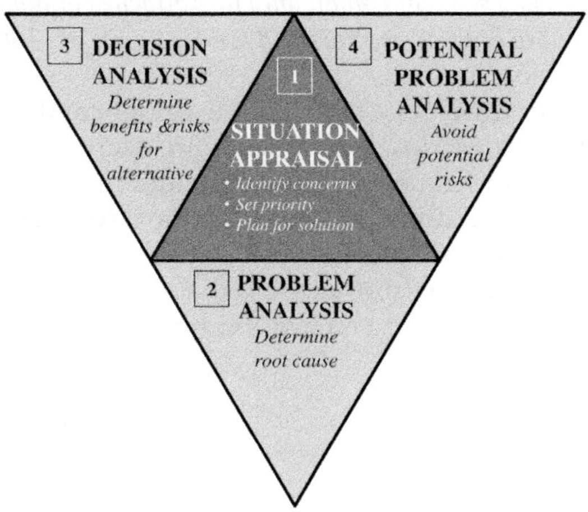

FIGURE 6.3
Scientific problem management by KT-model.

3. *Decision analysis* is identifying alternatives, benefits, and risks for each alternative—similar to "Hypothesis" and "Prediction" in Figure 6.1.
4. *Potential problem analysis* is scrutinizing potential problems and negative consequences and proposing countermeasure actions to minimize the risk—similar to "Test the prediction" and "Repeating process" in Figure 6.1.

Case Study: Internet Speed Problem

T-Company has built a second plant beside their existing plant. The second plant is intended to share the utilities as much as possible with the first plant. The employees working in the second plant begin to experience internet connection and speed problems, and the complaints escalate. Let's study how IT department engineers can analyze the situation and solve the problem by using the KT-method.

1. Situation Appraisal

The observation of problems should initiate an alert. In this step, we need to understand the contents of our concerns clearly. Then we need to set a priority on problems as discussed in Figures 4.8 and 4.9.

There are two types of initial reports by IT department engineers:

- *A-engineer says: The employees complain that the internet is slow.*
- *B-engineer says: The employees complain the internet is slow at any time of the day. When they send an E-mail with files attached or download attached files, it takes more than 30 seconds to download a file of ~100Mb size.*

There is a significant difference between the understanding of the issue by the two engineers. B's description is much more useful for the relevant people and managers to understand the situation and manage the concern. The concern can be narrowed down to a few specific issues rather than remaining entirely open for all possibilities.

The critical lesson from the above case is that we need to identify concerns in an understandable format by answering questions based on *what, when, how, where,* and *who.*

2. Problem Analysis

A primary purpose of problem analysis is to identify the true root cause of a problem, focusing on *"why."* Many techniques can be used, but more importantly, logical thinking and processes should be followed.

Let's revisit the previous example of the "Internet slowness issue." Let's see how IT department engineers can take step-by-step approaches to find the root cause of the problem to solve the internal customer complaint.

- *Problem: Sending and receiving files through the network is slow.*
- *The speed of receiving and sending files from a Wi-Fi–connected laptop is three times slower than wire-connected PCs for the same file size.*
- *There can be potential hardware and software problems: Internet speed, network communication software in a laptop, Wi-Fi router speed or signal strength, E-mail server problems, and so on as identified by using the cause and effect diagram shown in* Figure 6.4.
- *Wi-Fi–connected equipment has a more significant problem depending on the users' working locations; a possible cause is the signal strength of the Wi-Fi router.*
- *When the laptop was moved to within 10 meters of the Wi-Fi router, the problem disappeared—sending or receiving a file of ~100Mb size took 5 to 7 seconds instead of 30 seconds.*

Through this way of logical thinking and processes, we can identify *a potential root cause of the problem.*

3. Decision Analysis

The primary purpose of decision analysis is to identify what needs to be done to address the cause(s) found in the problem analysis. The decision

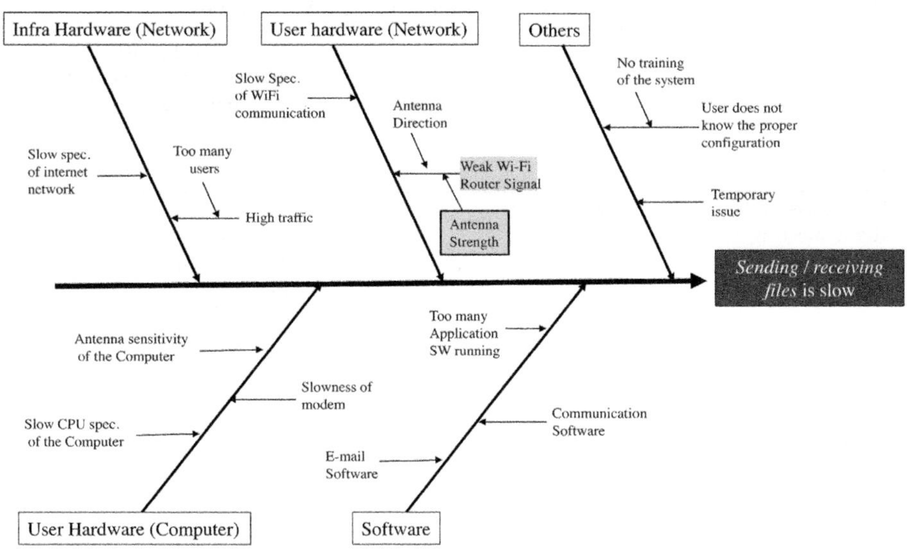

FIGURE 6.4
Cause and effect diagram for problem analysis.

TABLE 6.2

Determination of Critical Factors and Weighted Value

	Critical Factor	Weight
1	The signal should reach > 20m (across the open space)	"Must"
2	Signal accessibility ratio to meeting and testing rooms	10
3	Competitive total cost budget	8
4	Time of installing new routers	7
5	Data transmission speed	9
6	Impact on visual or environment concerns	7

analysis process should develop criteria for *a final decision for countermeasures* (e.g., Must and Want factors). *Several alternatives* should be evaluated systematically and logically to choose the best option to resolve the problem. The procedures are as follows:

- Determine *critical factors* for possible countermeasure options—Table 6.2.
- Assign a *relative weight value* for each alternative—Table 6.2.
- Calculate the *weighted score of each factor* for each alternative and add them up—Table 6.3.

TABLE 6.3

Calculation of Critical Factors and Weighted Value for Alternatives

		(Option-A) New Router: 802.11ax[6]			(Option-B) 30% More Routers of the Current 802.11.n		
Factor	**Wt.**	**Facts**	**Score**	**Wt. Score**	**Facts**	**Score**	**Wt. Score**
Signal accessibility ratio to meeting and test rooms	10	It can access more than 93% (current: 85%)	7	70	It can access more than 97% (current: 85%)	9	90
Competitive costs	8	~ $0.5M	9	72	~ 1.0M	7	56
Time of installation	7	Purchasing order to installation ~ 3 months	9	63	Purchasing order to installation ~ 5 months	7	49
Data transmission speed	9	Theoretical speed ~ 1300 (Mbps)	9	81	Theoretical speed ~ 450 (Mbps)	7	63
Impact on visual or environment concerns	7	No change compared with the 802.11n router	9	63	Having more routers makes them more visible	8	56
Total				349			314

TABLE 6.4

Possible Adverse Effect of the Best Alternative

	Possible Adverse Effects of Option-A	Degree of Impact (Risk Probability and Severity)
1	7% of rooms still will not have Wi-Fi signal	Reasonably acceptable
2	Some old computers may not employ the new Wi-Fi protocol	Old computers will phase out in two years

- From the analysis (Table 6.3), we are now able to compare each alternative numerically to determine the best alternative.
- Once we have determined the best score alternative from Table 6.3, which is Option-A, we need to list out *potential adverse consequences* for our choice alternative and evaluate probability and severity (Table 6.4).
- If the risk probability and severity are acceptable, we can make a final single choice—Option-A, in this case, to install a new protocol router, *802.11ax*.

4. Potential Problem Analysis

A decision is never a perfect decision, but an optimized one under a specific circumstances. The optimized decision can have potential weak points and risks. Therefore, identification of vulnerable areas of the decision can give an opportunity for management to prepare for a contingency plan or continual improvement.

Good managers always prepare a PPA (Potential Problem Analysis) plan in case of unexpected future problems. Regarding the PPA, the first thing managers should do is consider concerns for an uncertain future path, situation, or event while following a predetermined decision and action.

In Quality Engineering, we have a scientific tool that has been used in industry for a long time. The most popular and useful method for PPA is FMEA (Failure Mode and Effects Analysis) method. FMEA is one of the first highly structured and systematic techniques for failure analysis, as we discussed in Chapter 5.

By using the fundamental concept and procedures of FMEA, we can form a simple structure for PPA as shown in Table 6.5.

The last column, "Recommended Future Action," is the most important thing we need to pay attention to before we close this problem. In other words, we need to prepare any future plans and actions for continual improvement of the situation.

In this case, the countermeasure by a new Wi-Fi router of 802.11.ax may be suitable for overall signal speed improvement, but may not help much to penetrate the wall-enclosed meeting rooms. In the future they may build more wall-enclosed meeting or test rooms, so that the new protocol Wi-Fi may not

TABLE 6.5

Potential Problem Analysis by Using FMEA Concept

Original Failure Mode	Potential Effects of Failure	Current Decision Taken	Potential New Failure Mode	Recommended Future Action
The coverage area of the Wi-Fi router is limited	• Access to the internet and E-mail exchange can be slow due to weak signal • Some locations cannot access the network at all	• Higher speed Wi-Fi communication protocol (802.11.ax)	• Outdated Wi-Fi router (802.11n) cannot use a new protocol (802.11.ax) • The number of Wi-Fi routers installed is not enough	• Monitoring of the improvement ratio (speed and accessibility) • Installation plan (router locations and required number of routers) • Secure a budget for more routers for a 99% coverage ratio

help much to reach those rooms. Therefore, the IT department should consider installing more Wi-Fi routers or other options with a proper budget prepared. This is a kind of closed loop process, P-D-C-A, for continual improvement.

6.2 Problem Solving Methods and Tools

In the first section we focused on *general* scientific principles and methods of problem solving. In this section, our primary focus is on practical methods and tools to resolve actual quality problems that typically occur in the manufacturing industry.

6.2.1 Principles for Problem Management

Many types of quality problems are shipped to the market mainly because they were not managed in the system properly, whether they are technical or management problems. The causes of most problems occurring in the manufacturing industry are technical issues. However, once the problems occur, they become management issues. In this respect, we also need to use our scientific problem management methods focusing on the Type 2: Management issues in Table 6.1.

1. *Principle 1: Recognize the problem:* A primary reason for shipping the quality problem is because of (Type 2-1): Detection problem in Table 6.1. The cause of the problem is mainly due to a management issue—*no recognition of a problem. A problem recognition process is highly related to how we define something as a problem.*

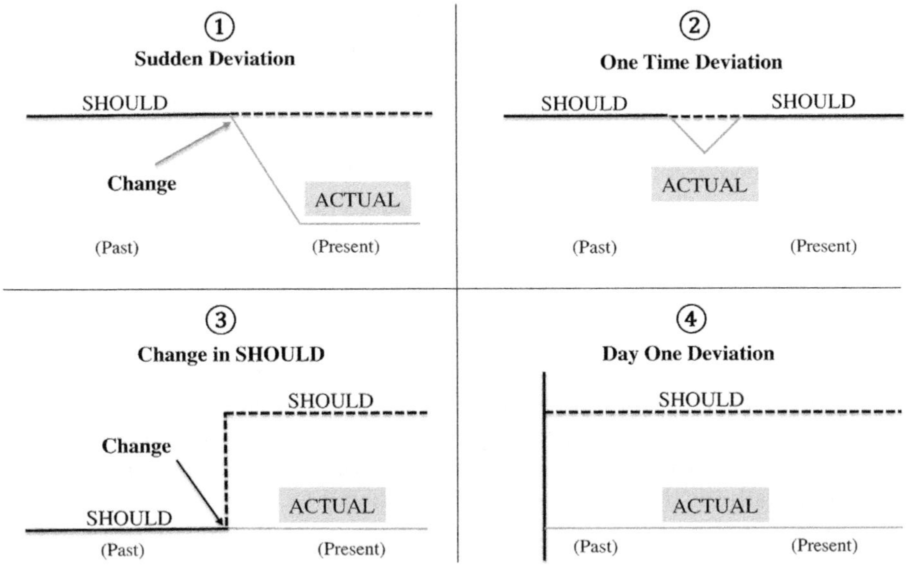

FIGURE 6.5
Recognition of four styles of problem.

Figure 6.5 explains four different styles of problems. All four styles have a deviation between what SHOULD be happening and what is ACTUALLY happening. Without a clear definition of what SHOULD be happening, we cannot recognize any gap, problem, or concern.

For example, consider an expensive new drilling machine installed in the manufacturing line. The management requests the machine operator that the machine should be running faster for higher ROI (return on investment). However, if the manager does not specify the target (SHOULD), e.g., 5 parts or 100 parts per hour, nobody can recognize a gap or problem.

2. *Principle 2: Record the problem:* Once we have detected problems, management should confront the reality *transparently*. Nobody likes bad news; however, hiding the problem or underestimating the concern can result in a loss of opportunity to solve the problem at the right time by the right method. Therefore, all issues must be *recorded in a problem tracking database with control numbers* like (b) in Figure 4.8.
3. *Principle 3: Prioritization of the problem:* Every project has its timeline to complete the milestone requirements to move to the next stage. Many quality problems can easily daunt management. As a result, critical issues sometimes slip away without introducing robust countermeasures, potentially resulting in a Type 2-2 problem as described in Table 6.1.

A fundamental principle of problem management is "first things first"—we need to solve important and urgent issues first. We discussed two methods of prioritization in Chapter 4: one is *"Prioritization Ranking by Business Needs,"* shown in Figure 4.8, and the other is *"Prioritization Ranking by Risk Levels,"* shown Figure 4.9.

We need to choose a proper ranking method for our project or event, typically by *"S," "A," "B," and "C" ranking*. Such a prioritization process is one of the necessary steps for scientific problem management.

4. *Principle 4: Rational management of the problem:* The fourth principle is a *critical thinking* process to solve the issues recorded in the tracking system by *rational management*. Intuition-based management, relying on experience and knowledge, has a risk in that our quick judgment can be biased, which can lead us to a wrong conclusion. It can result in a waste of time, opportunity, or resources. That is why we should minimize intuitive decisions and maximize the scientific processes and methods that we discussed in Section 6.1.

6.2.2 5P for Problem Solving

Efficient management of quality problems requires more than logical-principle approaches like Kepner-Tregoe. In the real world, situations are dynamically moving, complex, and changing. Much more practical processes and tools are necessary to manage daily issues promptly and efficiently.

Figure 6.6 explains 5P (5 Principles), which is used for problem solving by Honda and their suppliers. The basic concept and approach have a similar root and approach to the Kepner-Tregoe method.

The specific meaning of each principle is:

1. Problem definition *(What happened?)*
2. Identify the root cause *(Why it happened?)*
3. Put corrective countermeasures in place *(What will be done about it?)*
4. Confirm the countermeasures have worked *(Are the countermeasures working to resolve the problem?)*
5. Feedback/feedforward information *(How can we prevent any recurrence of the same kind of the problem in the future?)*

The principles alone are difficult to use in the field. There should be a standard process and format for everybody in the organization to understand and follow. Figure 6.7 presents a format necessary to apply 5P by responsible departments.

This is a one-page summary sheet; if additional information is necessary, it can be attached to this summary sheet. The format looks very simple; however, it is following the P-D-C-A closed loop principle as discussed in Chapter 4.

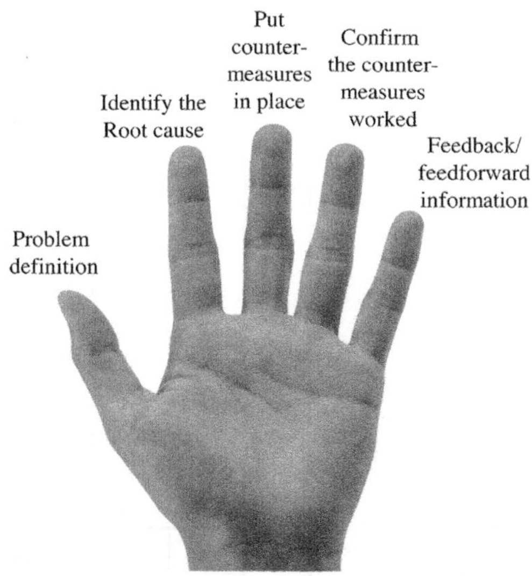

FIGURE 6.6
Five principles for problem solving.

First, "Problem Tracking" is a unique serial number following the code of a specific project or plant, and priority "Ranking" needs to be decided according to the company standard. And "Problem Statement" should be a simple title identifying the problem object and the content of the defect.

1-1. *Problem Definition* should describe the problem by using 5W2H principles: *who* found the problem; *which* part has the problem, including (object) model and type; *what* is wrong with it, describing the (defect) symptom of the problem; *where* the part was located when the problem was found; *when* the problem was first found, specifying the date and time; *why* it is a problem; *how* the problem was found, e.g., visual inspection or customer complaint; and *how* many parts or units have this problem.

1-2. *Problem Definition* describes the problem contents in detail by providing an in-depth analysis of the problem. It should include any past problem history, trends, other known variables, and containment measures taken. The detailed information can be used to identify true causes in Section 2-1.

2-1. *Identify Root Cause* deals with the root causes of two issues: first, "why was the defect made from the beginning," which is a more technical issue, and is a Type-1 problem as defined in Table 6.1, and second, "why was the defective product shipped," which is a management issue, a Type-2 problem as defined in Table 6.1. The questions should be answered in respect of both technical and

5-PRINCIPLES (Problem Solving Sheet)

Problem Tracking #:__________ Rank:__________

Resp:	ISSURER			RECEIVING DEPT.		
Task:	Prepared By	Checked By	Approved By	Checked By	Reported to	closed by
Name/Dept:						
Date:						

PROBLEM STATEMENT (OBJECT/DEFECT)

1-1 PROBLEM DEFINITION

Use the 5W2H to describe the effects and how thy were detected.

1-2 PROBLEM DEFINITION

Summary of analysis used to find and confirm true causes.

2-1 IDENTIFY ROOT CAUSE

(PROBLEM) WHY MADE	(WHY,WHY) Analysis of **Occurrence**
WHY SHIPPED	(WHY,WHY) Analysis of **Detection**

2-2 IDENTIFY ROOT CAUSE

Root Causes to be C/M and Why Selected

Root Causes to be C/M and Why Selected

3 CORRECTIVE COUNTERMEASURE(S)

Countermeasure Implementation	Responsibility	Date

4 CONFIRM COUNTERMEASURE(S)

How	Results	Confirmed by	Date

5 FEEDBACK/FEEDFORWARD

Feedback to	Method	Who	Date

FIGURE 6.7
Five principles format for problem solving.

management viewpoints—both Type-1 and Type-2 issues. In this process, a multiple whys approach, Why-1, Why-2, Why-3, Why-4, and Why-5, should be taken for thorough investigation of the issue and identification of the true root causes.

2-2. *Identify Root Cause* is to select the final root cause from 2-1 with reasons for the selection. Also, this section should identify the cause that needs countermeasures in later sections.

3. *Corrective Countermeasure(s)* is to identify the temporary and permanent countermeasures to take. Each countermeasure should state the name of the responsible person who will deploy the countermeasure with a clear description of the schedule.

4. *Confirm Countermeasure(s)* is to identify the effectiveness of the temporary and permanent countermeasures taken. This section should clarify the confirmation methods linked with the root cause identified in Section 2.2. The confirmation results need to be described in detail with the information of the person in charge and the date of confirmation.

5. *Feedback/Feedforward* is recurrence prevention activity by a systematic process. The methods of feedback/feedforward should be clarified with the responsible person deploying the activities. Most of the standard methods are registering the problem into PPH (past problem history) database and reflecting them into FMEA activity, PQCT (process quality control table), and operation standard, followed by training the relevant workforce.

Since the 5P methods is a simple document, a small misunderstanding or a lack of understanding of each section can result in ambiguity and confusion regarding the problem and solution. Therefore, widespread training followed by a good habit of using the sheet is mandatory for successful results.

Figure 6.8 is an example case to show how we can use the 5P format and procedures to solve real-world problems.

The problem case in Figure 6.8 is "creaking noise from the cup holder cover when opening and closing." This looks like a relatively simple case: the root cause of the problem is the missing of the grease application in the opening gear mechanism. However, every problem at the beginning is difficult to analyze due to many potential causes. A fundamental approach should be similar to forensic methods. The truth lies with the part and conditions in which the part is being used. The service department needs to listen to the customers' usage conditions and complaints very attentively. The evidence lies in the customers' vehicle from which the complaints are arising. Using the 5P method, we can conveniently follow scientific problem-solving procedures.

In this particular case, the noise level is not very noticeable at normal temperature conditions. Therefore, the standard procedure has difficulty detecting the specific noise in a typical manufacturing environment due to the high background noise. The lesson in this analysis is that this kind of robustness issue should be addressed as a critical control point in the design and process stages. The potential failure mode

5-PRINCIPLES (Problem Solving Sheet)

Problem Tracking #: FP-001 **Rank:** B-rank

Resp:	ISSUER			RECEIVING DEPT.		
Task:	Prepared By	Checked By	Approved By	Checked By	Reported to	closed by
Name/Dept:		K-Suppliers			Part Quality Dept.	
Date:						

PROBLEM STATEMENT (OBJECT/DEFECT)

FP5 Model Cup-holder Creaking Noise during operation

1-1 PROBLEM DEFINITION

- Who found the problem? Customer
- What part has the problem? Cup holder cover
- What's wrong with it? Creaking noise during the opening
- Where was it found? Market
- When was the first problem found? January 15, 2019
- How many parts had this problem in the market? n ≈ 60 parts

1-2 PROBLEM DEFINITION

The customer noticed there was a creaking sound in the morning when opening the cup holder cover. The service man identified there was grease application missing in the opening gear mechanism.

2-1 IDENTIFY ROOT CAUSE

	Why-1	Why-2	Why-3	Why-4	Why-5
WHY MADE?	Application of grease is missing in assembly line	Line workers missed the job	Newly hired manpower	No training of manpower	No standard requirement of training
WHY SHIPPED?	No noise was detected in line	Background noise is high	Room temperature does not make big creaking sound	No sound proof chamber in assembly line	

2-2 IDENTIFY ROOT CAUSE

Training and checking procedure were not in the quality management system.

In-line quality checking system and method are not practical.

3 CORRECTIVE COUNTERMEASURE(S)

Countermeasure Implementation	Responsibility	Date
•1000 parts were checked @ factory. 10 parts were found N/G	- Rick, Assy	•Feb. 1, '19
•Update of QMS of training	- Jack, Assy	•Feb. 20, '19
•Installation of sound proof in-line test facility for sampling check	- Mike, Mfg. Eng'ing Dept.	•Mar. 1, '19

4 CONFIRM COUNTERMEASURE(S)

How	Results	Confirm by	Date
•Check QMS Manual	-Updated	Kim, Q-dept.	•Feb. 22,'19
•Interview of workers	-Training is good	Kim, Q-dept.	•Mar. 15,'19
•Check the new facility for noise detection capability	-Can detect the noise	John, Assembly	•April. 2,'19

5 FEEDBACK/FEEDFORWARD

Feedback to	Method	Who	Date
•Emily, NM Dept.	- Meeting, PPH DB, FMEA	Jack	•Feb. 26, '19
•Ross, ME Dept.	- Meeting, PPH DB, PFMEA	Mike	•Mar. 20, '19

FIGURE 6.8
Problem solving by using 5P. (Example: Cup holder noise.)

should be in the PPH database and FMEA's potential failure modes during the development. Also, the OEM company should ask their suppliers to identify the control mechanism both in design and in their manufacturing process control.

The 5P method is a simple but very comprehensive and convenient tool that can be used by different levels of people from managers to assembly line workers.

6.2.3 8D Report for Problem Solving

Eight Disciplines is a problem-solving method that was used during World War II by the U.S. government. In 1974 it was first described by the U.S. Department of Defense in a military standard for the management of nonconforming materials by their contractors.[7] Later, the method was further developed and used extensively in the late 1980s by the Ford company.[8] This systematic method has been used to resolve product and process problems. This method has now become prevalent in the automotive industry, and today many automotive companies and auto parts suppliers are using this method.

The primary purpose of 8 Disciplines is to identify, correct, and eliminate recurring problems based on teamwork. Its basic concept and approach are similar to the 5 Principle method to solve problems scientifically. It starts with a proper description of the problem, identifies the root cause, applies countermeasure(s), confirms the effect of the countermeasure(s), and deploys a feedback/feedforward method. However, it explicitly adds team structure and congratulations on the output recognized by the organization, whereas in 5P, these factors are implicit.

8D follows the logic of the P-D-C-A cycle in the format shown in Figure 6.9. The 8 Disciplines are:[9]

D1: Use a Team: Establish a team of people with product/process knowledge.

D2: Describe the Problem: Specify the problem by identifying it in quantifiable terms using 5W2H—who, what, where, when, why, how, and how many. *This is the same concept as Principle-1 of 5P.*

D3: Develop an Interim Containment Plan: Define and implement containment actions to isolate the problem from the customer. *In the 5P, this is combined into Principle-3.*

D4: Determine and Verify Root Causes and Escape Points: Identify all relevant causes that could explain why the problem has occurred. Also, identify why the problem was not detected at the moment it occurred. All causes must be verified or proved. One can use five whys or cause and effect diagrams to map out causes against the effect or the problem identified. *This is the same as Principle-2 of 5P.*

D5: Verify Permanent Corrections for the Problem: Design a proper way of verification in the same way that the problem occurred and might occur in the customer's usage conditions. *This is the same as Principle-4 of 5P.*

D1: Team members
Who are your team members and who is going to be the team leader? What are the skills and functions? Create your problem solving team.
D2: Describe the problem
What is the problem? When did the problem occur? What is the problem history? Define the problem as objectively as possible. If possible, conduct a 5W2H Analysis as an addition to define a clear description of the problem.
D3: Interim containment action
What containment actions can be taken? Which (temporary) fixes can be integrated for preventing a problem from getting worse until a permanent solution is implemented?
D4: Identify the root cause
Identify all possible root causes that could explain why the problem occurred. Methods such as the fishbone diagram or the 5 whys method can help. Why was the problem not noticed at the time it occurred?
D5: Developing permanent corrective actions
What permanent corrective actions can be developed to solve the problem(s)? What are the deadlines? What is the proof of effectiveness?
D6: Implementing permanent corrective actions
Which permanent corrective actions will be implemented and who is responsible for the implementation? What are the deadlines?
D7: Preventive measures
Prevention is the best cure. Which additional measures need to be taken to prevent similar problems? What lessons are learned and could be applied to other problems?
D8: Congratulate the team
Congratulating the team for the results will boost the team motivation. How could the team be rewarded and how could this help fixing problems in the future?

FIGURE 6.9
8D format for problem solving.[10]

D6: Define and Implement Corrective Actions: Define and implement the best corrective actions. *This is the same as Principle-3 of 5P.*

D7: Prevent Recurrence/System Problems: Modify the management systems, operation systems, practices, and procedures to prevent the recurrence of this or similar problems. *This is the same as Principle-5 of 5P.*

D8: Congratulate the Main Contributors to Solving the Problem: Recognize the collective efforts of the team. The team should be formally thanked by the organization.

The same case study we used for 5P in Figure 6.8 can be applied to the 8D Report format in Figure 6.9. Figure 6.10 shows that both of the concepts and structures of the 5P and the 8D reports are highly comparable.

6.2.4 QC-Story

There are other types of problem management tools and formats that can be used for scientific problem management. For example, N-Company

8D REPORT

Title	FP Model Cup-holder Creaking Noise	**Date Opened** Jan. 15, 2019	**Last Update** Mar. 20, 2019
Product/Process Information	FP5 Model Cup-holder	**Organisation Information**	K-Supplier
D0 Symptom(s)	When the customer Opening the cup-holder, it makes noise in the moring.	Sketch	
D0 Emergency Response Action(s) Stock and in-transit Quantity)	Checked all stocked parts in the supplier site	% Effective 99%	Date Implemented Feb. 1, '19
Verification/ Validation	Operation noise check in a sound proof room		
D1 Team (Name, Dept. Phone)		**D2 Problem**	
Team Leader	Kate, Assembly Dept	Problem Statement:	FM Model Cup-holder Creaking Noise
Team Members	Rick, Assembly Dept Jack, Assembly Dept. Mike, Mfg. Engineering Dept.	Problem Description:	The customer noticed there was a creaking sound in the morning when opening the cup holder cover.
D3 Interim Containment Action(s)	1000 parts were checked in house @factory, 10 were found NG (noise)	% Effective 100%	Date Implemented Feb. 1, '19
Verification / Validation	100% Operation Noise check in a sound proof room before shipping	Very Effective	Mar. 1, '19
D4 Root Cause(s) and Escape Point(s)			
Occurrence Root Cause:	Grease was not applied due to the lack of training.	% Contribution	100%
Verification:	With the application of grease on gear linkage, the noise disappered	Date	Jan. 30, '19
Escape Root Cause:	The noise was diffcult to detect in the plant's nosy environment.	% Contribution	
Verification:	Checked the noise detection of the problem part -very diffciult to detect in normal manufactuing environement condition	Date	Jan. 30, '19
D5 Chosen Permanent Corrective Action(s)			
Occur Permanent Corrective Action(s):	Double checking of the grease application by secondary check process	% Effective	99%
Verification:	Checked the secondary process and marked by white pen	Date	Feb. 20, 2019
Escape Permanent Corrective Action(s)	Noise level double confirmation in a temporary sound proof room	% Effective	97%
Verification	The noise detection level was good (29 NG deteced out of 30 NG parts)	Date	Feb. 25, '19
D6 Implemented Permanent Corrective Action(s)	1. Update of QMS (Quality Management System) Manual for training and bi-monthly audit of the system 2. Installation of secondary check process for grease application 3. Instlaltion of a sound proof in-line test facility	Date Implemented	Feb.25, '19
Validation:	Confrimation of the manual and the facility usage	Date	Mar. 1, '19
D7 Preventative Action(s)	1. Add the problem in PPH (Past Problem History) data base of New Model and Manufactuing Engineering Dept. 2. Update of D-FMEA and P-FMEA	Date Implemented	
D8 Team and Individual Recognition	Quality Chief (Allan Muller)	Date Closed Mar 20, '19	Reported by Jack

FIGURE 6.10
Problem solving by 8D. (Case study: Cup holder noise.)

uses a "QC-Story" format. The basic structure of a QC-Story is composed of nine steps:

1. Title (problem)
2. Reason for the title (purpose)
3. Phenomenon confirmation (situation analysis)
4. Target
5. Root cause analysis
6. Countermeasure (CM)
7. CM effect
8. Standardization
9. Next action plan (Capitalization)

Problem management may require more than just one "tool," just as a carpenter uses more than a hammer to build a house. Some companies may want to use much simpler tools and methods. No matter what formats the company decides to use, the scientific principles and structures should be maintained: clarification of the problem, implementation of outflow prevention activity, root cause analysis, study of countermeasures, confirmation of the effectiveness, and recurrence prevention activity. The overall mindset and critical thinking approach must be instilled into the workers by utilizing the *learning by doing principle* with coaching by experts. Once a specific tool has been chosen, everybody in the organization should be trained to follow and respect those procedures.

Research shows that scientific problem management tools significantly reduce the organization's quality issues and financial cost by addressing the quality issues promptly and preventing the defects from recurring.[11]

6.3 Preventive Quality Management

In our daily lives, we always experience some problems or issues due to unexpected situations. Some are minor nuisances, but some are critical problems endangering our business operations or our safety. We may want to ask two fundamental questions on the issue: Are the problems inevitable because they occurred abruptly? Alternatively, could some problems have been prevented if we had taken some actions before they occurred?

The answer is *yes* in most cases. We know from our experience that we could have avoided some problems or at least minimized the impact of the problems if we had done something in advance.

TABLE 6.6

Quality Problems by Modes of Failure

Failure Mode Types	Characteristics	Examples
Mode-1 Failure	The failure could have been prevented, or the risk could have been mitigated.	• Most Type-1 and Type-2 problems in Table 6.1 • Many quality failures and historical recalls
Mode-2 Failure	The failure is difficult to prevent or avoid.	• Some Type-2 problems • Extreme weather or unusual usage conditions
Mode-3 Failure	The failure is intentionally created by humans (e.g., situations of moral hazard).	• VW Emission scandal in 2016 • Mitsubishi Motor's fuel economy scandal in 2016 • Wells Fargo scandal in 2018[13]

In the previous section, we discussed how to manage quality problems scientifically. Many types of quality problems could have been avoided if we had done something better in advance, e.g., implementation of *a robust product design and stable control of the process, thorough training of the workforce, better control of the suppliers' quality, better problem detection method and management, etc.*

From a product quality viewpoint, there are three modes of problems or issues as summarized in Table 6.6.

The first mode of failure is the problem that we could have prevented or at least could have minimized the impact of. This mode of failure is the case in most Type-1 and Type-2 problems as shown in Table 6.1 for technical and management problems. Some examples are the Space Shuttle Columbia's disaster in 1986 and many historical recalls of automobiles.

The second mode of failure is an unexpected problem. Therefore it is impossible or very difficult to prevent it from occurring. For instance, functional defects can occur due to a natural disaster or unexpected product usage by customers. Some examples are water penetration into cars by sudden flood or using smartphones in extremely cold conditions.[12]

The third mode of failure is the human-made problems that we created intentionally with our knowledge. The emissions scandal in 2015 by VW and the fuel economy issue in 2016 by Mitsubishi Motors are some examples where the problems were created by the companies with their knowledge.

Table 6.6 shows a summary of the failure mode types and examples.

Mode-2 failure is very difficult to forecast or prepare for. A solution to prepare for extreme weather conditions or every usage condition is too costly or beyond engineering capabilities—there is necessarily a trade-off between technology and cost.

Mode-3 failure is a moral hazard issue as discussed in Chapter 4. In the automobile industry, there have been many similar scandals like the VW emissions scandal. For example, in 2016, Mitsubishi Motors' president stepped down due to the company's falsification of fuel economy data[14] followed by Nissan's inspection scandal in 2017.[15]

Now the remaining category is *Mode-1 failure*. Most of the Type-1 and Type-2 problems in Table 6.1 are problems that could have been prevented, or for which the impact could have been mitigated. Most quality problems belong to this Mode-1 category, and this is the target category for preventive quality management.

Problems do not occur suddenly. Usually, there are symptoms of risks beforehand. Most risks arise directly from apparent problems, but some arise from the weaknesses in the system, management, and engineering. Risk detection and risk management are critical in preventive quality management.

6.3.1 Risk Management

The key strategy of preventive quality management is *"risk management." "Risk" and "problem" are different.* A *risk* is an uncertain future event that could have an adverse effect (threat). However, a *problem* describes a 100% certain condition that exists now, threatening customer satisfaction.

Typically, risks come before problems. Generally, there are some symptoms of risks before problems occur. At an earlier stage, a risk can be detected only by an expert, but a problem can be recognized easily by the general public. If risks are not appropriately managed, they develop into problems.

The purpose of risk management is to develop a countermeasure plan or to mitigate or eliminate the potential consequences of risks.

There are four steps in risk management:

1. Categorization of potential risks
2. Risk detection and analysis

3-1. Risk management (proactive)

3-2. Risk management (reactive)

Figure 6.11 shows the risk management process in the P-D-C-A flow. Based on a product and a business type, *expected risk types can be predefined in the early project planning stage.* Then risk detection activity needs to be deployed during the project—*knowledge and experience are essential for risk detection skills.* Once potential risks have been detected, necessary risk management actions need to be followed in the form of *"3-1. Proactive risk management"* and *"3-2. Reactive risk management." Proactive risk management is identifying potential risks in the same or other projects, and reactive risk management is mitigating the impact of the risks identified.* Both types of risk management require countermeasure activities as defined in Table 6.7, and the effects should be re-verified through "2. Risk Detection."

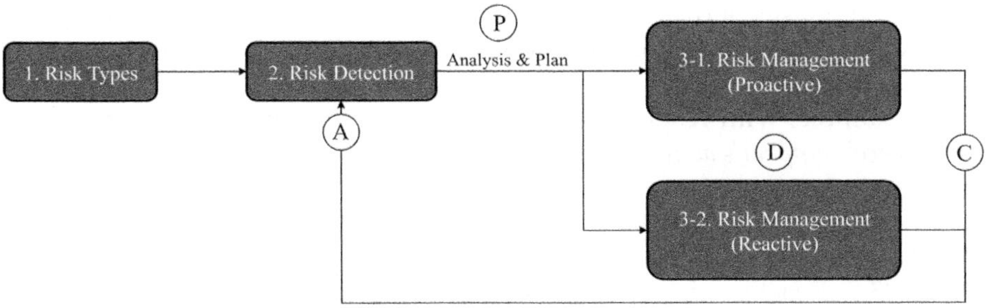

FIGURE 6.11
Risk management flow.

Table 6.7 shows examples of risk types, risk detection by symptom, and risk management methods. These risks are the typical cases that most manufacturers experience during a new product's development. The risk detection goes by the symptoms that developers (designers, test engineers, manufacturing engineers, etc.) can experience during the verification and

TABLE 6.7
Potential Risk Types, Detections, and Management

1. Risk Types	2. Risk Detection (Symptoms)	Potential Problems (Examples)	3. Risk Management
• Design is not robust	• Product performance is below expectation • Function failures during the validation test	• Customers complain about bad fuel economy • Navigation display freezes	• D-FMEA/FTA • CAE • Software Bench Testing
• Supplier Quality Management is not robust	• Parts do not fit well • Functional defects found during pilot production • Datasheet provided by the supplier is not accurate	• Parts are difficult to assemble in-line mfg. • High warranty issues in market	• APQP/PPAP • Supplier's FMEA/FTA • Supplier Process Capability Check and Improvement • Firewall Activity
• In-house Process is not stable	• Many assembly mistakes during the pilot run • Body accuracy variance between products	• Customers complain about abnormal squeak noise from the seat areas • Wind noise comes from the front left side door.	• P-FEMA • PQCT • Training • Process Capability Check and Improvement
• Repair and Service are not well prepared	• Service and maintenance methods are not economical • Service parts flow is not established	• Customers complain about long waiting times and high service fees • Reentry for the same repair is high	• Maintenance friendly design • Proper repair tools and training • Forecasting of necessary initial service parts

validation phases of the design and the product. The risk management column shows several methods and tools that are deployable at the beginning of the project.

For example, with respect to *supplier quality risk management, it is imperative to check supplier's activity of quality planning, control, assurance, and improvement, deploying necessary quality engineering methods and tools before a new model launch.*[16] The technical details of each method (D-FMEA, P-FMEA, APQP, PPAP, etc.) were explained in Chapter 5.

The "3-2. Reactive risk management" process defined in Figure 6.11 is a critical part of risk management. If the apparent risks are well managed, they are *less* likely to become problems. However, it is less practical to eliminate all level of risks because not all risks become problems. Therefore, a proper *risk assessment* is an essential step to prioritize reactive actions for risk management. We can use the risk criteria predefined in Chapter 4.

> *Let's say that there are two kinds of risks identified during new product development.*
>
> *(Risk-A): "The noise level of the air cleaner is 2 dB higher than the previous model."*
>
> *(Risk-B): "During the field testing of a new model, the test team experienced the digital instrument panel turning off one time."*
>
> *From a quality viewpoint, both of these risks can be quality problems disappointing customers' expectations. However, the risk levels are very different.*
>
> *The first one, (Risk-A), can generate some uncomfortable feelings for noise sensitive customers. However, they will be less likely to ask for a vehicle exchange or ask for a recall. Furthermore, not all customers will be able to catch the 2 dB difference.*[17] *So this issue can be graded as a low-risk item, which belongs to "C-rank" category by the definition in Figure 4.9.*
>
> *However, the second one, (Risk-B), is a functional issue from the beginning. The instrument panel is supposed to display driving information to a driver without interruption. If the driver cannot get the right information, the driver would have difficulty making a sound judgment of his driving conditions, which means it can be a potential safety issue. Furthermore, proper display of driving information is a regulation item in most countries. Even though the problem occurred only one time during the test driving, the criticality is very high. Therefore, this issue should be designated as "S-rank" by the definition in Figure 4.9.*
>
> *Now priority of risk management is clear between (Risk-A) and (Risk-B), and we should focus more on (Risk-B) rather than (Risk-A).*

6.3.2 Preparation

Despite all the scientific risk management, unexpected risks without warning may happen. It is like a sudden rainfall without any weather forecast of rain. The people living in rainy regions during monsoon season always carry an umbrella just in case. For proactive risk management as defined in Figure 6.11, we need this kind of mentality—*preparation just in case.*

During new product development, not all risks are identified. Products are becoming more complicated, product variety is increasing, and the product development cycle is becoming shorter. According to a survey, time-to-market has dropped by 12% over a three-year period (2014–2017).[18] Therefore, companies need to handle multiple product development projects simultaneously. The development speed of new products is increasing, and less time and budget are allocated for quality validation. In such a highly competitive business environment, it is too idealistic to think we can launch a perfect quality product.

Therefore, *proactive risk management* with a prepared back-up solution is also an essential part of preventive quality management.

> *In 2016, the Samsung Galaxy Note 7.0 had a battery problem. The company eventually found the root cause of the battery melting and fire. Samsung's reaction was neither fast nor accurate.[19] Typically, new product development has a design review, prototype build event, verification and validation test, and a pilot-production event. During the development, some potential risk issues should have been discovered. In the case of Samsung's Note 7.0, it had a severe design risk that compromised the safety space margin in the battery design.[20] The company should have noticed the risky design so that they could have prepared an improvement solution as proactive risk management.*

There are several management tools and methods for preparation as a part of preventive quality management: preparation of a back-up solution; firewall and risky parts management; new product launch team; one-voice communication manual.

- *Preparation for a Back-up Solution:* Despite preventive quality management activities, there could be some remaining issues at the time of a new product launch. The remaining issues can range from high-risk grade to low-risk grade items with a low occurrence probability. One of the best ways of risk management is to prepare a *back-up solution in advance.* If any similar symptoms of the problem occur in the market after the product launch, the manufacturer has an option to promptly apply the prepared back-up solution or start from the back-up solution ideas that were prepared.

  Such preparation is proactive risk management that makes prompt reactive risk management possible.
- *Firewall and Risky Parts Management:* Some suppliers have an inherent weakness in their quality management or process capability. We can detect risks through their mass production quality of the current products or pilot production of new products. *Risky parts suppliers need to be selected for firewall activity that prevents bad parts from being shipped from their plants.* The client plant needs to request that

the suppliers have special control of potential bad parts that might impact its quality.

Firewall management includes but is not limited to 100% of additional function check, 100% check by gauge, 100% of dimensional measurement, and/or 100% of visual inspection, etc. This is a special temporary activity that can be requested to risky suppliers, but it should continue until the supplier's quality becomes stable enough to guarantee quality to the client.

Such activities are painful for the suppliers, but through this special activity they can give a minimum amount of quality assurance to the client's plant and their customers.

- *New Product Launch Team:* No matter how perfectly we prepare for a new product, there may be surprises in the market after a new product launch. This is because our validation process cannot mimic exactly the way our customers use our products and because our validation samples cannot entirely represent the total number of the products in the market.

  Therefore, it is wise to prepare for surprises after a new product launch. When there are surprises there should be a contact point through which we can receive critical information accurately and promptly. Traditionally, the customer service team and the call centers are the primary information source, but nowadays, more accurate information can be collected from an online portal or social media. A special mission team composed of cross-functional experts is very effective in monitoring surprises and reacting accordingly. Figure 6.12 shows a simplified structure of a new product launch team.

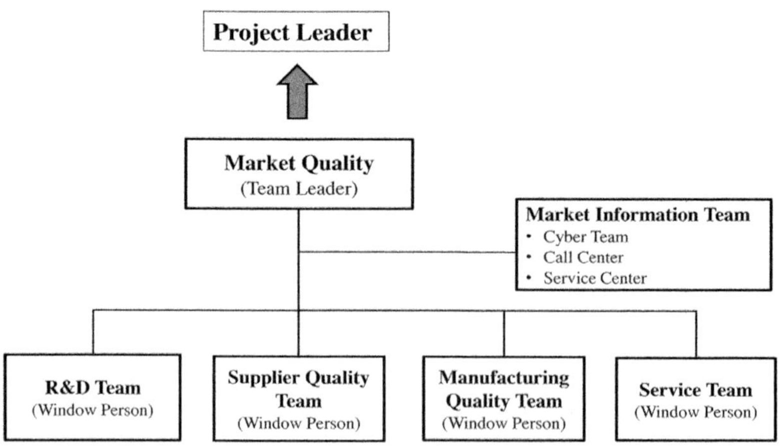

FIGURE 6.12
Organization of new product launch team.

In the team structure, the role of the market information team is critical. The traditional ways of collecting data directly from customers are old-fashioned and slow. Much more information, called "Big Data," can be collected from social media where customers upload and share the good and bad points of the product.

Manual analysis of Big Data is very challenging for a prompt and accurate reaction. With the help of Big Data processing technology, the team can efficiently search, process, and analyze Big Data (see section 11.2.5). The AI can alert the team daily and hourly for critical issues.

Risks can be broken down into functional categories and allocated to each functional group for in-depth analysis and countermeasure planning. Speed and accuracy are the core elements for a successful operation of the team.

- *One-Voice Communication Manual:* Usually, a new product is launched with new features, functions, and technologies. Customers may not be used to these new things at the beginning and can even complain about the awkwardness of the new functions and technologies. In such a case, the new product may receive mixed reception in the market. Typically, service manuals cannot cover all possible misunderstandings or misusage of the new product by customers.

  Before the launch of a new product, the concerns of new features and functions need to be pre-identified through customer monitoring or clinic activities. Then the identified concerns need to be addressed through a *one-voice manual* format (see Figure 6.13).

  This *one-voice manual* needs to be communicated to the relevant people, such as the cyber team, call center, and service center so that when customers have concerns or complaints the teams can react properly with *one consistent voice.*

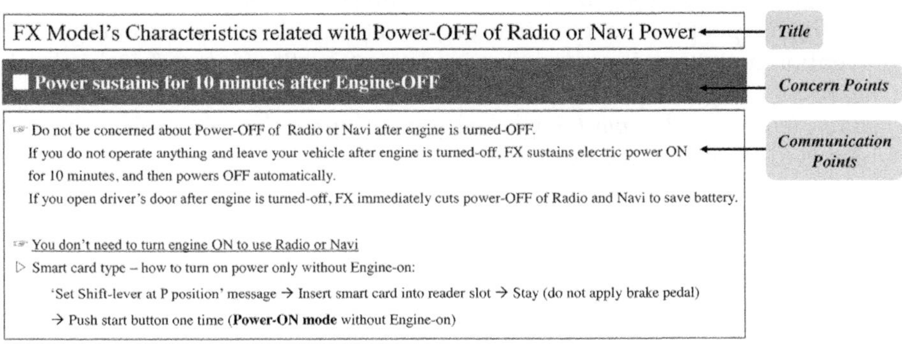

FIGURE 6.13
Example of "one-voice communication" manual.

6.3.3 Audit for Vigilance

Whatever robust system or method we set up, if the organization is not vigilant, the systems or methods will not be respected by the employees. Many kinds of risk warning messages can be neglected or ignored. As an analogy, let's say that highly sensitive radar is installed in the air defense system. If the operator of the equipment does not pay attention to warning signals, the radar becomes useless. In the same way, the risk management and preparation system can be useless if the organization does not respect the quality procedures.

Therefore, it is crucial that the organization be vigilant in its monitoring of risk signals. To instill this kind of culture and spirit into the organization, one efficient management method is to deploy regular *audits*. There are two kinds of audits; one is an external audit, and the other is an internal audit. The external audit usually focuses on system audits and procedures like an ISO 9001 audit. As we discussed in Figure 3.7, many internal non-conforming practices are intentionally hidden from the external audit. Therefore, internal audits should focus on *actual practices* of the systems, processes, and management.

The major functional categories of actual management practices to check for are:

- R&D Quality Management
- Supplier Quality Management
- Manufacturing Quality Management
- Service Quality Management

The way of deploying an audit is important as well. The audit team should focus on the *contents* of quality management of real cases and projects. *The auditing should be more than the judgment of good or bad, but to identify weak points for improvement.*

Preventive quality management may sound like too much work. However, as we discussed in Chapter 2 and showed in Figure 2.12, upstream problems cause *"10 times more cost"* in the downstream processes.

Intelligent people and effective organizations always pour energy into prevention activities. Juran emphasized the "Quality Trilogy" in his book, *Quality By Design*[21]: quality planning, control, and improvement. *Scientific quality planning is the most important part of preventive quality management,* which is discussed in the next chapter.

Notes

1. Russell, Bertrand. *The Impact of Science on Society,* Unwin Hyman, 1976, Chapter 6. Science and Values.

2. Kahneman, Daniel, *Thinking Fast and Slow*, Farrar, Straus and Giroux, 2011.
3. Barker, Thomas, *Engineering Quality By Design*. Marcel Dekker, Inc., 1990.
4. Kepner, Charles and Benjamin Tregoe. *The New Rational Manager*, Princeton Research Press, 1981.
5. Kepner-Tregoe, www.kepner-tregoe.com/about-kt/company-overview/benjamin-b-tregoe-phd-our-founder/ (accessed January 22, 2019).
6. Wen, Howard. *Network World*, October 1, 2018, "802.11ax preview: Access points and routers that support the Wi-Fi 6 protocol on tap," www.networkworld.com/article/3309439/mobile-wireless/80211ax-preview-access-points-and-routers-that-support-the-wi-fi-6-protocol-on-tap.html (accessed January 24, 2019).
7. Eight Disciplines Problem Solving, https://en.wikipedia.org/wiki/Eight_Disciplines_Problem_Solving (last modified March 1, 2019).
8. Krajnc, Marjanca. "With 8D method to excellent quality," *Journal of Universal Excellence*, October 2012, Year 1, No. 3, pp. 118–129.
9. Eight Disciplines of Problem Solving (8D), https://quality-one.com/8d/ (accessed January 25, 2019).
10. 8D Report Template, www.toolshero.com/problem-solving/8d-report/ (accessed January 25, 2019).
11. Eight Disciplines Problem Solving, https://en.wikipedia.org/wiki/Eight_Disciplines_Problem_Solving (last modified March 1, 2019).
12. Patricia, November 29, 2014, "HOW TO HANDLE IPHONE EXTREME TEMPERATURE WARNINGS," www.iphonetricks.org/how-to-handle-iphone-extreme-temperature-warnings/ (accessed January 25, 2019), iPhone's a non-operating temperature interval ranges between -4° *to 113° F*, that's -20° to 45° C. *If your device's temperature drops below or rises above its operating limits, iOS will prompt you with a warning message and ask you to warm it up or cool it down.*
13. Flitter, Emily. February 2, 2018. "Federal Reserve Shackles Wells Fargo After Fraud Scandal," www.nytimes.com/2018/02/02/business/wells-fargo-federal-reserve.html (accessed January 15, 2019).
14. BBC News, "Mitsubishi to stop selling eight models in fuel scandal," August 30, 2016, www.bbc.com/news/business-37218510 (accessed January 25, 2019), *"The fuel economy on eight models were as much as 8.8% and on average 4.2% lower than advertised," the transport ministry said in a statement.*
15. "Nissan chief to return part of salary following inspection scandal," *Financial Times*, November 17, 2017, www.ft.com/content/27d9dc68-cb87-11e7-ab18-7a9fb7d6163e. *The scandal centers on the habit of allowing uncertified inspectors within Nissan to sign off on "final inspection" tasks conducted on behalf of the Japanese government.*
16. Toyota and Its Component Suppliers—Case Study, https://elsmar.com/pdf_files/Toyota%20and%20Component%20Suppliers.pdf (accessed January 25, 2019).
17. Sound Control, www.metroglass.co.nz/catalogue/100.aspx (accessed January 25, 2019), Under typical field conditions, the ear cannot detect a change of 1–2dB.
18. Dirlea, Violetka, Ram Kidambi, Stephan Krubasik, and Christine Sachseneder, "Quality 4.0 Preventive, Holistic, Future-Proof," www.atkearney.com/industrial-goods-services/article?/a/quality-4-0-preventive-holistic-future-proof (accessed January 25, 2019).

19. Chen, Brian X. and Choe Sang-Hun, "Why Samsung Abandoned Its Galaxy Note 7 Flagship Phone," *The New York Times*, October 11, 2016.
20. Shedletsky, Anna-Katrina. December 01, 2016 "Aggressive design caused Samsung Galaxy Note 7 battery explosions," www.instrumental.com/blog/2016/12/1/aggressive-design-caused-samsung-galaxy-note-7-battery-explosions (accessed January 26, 2019).
21. Juran, J. M. *Juran on Quality by Design*. The Free Press, A Division of Simon & Schuster, Inc., 1992.

7

Scientific Quality Planning

Most companies experience a significant gap between their quality targets and actual achievements. New products do not sell as well as expected, and unexpected quality issues and crises occur in the market. The companies then wonder what is wrong.

New products with new features and technologies do not necessarily increase sales. Certificates of the company like ISO or CMMI (Capability Maturity Model Integration)[1] do not warrant good quality outcomes.

Joseph Juran emphasized quality planning as a starting point for good quality outcomes. Without scientific quality planning, it would be difficult to trace the reasons for the gaps. Through the scientific and systematic deployment of quality, we can reflect weak processes or activities into our improvement activities to achieve our targets better.

The scientific quality planning process respects systematic, logical, and objective approaches from the planning stage. The planning should match the needs of the customer as well as the company. Figure 7.1 represents how a company's goal and strategy can lead to action at the bottom, and also how bottom-up activities can be aligned with the strategy and goal.

- A **goal** is a broad primary outcome of the business operation.
- A **strategy** is an approach we take to achieve a goal.
- A **tactic** is about how to deploy the strategy to achieve the goal.[2]

Business strategy is the firm's approach to achieving its vision and prioritizing objectives considering the business's environmental factors and organizational capabilities to succeed. The strategy needs to be cascaded and deployed to functional groups logically in the form of *"tactics."* Given the tactics, each functional group needs to take appropriate *actions*. These actions are real effort coming from an actual workplace, *Genba*, building quality inside products.

From this perspective, in this chapter we have an in-depth discussion of the following topics:

1. Quality Follows Business Strategy
2. Build Quality as Fundamental for Business Plan
3. Quality Goal Setting Process
4. Quality Strategy Setting
5. Specific Action Plan
6. Quality Planning of New Product (Case Study: Automobile)

FIGURE 7.1
Logical approach for organization activity.

7.1 Quality Follows Business Strategy

In earlier chapters, we discussed that *"Quality for Quality" has no meaning.* Quality should bring about value to customers as well as benefits to company business. *Quality should be reciprocal in the sense of customer satisfaction and the success of a company.*

Alfred Chandler said, "structure follows strategy."[3] By the same principle, *quality should follow a company's structure and business strategy.* Quality organization is a subset of the corporate business organization. We have to respect the company's business structure and strategy. In other words, *we need to align the quality strategy to the business strategy.*

Figure 7.2 explains how quality strategy and tactics need to be aligned with business strategy. Every organization and business has constraints due to business environments and organizational capabilities. That is why we need to devise a strategy—*business strategy leads to quality strategy.*

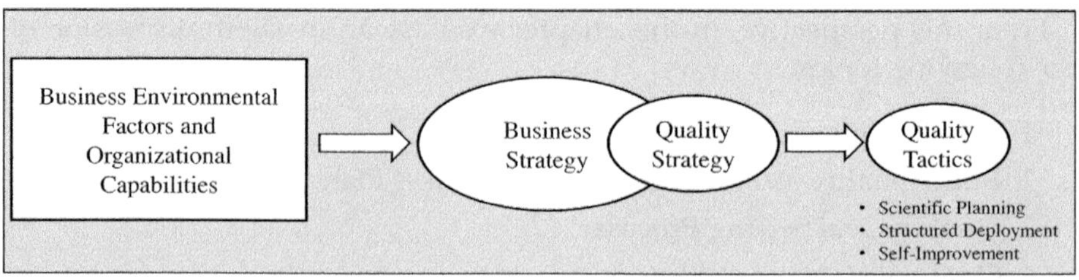

FIGURE 7.2
Quality follows business strategy.

The quality strategy is about building quality to support business strategy. For instance, "Number one market share in MRI (magnetic resonance imaging) equipment in the medical industry" can be a business goal. The relevant business strategy can be "aggressive market expansion in emerging economy countries by providing superb service quality." Here the business strategy dictates the quality strategy and tactics to be followed. Therefore, a relevant quality goal can be "Number one customer satisfaction in service quality by 2020." To achieve this quality goal, the relevant quality strategy can be "build after-service infrastructure in BRICS (Brazil, Russia, India, China, and South Africa) countries and provide speedy and highest quality service supported by locally trained, skilled service people."

7.2 Build Quality as Fundamental for Business Plan

Unfortunately, many business leaders take quality for granted until they encounter a quality crisis in the market. In such a crisis circumstance, regardless of any recovery efforts, such as a new *product*, a *promotion, a discounted price, and more distribution channels*, sales will continue to drop.

Why? Quality is a matter of trust in the company, the brand, and its products. Without trust, a good product is not a good product anymore, and a good price does not mean good value. Building quality as fundamental for business is the best strategy for successful business operations.

In Chapter 3, we discussed three quality pillars: *Quality System, Quality Management,* and *Quality Engineering*. Building these three pillars companywide inside functional groups vertically and horizontally is a critical quality activity for leaders to focus on. These three pillars also need to be raised up inside the organization through concurrent *Quality Planning, Deployment, and Improvement* as represented in Figure 7.3.

7.3 Quality Goal Setting Process

Good planning is a necessary condition for success. A good plan requires good forecasting. *Forecasting ability is the organization's capability,* which requires knowledge and experience. Good quality planning follows certain principles: data collection, information analysis, and decision making using a logical approach. We can call this *"scientific quality planning."*

Scientific quality planning has foresting capability: what actions and resources are required to achieve quality goals? When something is wrong, we should be able to identify what observation or hypothesis was wrong

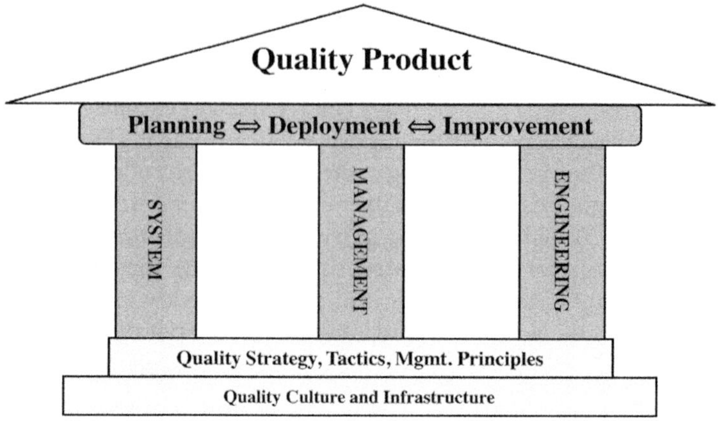

FIGURE 7.3
Build three pillars of quality.

from the planning stage. Good planning starts with a goal setting process. Well-defined, challenging goals enhance an organization's productivity. According to the Edwin Locke's Goal Setting Theory, "hard goals drive performance more effectively than easy goals, and specific hard goals produce a higher level of output than vaguely worded ones."[4]

Scientific quality planning needs to follow key principles and approaches in a logical way:

- set up quality goals and strategies required by business needs
- set up a quality organization, roles, responsibilities, and workflows
- make a specific action plan in the timeline

As stated earlier, quality goals should follow business requirements. *We should have a clear link between business goals and quality goals.* When setting business goals, the most critical parameter is the market where customers are buying or will buy the product.

By the same principle, when setting quality goals, we need to understand the market. The market represents reality, which is both a testing ground and a battlefield, showing evidence of customers' needs and what competitors are doing. Figure 7.4 explains the methodology of a market situation analysis for the purpose of quality goal setting. *The critical viewpoints are customers' needs, competitors' positions, the market trend, and the technology trend.*

Here, *customers* represent current and target customers or potential buyers. *Competitors* are the current or potential contenders to compete against and/or to learn from. The *market* represents an opportunity ground where the products will compete. *Technology* is the innovative features or performance required to make new products attractive.

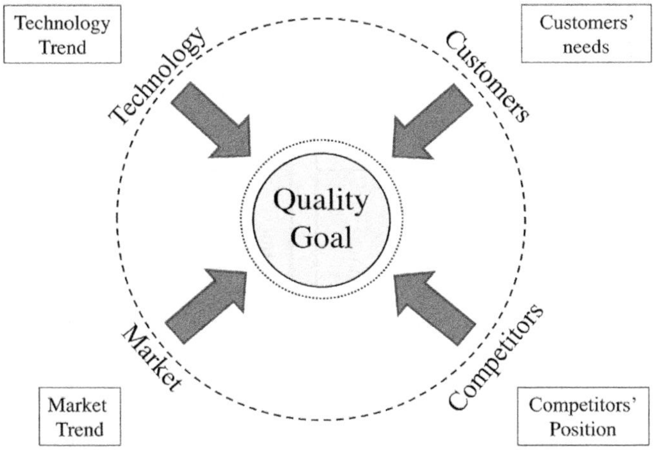

FIGURE 7.4
Market analysis methodology for quality goal setting.

From these viewpoints, we need to focus on the following areas: 1. *Who are our target customers, and what are we targeting for?* and 2. *How are our competitors doing in the market, and what is the trend of the market and technology?*

7.3.1 Defining Target Customers

Not every person is our current or even prospective customer. Our target customers are the people we are aiming for. First, we need to understand the current customers. With an excellent analysis of current customers we can identify potential customers who may be interested in purchasing our new products in the future.

The primary starting point is to identify the demographic distribution of our current customers by age, gender, income, education, and jobs, as shown in Figure 7.5. Such a demographic analysis is one of the most effective methods we can use to understand *our current and potential future customers.*

7.3.2 Benchmarking

Civilization has evolved through competition. Darwin's evolutionary theory, natural selection, is about survival through competition in nature. "Nature gives more lives than she can support."[5] From that perspective, we can paraphrase that quote as, *"the market gives more competitors than she can support."* We are competing in the market on price, technology, performance, quality, delivery, trust, and sales. Competition is a natural law to make things to move faster, smarter, more trustworthy, and more advanced.

Companies learn from other companies, just as athletes learn from other, better-performing athletes. Consciously or unconsciously, we compare

Demographics of Customers

[Sample(n) = 5432]

Age	Total
20's	18.3%
30's	44.6%
40's	25.9%
50's~	11.2%
Average	37.25yrs

Marital status	Total
Single living in Own household	6.8%
Single living with parents/relatives	23.5%
Married couple living with children	54.5%
Married couple Not living with children (including No child)	11.6%
Married living separately with a partner	2%
Others	1.6%

Income/month	Total
~ $4,000	17.4%
$4,000 ~ 6,000	45%
$6,000 ~	37.6%

Gender	Total
Male	59.5%
Female	40.5%

Region	Total
West Coast	43.9%
East Coast	24.9%
Northern States	12%
Sothern States	12.8%
Others	2.9%

Education	Total
Below middle school	0.5%
High school	13.9%
University	73.4%
Master's degree	12.3%

FIGURE 7.5
Demographic study of the current customers (Example: K-Company SUV Model).

ourselves with what others are doing and try to imitate better ideas. Idea diffusion has the most significant role in human civilization—even more so than independent inventions.[6] Benchmarking activities is part of idea diffusion—learning ideas and evolving them. Benchmarking is a fundamental learning process from competitors. The essence of benchmarking is the process of identifying the highest standards and *best practices* of excellence for products, services, or processes, and then making the improvements necessary to reach or exceed those standards.[7]

The major viewpoints of benchmarking activity should be:

- *where* are their unique strengths?
- *why* are they better?
- *how* are they doing that?

With this perspective, we need to set up a standard benchmarking process and expected outcomes. Figure 7.6 shows a quality benchmarking process adapted to meet our purpose of setting quality goals and strategies to be competitive in certain characteristics of the products, benchmarked with other better performers in the market.

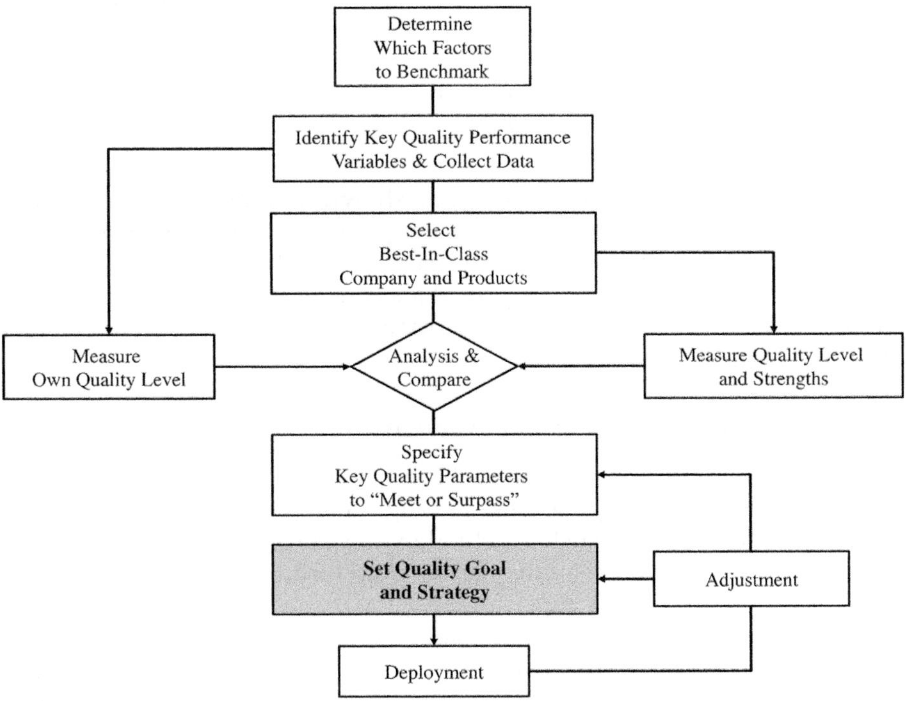

FIGURE 7.6
Benchmarking process. (Adapted from Bateman, 1989.)[8]

TABLE 7.1

Types of Benchmarking and Parameters (Examples)

Benchmarking Type	Viewpoints	Parameters (examples)
Company Strategy	The firm's approach or methods for achieving its vision & goals and prioritizing objectives	• Company tactics, management principles, etc.
Organization	Organization structures to perform specific activities	• R&D organization, manufacturing structure, etc.
Marketing	Special or unique marketing strategy and process for products and services	• Network set up strategy or process • Promotion strategy or process
Productivity	Performance of productivity	• Number of vehicles per workforce • Number of hours per vehicle, etc.
Product Cost	Cost structure and competitiveness of the product	• Component cost, selling cost, etc.
Product Features	Special features provided or options for customers to choose from	• Safety features of the vehicle, camera options, expandable memory, etc.
Product Performance	Special attributes that customers appreciate	• CPU speed, booting time, fuel economy, engine power, etc.
Product Functions	Special technical functions that make the product competitive	• Inductive charging, auto braking, auto cruise, etc.
Product Quality	Key quality factors that customers can perceive as significant value	• Initial and durable quality, and appeal quality, etc.

During the benchmarking activity, as shown in Table 7.1, we can measure the values of the primary interesting parameters like the examples in the third column of the table and compare them with our values. Then we need to determine and analyze critical success factors of the competitors. From that result, we can set the measurable, specific values as our quality goals and targets.

7.3.3 Understanding the Current Products

The customers who buy or leave a brand or its products can provide a detailed story of their experiences. Through a simple survey, we can find out valuable information, such as why they buy, why they do not buy, what they like, what they hate, why they leave, etc. This kind of study focuses on *purchasing behavior.*

Through a survey like the one in Table 7.2, we can identify the strengths and weaknesses of our current products. Then we can incorporate the study results into our future new product planning. We can focus on quality related items for quality goal and target setting, such as the italicized items in Table 7.2.

TABLE 7.2

Study of Purchasing Behavior

Reasons for purchasing	What do they like?
- *Safety*	- Brand Image
- Size	- *Driving performance*
- Etc.	- Etc.
What don't they like?	**Reasons for not purchasing**
- Styling	- *Quality*
- *Perceived quality*	- Price
- Etc.	- Etc.

7.3.4 Understanding the Market

Market research can show an overall landscape of the potential growth field—the position of our products versus competitors and the market trend. We can also get specific information about where we need to focus our products and quality efforts.

One useful method is to use perceptual mapping. *Perceptual mapping is a diagrammatic technique to display the perceptions of customers or potential customers visually.* A perceptual map shows an overall landscape of how the market is behaving. In this map, the position of a company's products, product line, or brand is displayed relative to the competition.[9]

Figure 7.7 shows an example of a market study explaining the overall trend of the target products as well as a growing segment in the market. This example explains that there is a growing trend in functional and quality premium areas. Therefore, we can focus on building more functional features and technologies as well as setting more aggressive quality goals to appeal to premium customers.

7.3.5 Product Audit

There is always a gap between our conceptual understanding of our product quality vs. the actual situation in the market. We tend to favor our own created products over others because of our propensity for "in-group favoritism." Sometimes it is not easy to identify the problems that customers experience in the field. Many VOC (Voice of the Customer) complaints seem to be less understandable than warranty claims.

One of the best ways to understand the reality of our products is to regularly audit and monitor the actual quality of the products being shipped. Figure 7.8 represents a simple and clear product audit process that we can apply to understand the quality situation of our products in the market. The audit result can then be used to set our quality goals for a new product.

During this process there are three essential activities: preparation of a checklist based on the market information, random sampling of final shipping products, and objective evaluation by experts. The output is the problem or concern list reflecting the market situation.

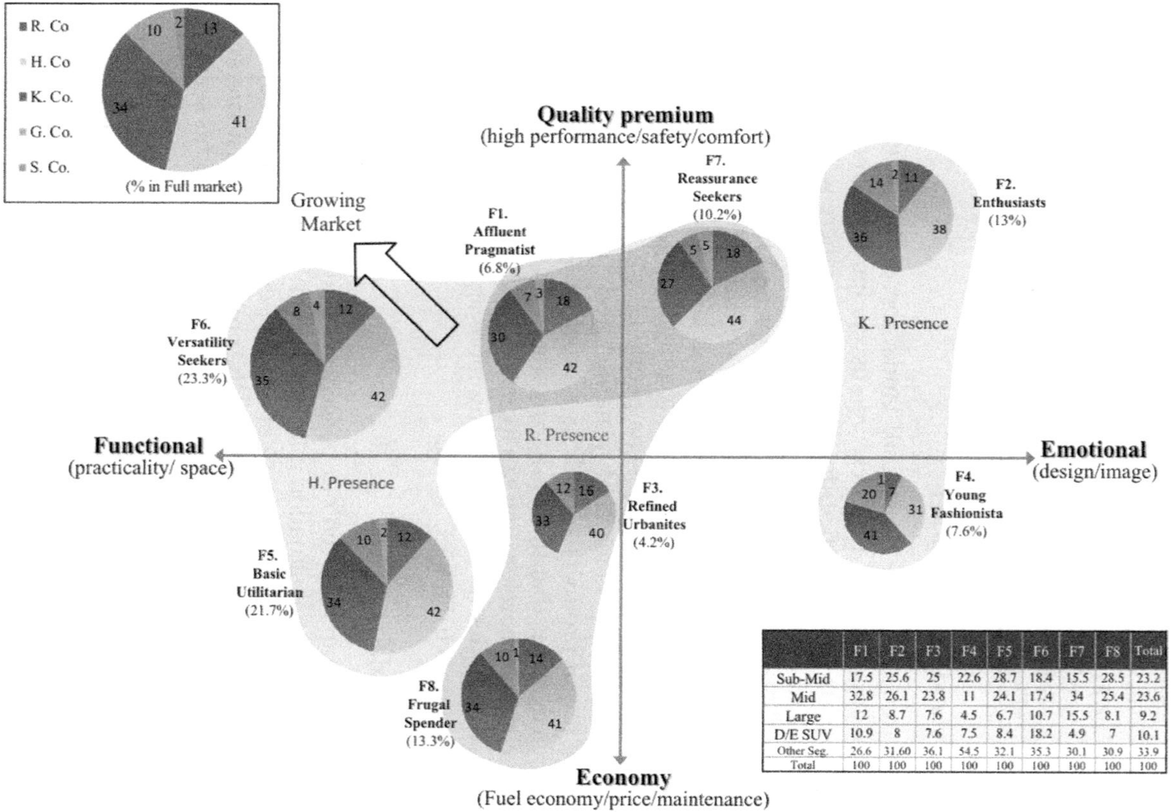

	F1	F2	F3	F4	F5	F6	F7	F8	Total
Sub-Mid	17.5	25.6	25	22.6	28.7	18.4	15.5	28.5	23.2
Mid	32.8	26.1	23.8	11	24.1	17.4	34	25.4	23.6
Large	12	8.7	7.6	4.5	6.7	10.7	15.5	8.1	9.2
D/E SUV	10.9	8	7.6	7.5	8.4	18.2	4.9	7	10.1
Other Seg.	26.6	31.60	36.1	54.5	32.1	35.3	30.1	30.9	33.9
Total	100	100	100	100	100	100	100	100	100

FIGURE 7.7
Study of market. (Example: Vehicle market study.)

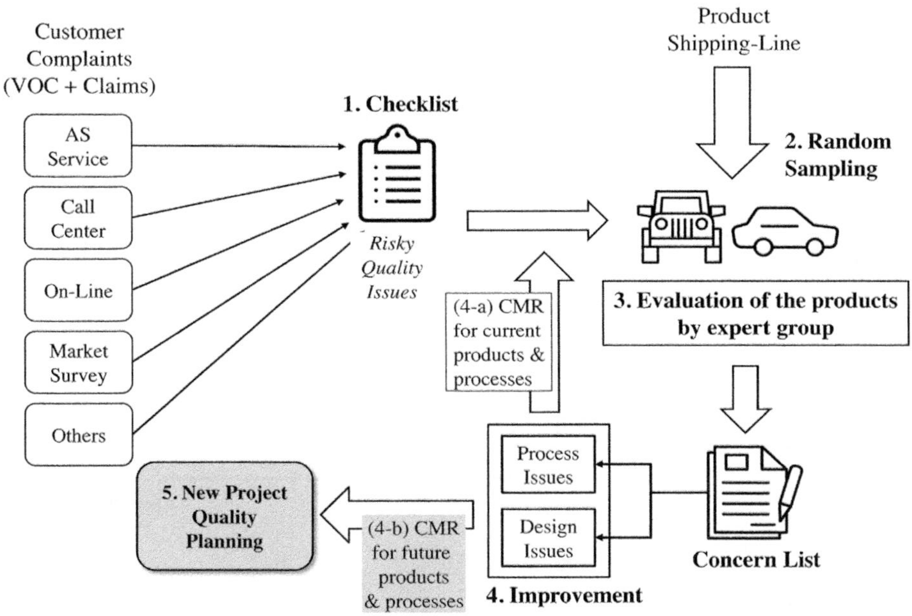

FIGURE 7.8
Product audit to quality planning process.

Usually, there are two types of problems: one is product design problems, and the other is process problems. First, we apply CMR (countermeasure) for the current products and processes (4-a), and apply long-term CMR to future products and processes (4-b) as shown in Figure 7.8. Such long-term planned improvement activities should be reflected in the *quality planning* of a new project including quality goals and targets.

7.3.6. Process Audit

QMS (Quality Management System) procedure dictates what and how things should be done in each business unit. In manufacturing process, we frequently find out some gaps between preset rules and actual practices, as shown in Figure 3.7. Such a kind of concern or risk can be identified through a process audit in the actual workplace, *Genba*. The major purpose of the process audit is to find out non-conformity issues that are not following predefined rules and procedure.

Figure 7.9 explains how we can identify concerns and rectify them in the process.

The concerns identified may have two issues: the *Genba* does *not* follow the procedure in QMS, or the procedure is outdated, *not reflecting* the actual process in *Genba*. When the concerns are identified, two types of corrective actions need to take place: 1. Update the QMS procedure,

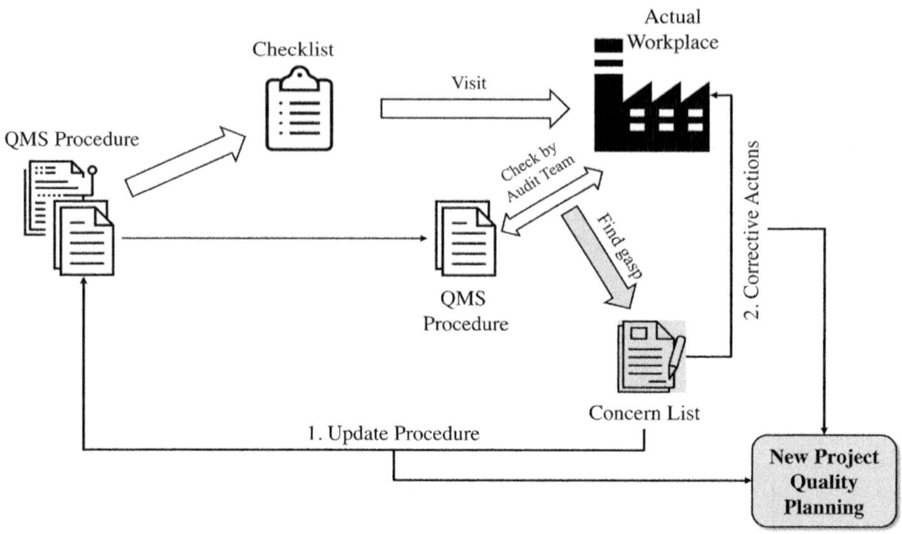

FIGURE 7.9
Process audit flow.

and/or 2. Correct actions in the actual process as shown in Figure 7.9. Moreover, these corrective actions need to be reflected in the quality planning.

7.3.7 Quality Goal/Target Setting

In the previous sections, we discussed the overall processes and activities required for situation analysis of the market, products, and processes for quality goal setting. From the outcomes of these activities, we can focus on the critical information and parameters for setting specific quality targets of each critical parameter for the success of the products.

Figure 7.10 is the summary of the link between the situation analysis studies and quality goal setting. Each type data and information has a unique purpose as an input to the quality goals and targets. Two major categories are product and process, where most of the product issues are relevant to product engineering department responsibilities, and process issues are mostly related to manufacturing responsibilities. The detailed target breakdowns should reach from level-1 (the corporate level) to team level targets.

Business leaders may have a concern that customers may not recognize some types of quality activities. In this perspective, when setting a "quality goal," *we should focus on delivering tangible outputs that the customers want and can perceive.*

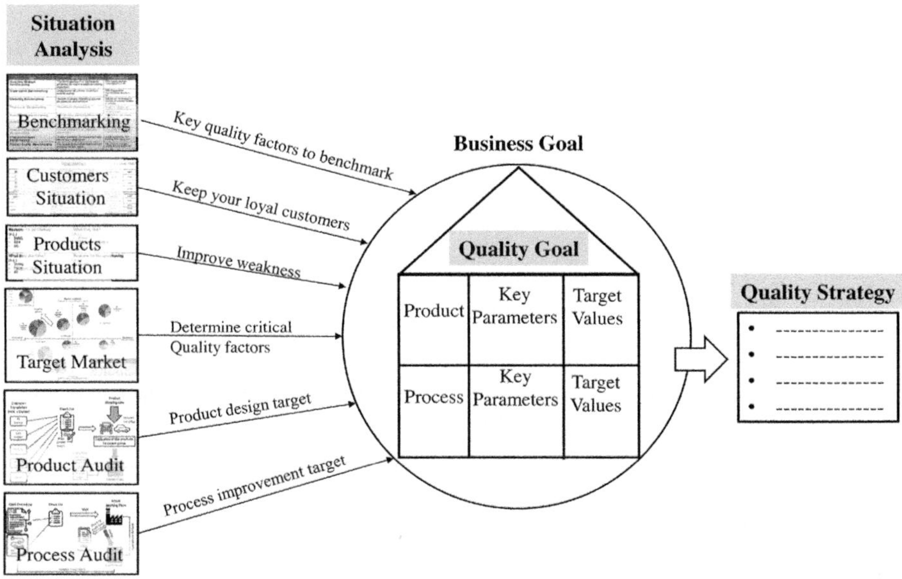

FIGURE 7.10
Quality goal setting process.

To enable customers to recognize the quality differences between products, we should focus on differentiation of quality characteristics compared to our competitors.

When setting quality goals, one of the common principles we can use is SMART:[10]

- **S**pecific: Key quality parameters should be identified.
- **M**easurable: How much or how many need to be set?
- **A**chievable: Is the organization capable of achieving the goal?
- **R**elevant: Is it consistent with higher-level goals, e.g., business goals?
- **T**ime-based: When should it be achieved?

Table 7.3 shows simple example cases on how to set quality goals by using a logical approach that *links the quality goals to the business goals*. In reality, quality goals are much more detailed and complex. However, these example cases are quality goals at a corporate level, that is, Level-1 quality goals. Then the more detailed and specific goals need to be set at the team level by each functional group through a quality deployment activity that we are going to discuss in later sections.

TABLE 7.3

Logical Approach for Quality Goal Setting (Level-1 Target)

Case (Example)	Business Goals	Quality Goals / Targets
1	No. 1 M/S (market share) in MRI equipment by 2021	• No. 1 Service Quality Satisfaction by 2020
2	A top-3 bestseller in SUV Segment by 2021	• Achieve Top-1 IQS Quality by 2020 • Achieve Top-3 Appeal Quality by 2020
3	No. 1 M/S in MRR (Mid-Range Radar) market for an autonomous vehicle by 2021	• Develop 100% weatherproof QA (quality assurance) by 2019 • Develop the best EMI proof quality in the market by 2020
4	Make new Model-X a category leader in the smartphone market by sales revenue in 2022	• No. 1. CSI (Customer Satisfaction Index) in 2021

7.4 Quality Strategy Setting

A goal is a broad primary outcome we target to achieve. A strategy is an approach we take to achieve the goal. Previously in section 7.1, we discussed how a business strategy is the firm's approach to achieving its vision and prioritizing objectives considering business environmental factors and organizational capabilities to succeed in business.[11]

Constraints in the business environment require "business strategy," business strategy requires quality strategy, and quality strategy requires quality tactics. This is the chain reaction process. Quality action follows business needs.

7.4.1 SWOT Model for Quality Strategy

In a broader perspective, when setting quality strategies, we need to consider both constraints and strengths of the organization. The constraint factors require our way of thinking and management to devise a method to avoid or overcome those constraints. Strength factors give us an opportunity to fully leverage our abilities to perform better and differently from our competitors.

In the semiconductor industry, Intel and Samsung have different types of constraints and strengths. Therefore, they have different business and quality strategies. Intel has a long history and powerful knowledge of CPU design. Intel's primary focus is on innovative CPU and security design for the smartphone market, leveraging Intel's competencies in technology and quality.[12] *On the other hand, Samsung has been the world's biggest chipmaker since 2017.*[13] *Samsung's strength is in producing memory and flash storage. Samsung's business strategy is to secure the strongest patents and technology portfolios among*

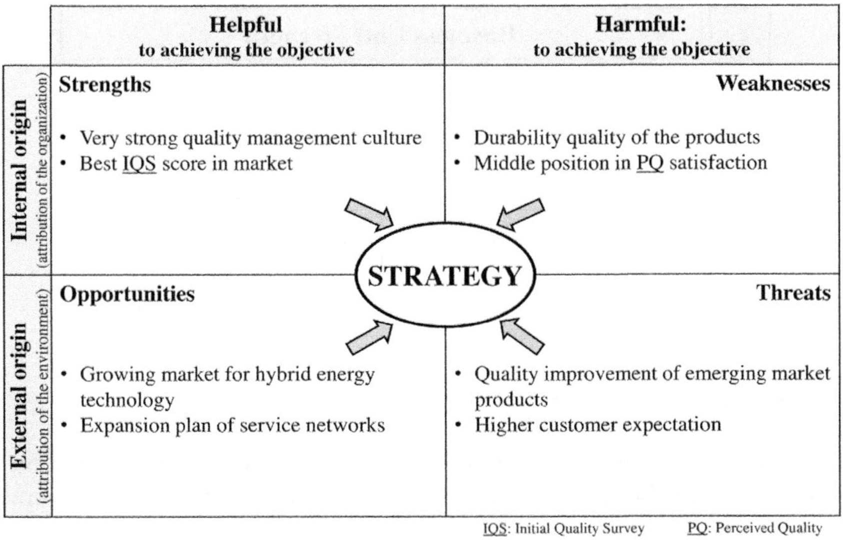

FIGURE 7.11
SWOT model for quality strategy setting.

> *technology companies,*[14] *and the quality strategy is to significantly improve the chip's yields by more than 70% on the lead products like 14-nanometer technology.*[15]

In setting business and quality strategies scientifically, the SWOT (Strength, Weakness, Opportunity, and Threat) model approach is simple and effective. Figure 7.11 shows a SWOT model for quality strategy setting.

The majority of the information in the SWOT analysis comes from the situation analysis of products in the market as discussed in the previous section. Once the information has been broken down into the SWOT analysis, quality strategies can be appropriately set according to the priorities linked with the business strategies.

7.4.2 GE McKinsey Matrix for Quality Strategy

When we have a complex product portfolio it is difficult to make a decision about which products and which areas we need to focus on. This is because each product has different demands, requirements, strengths, and weaknesses.

Since we have limited resources in the company, we have to decide which products we need to ramp up the strengths of by inputting more resources, or which products we simply maintain the status quo for. Such strategy setting can be done by using the GE McKinsey matrix strategy.[16]

The GE–McKinsey matrix offers a framework for a systematic approach for the multi-business corporation to prioritize its business strategies.

Market Attractiveness	Business Unit Strength: High	Medium	Low
High	Challenging Q-Strategy	Challenging Q-Strategy	Improvement Q-Strategy
Medium	Challenging Q-Strategy	Improvement Q-Strategy	Current Q-Strategy
Low	Improvement Q-Strategy	Current Q-Strategy	Current Q-Strategy

FIGURE 7.12
Quality strategy setting. (Modified from GE McKinsey matrix.)

The GE McKinsey matrix has two main variables which are plotted on the *x*- and *y*-axes of the matrix. The *y*-axis represents "market attractiveness," i.e., market size or growth prospect. The *x*-axis represents "business unit strength" of the organization, i.e., different product groups. The *y*-axis, market attractiveness, can be decided based on information from market surveys as we discussed in the previous sections. There are three main strategies in this method: investment/grow, hold, and harvest/sell.

Since our quality strategy goes with business strategy, we can use the same principle and approach of the GE–McKinsey matrix for our quality strategy matrix. The model can be modified for a quality strategy setting, as shown in Figure 7.12. There are three main quality strategies, such as *"challenging quality strategy," "improvement quality strategy,"* and *"current quality strategy." The "Challenging Q-strategy"* requires us to input the best and most of our resources to achieve the aggressive quality goal. The *"Improvement Q-strategy"* means we input reasonable resources to improve the quality. The *"Current Q-strategy"* means we keep the current quality strategy without additional resources.

Table 7.4 shows some practical examples of quality strategies to achieve quality goals as discussed in Table 7.3. The next step is to set up specific action plans to support the implementation of those strategies.

7.5. Specific Action Plan

After we have set up the quality goal and strategy, we should make a plan on how to deploy the strategy. Our plan should include required activities by each functional group and the specific timeframes required by business and quality needs.

> *As examples, the grand projects of the Egyptian pyramids and the Great Wall of China must have been managed based on a specific action plan. The construction*

TABLE 7.4

Logical Approach for Quality Strategy Setting (Examples)

Case	Business Goals	Quality Goals / Targets	Quality Strategy
1	No. 1 M/S (market share) in MRI equipment by 2021	• No. 1 Service Quality Satisfaction by 2020	• Set up local service infrastructure by 2019 • 100% training of local serviceman for self-independence by 2019
2	A top-3 bestseller in SUV Segment by 2021	• Achieve Top-1 IQS Quality by 2020 • Achieve Top-3 Appeal Quality by 2020	• Set up IQS forecasting Model by 2018 • Set up PQ (perceived quality) Joint TF Team (Design + Mfg.) • 3rd party IQS and PQ evaluation
3	No. 1 M/S in MRR (Mid-Range Radar) market for an autonomous vehicle by 2021	• Develop 100% weatherproof QA (quality assurance) by 2019 • Develop the best EMI proof quality in the market by 2020	• Securing QA through innovative weatherproof technology and lab/field test • 50% better EMI performance than the best in the class model
4	Make new Model-X a category leader in the smartphone market by sales revenue in 2022	• No. 1. CSI (Customer Satisfaction index) in 2021	• Apply the best resolution display in the segment • Develop the lowest energy consumption in the market • 100% defect-free at launch timing

> *of a single pyramid occupied more than 100,000 workers for 20 years. Who told each worker what to do and by when? Certainly, someone had to plan what was to be done, organize the people and the materials to do it, lead and direct the workers, and impose some controls to ensure that everything was done as planned.*[17]

In human history, the oldest and most efficient organizational structure is still a pyramid style of hierarchical organization—it is a top-down, multi-layer organization where a few top people govern a larger number of people at the middle and at the bottom. The pyramid organizational structure largely dictates to whom each employee needs to report. Managerial hierarchy is the most efficient, most natural structure.[18]

The plans should be detailed and specific, which we can call "SAP"—Specific Action Plan. Here, the SAP also has hierarchy from top to bottom. It is not efficient for top management to define every single activity required at the bottom of the organization. The primary responsibility of the top management is to define the top level of activities, e.g., at the corporate and divisional levels. Then each function group needs to take their portions of responsibilities to meet the top level's action plan regarding their requirements and timeframe.

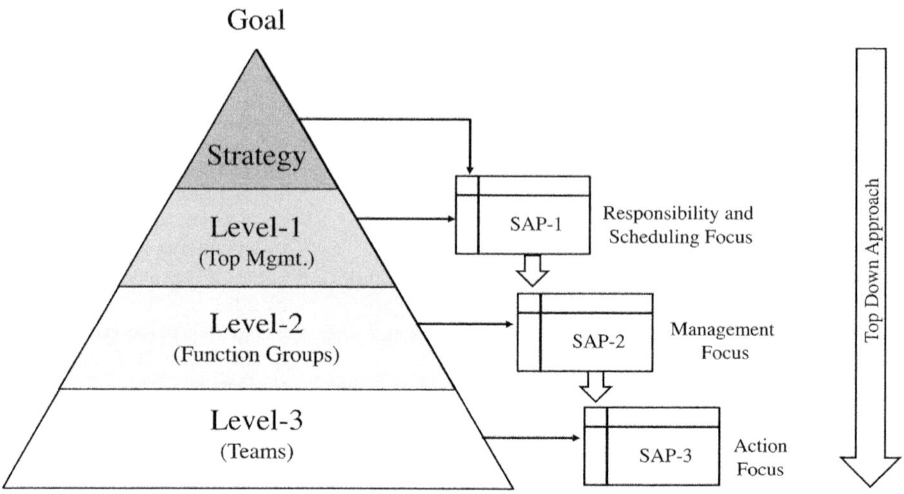

FIGURE 7.13
Top-down approach for specific action plan.

Figure 7.13 shows the structure of the top-down approach of SAP. Each level of SAP should take responsibilities requested by a higher level and define its own jobs as well as cascading responsibilities to the next lower level of the organization.

Each level of SAP has its focuses and details. For instance, "SAP-1" is a top level's SAP, taking quality strategy and transforming it into specific actions focusing on defining functional groups' responsibilities with major event schedule. The lowest level of SAP should be a working team level's SAP, concentrating on specific actions in the actual work to achieve the quality goals.

7.5.1 Group Planning

Planning requires good knowledge and experience. We have to forecast the actions required in the future to accomplish the quality goal. That is why most companies want to hire smart and experienced people for the planning group in the company. However, a single person's knowledge and experience are limited, weak, and even dangerous. We need to have critical thinking from different viewpoints. We also need a good imagination to expand the horizons of our thinking.

That is why we need group thinking and planning. *Brainstorming* is a mandatory process for group planning. We can invite a relevant group of people and have intensive open discussions for improvement and modification of the plan.

There is also another type of group thinking other than brainstorming—*"hybrid thinking."* Hybrid thinking is done by *diversified people who have different backgrounds and expertise.* When we plan for a journey in uncharted territory, hybrid thinking can be a strength. Hybrid thinking is a conscious blending of different fields of thought to discover and develop opportunities that were previously unseen by the status quo.[19]

Group planning by *brainstorming* complemented by *hybrid thinking* is the right approach for a solid plan for *an entirely different product category.*

7.5.2 Concurrent Planning

Quality activities for new product development require a substantial amount of time. Planning by sequential methods seems to be thorough but takes a long time. Sequential planning is not efficient with respect to cost and time. In modern industry, speed is a critical factor for business and industrial competitiveness.

Quality planning requires many different functions' inputs, feedback, and corrective actions. If we use sequential development, the feedback and corrective action cycles take a long time. At the same time, large scale re-design or re-tooling activities become necessary because the upstream work, such as concept, design, tooling, and equipment order, have been fixed already.

With respect to engineering planning, concurrent engineering has different approaches and processes compared to sequential engineering. In concurrent engineering, all relevant functions participate in the development activities together. Such a management and engineering philosophy can improve quality and reduce costs and lead time from product conception to product development for new products and product modifications. From the outset, this approach is intended to help the developers consider all elements of the product lifecycle from conception to service, including quality, cost, schedule, and user requirements.[20]

Concurrent engineering has shorter cycles of feedback and corrective actions. Figure 7.14 shows the SAP process using concurrent engineering—all relevant functional groups participate concurrently in the development. As a result, each step of the development process is completed by shorter "P-D-C-A" cycles in a shorter time.

For instance, when a product concept has been conceived, an early feasibility study participated in by all relevant functions can give developers different viewpoints, concerns, and new ideas. With such information, the developers can embed the right features and competitive performance into the design of a product. This approach can also minimize unnecessary modifications during the downstream industrialization stage that might happen due to issues associated with manufacturing equipment, processes, repair tools, service, or maintenance.

After V-Company had experienced many failures of new product launches in its history, it established a strong product planning process by writing constitutions in three books: Red book for technical teams' activity, Blue book for commercial teams' activity, and Green book for Business teams' activity.

The books dictate each team's role and responsibility clearly starting from the planning to sales stage by working concurrently. The core concept of the rules is to enforce each team to start and work together from the beginning. Since then, the new models were developed with proper funding and resources, and were launched in time without critical late design changes. As a result, the number of sales hit the target accurately.

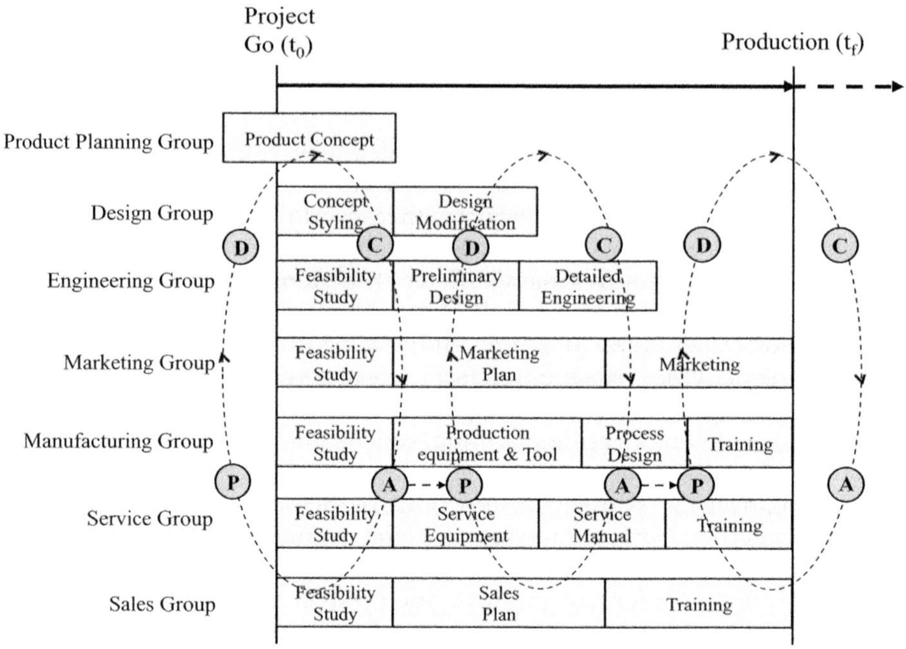

FIGURE 7.14
Specific action plan based on concurrent engineering.

> *Later, the same methods and processes have been benchmarked by G-Company, which also made a huge leap forward success in its product development and marketing in China.*

7.5.3 Linking the Activities with Dependencies

Quality activities are not standalone operations. They are tied to and aligned with a bigger scheme of activities within the product development schedule: tooling, testing, and so on. Quality tasks require inputs and expect outputs. The outputs of one task could be the inputs to initiate other relevant tasks. Such relationships should be explicitly mentioned in the management schedule.

Planning requires project management tools; every task should be linked with responsible functions and people and should be aligned with a time schedule. *Tasks and time go together.* In industry, the most popular project management tools are *"PERT"* and *"Gantt" charts.*

PERT (Program Evaluation Review Technique) is a methodology developed by the U.S. Navy in the 1950s to manage the Polaris submarine missile program. Figure 7.15 shows SAP by using a PERT chart; it illustrates project management tasks and their schedules by a network diagram consisting of numbered nodes (either rectangles or circles) representing events, or milestones in the project linked by labeled vectors (directional lines) representing tasks

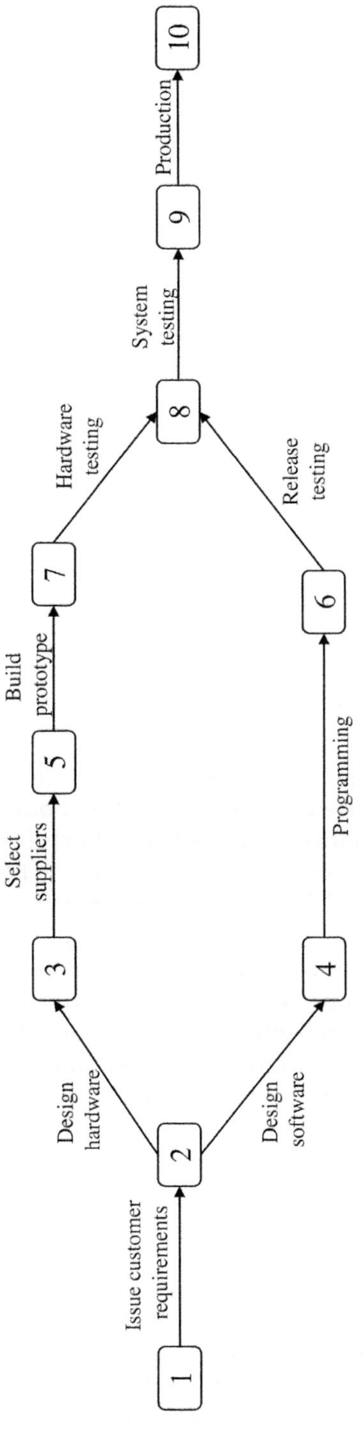

FIGURE 7.15
Specific action plan by PERT chart.

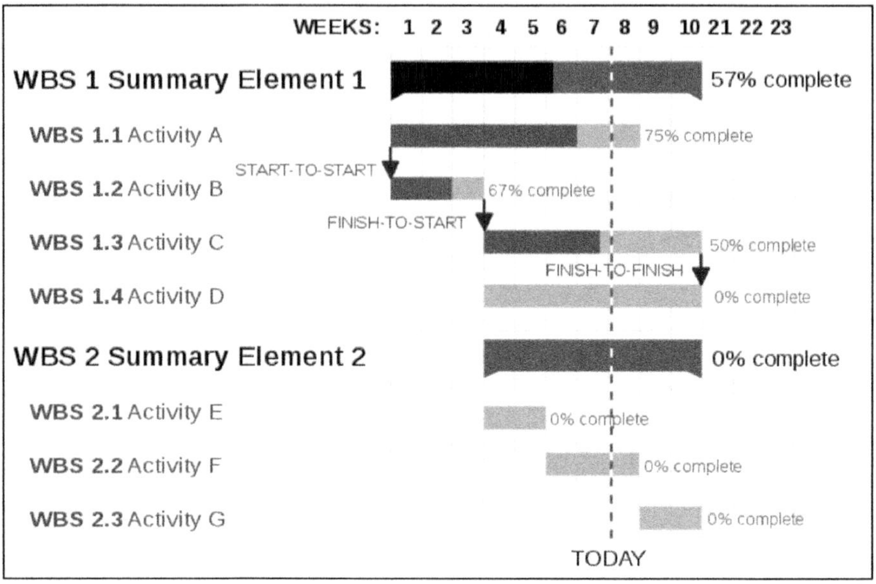

FIGURE 7.16
Specific action plan by Gantt chart. (Image: Public.)[23]

in the project.[21] The PERT chart illustrates task dependencies: some tasks are sequential, and some are concurrent.

The other project management tool used in the industry is the Gantt chart. *Gantt charts* can be easily used and interpreted without much training—refer to Figure 7.16. A Gantt chart is a type of bar chart that illustrates a project schedule, named after its inventor, Henry Gantt, who designed the chart around the years 1910–1915. Modern Gantt charts can easily show the dependency relationships between activities and the current schedule status.[22] Typically, the first column represents the main tasks to be completed in the project, and the first row represents a time schedule. Every activity is represented by a bar that stretches from the start date to the end date of that activity. Moreover, we can also insert much more detailed information inside the chart. Gantt charts can be conveniently drawn by using Microsoft Excel or other similar software.

7.6 Quality Planning of New Product (Case Study: Automobile)

In the previous sections, we discussed a general scientific process of quality planning. In this section, we will discuss how we can practically apply the procedures to the quality planning of real products in the workplace.

Quality planning can be applied to improve the overall quality of the company, the current mass-produced products, or new products. However, such thorough scientific quality planning can be particularly effective when applied to new product development.

There are ten major steps we can apply for new product quality planning (see Figure 7.17):

1. Initial Product Concept: proposal for a new product concept for the market and business
2. Market Research: the trend analysis of customers behaviors and the market
3. Technology Research: the trend analysis of new technologies and the company's capabilities
4. Feasibility Study: the feasibility (design, industrialization, marketing) study of a new product concept considering the company's capabilities
5. Product Concept: a final direction of the new product concept
6. Resource Set-up: secure the necessary resources
7. Target & Strategy Setting: clarification of the project target and strategy
8. Project Organization: workforce organization for the project
9. Project SAP: specific action planning for the project
10. Project Kick-off: deployment of the project to all relevant functions

Figure 7.18 shows an example of market research based on current customers. From this market study we develop a good understanding of the demographic distribution of our customers and their overall opinions on the current product's characteristics and quality. From the analysis, we need to focus on keeping what our customers like and improving the weak points of the current products for the new product.

Figure 7.19 shows the overall market trend in the luxury SUV (LSUV) market. This market research shows the trend of several different segments. From this trend analysis we can foresee an opportunity that Premium & Performance LSUVs will grow faster.

Technology research is the primary part of vehicle development. Which technology and features we should develop into a new vehicle is a critical decision-making factor for customer satisfaction. For this purpose, we can use the QFD (Quality Function Deployment) method discussed in Chapter 5.

Figure 7.20 is an application of QFD to assess which technologies or features we should use to develop an autonomous vehicle from the customers' standpoints. To create an autonomous vehicle, different types of sensors and technologies are required. We can start with the many available features and functions and decide the most effective ones for our potential customers.

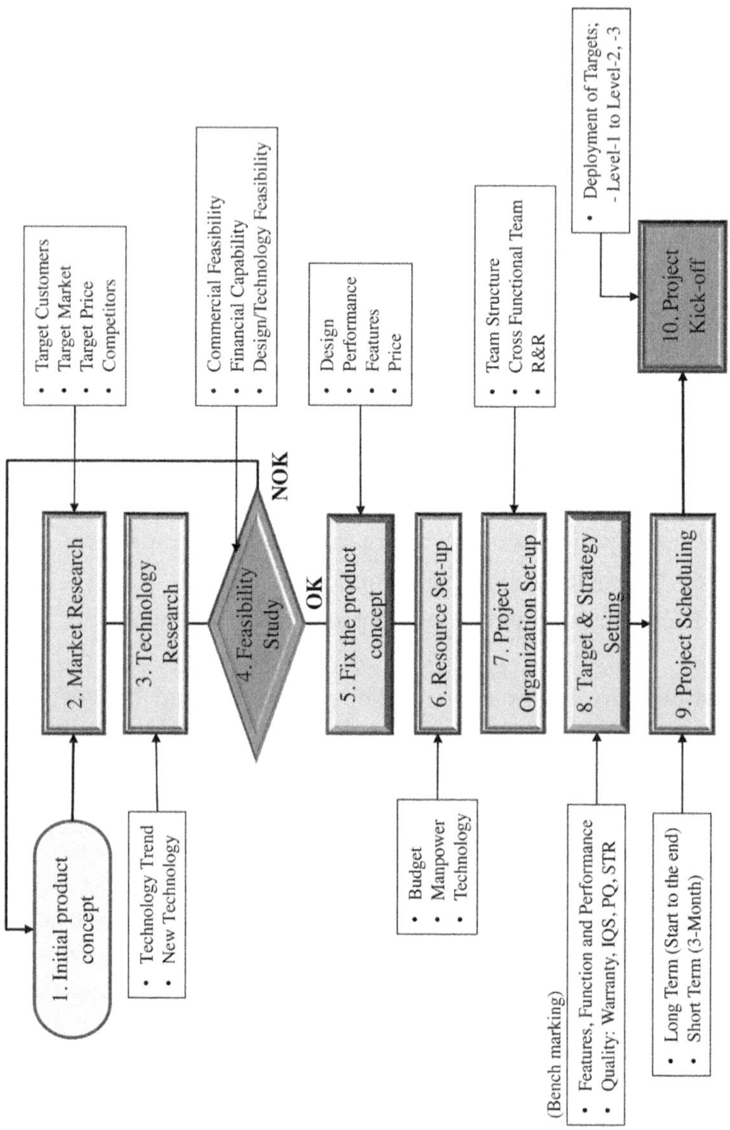

FIGURE 7.17
New product development planning (process flow).

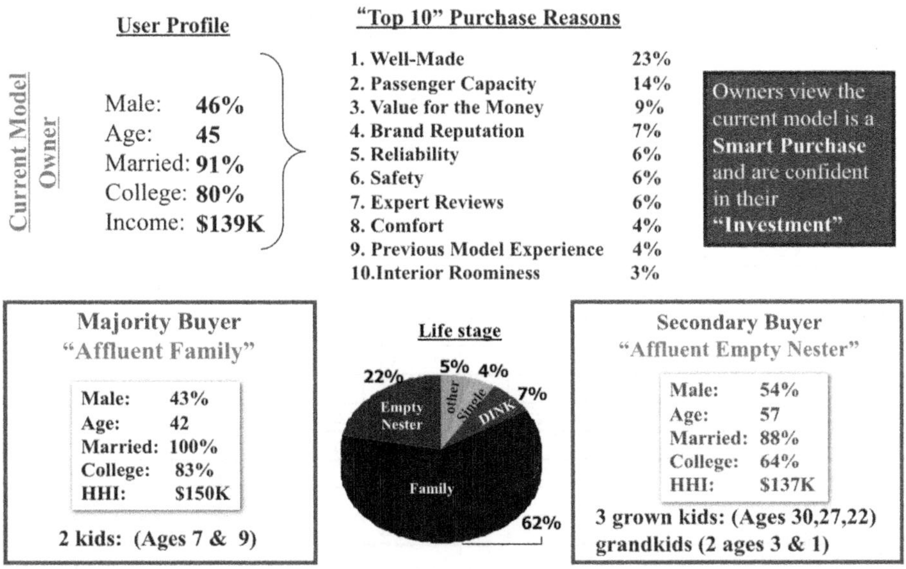

FIGURE 7.18
Market research of the current product buyer. (Example: SUV model.)

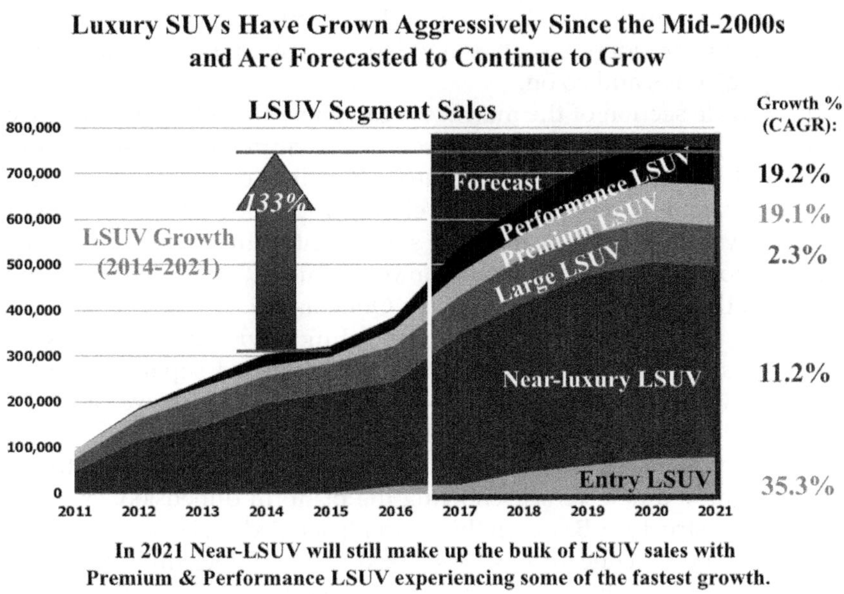

FIGURE 7.19
Market trend research of LSUV market.

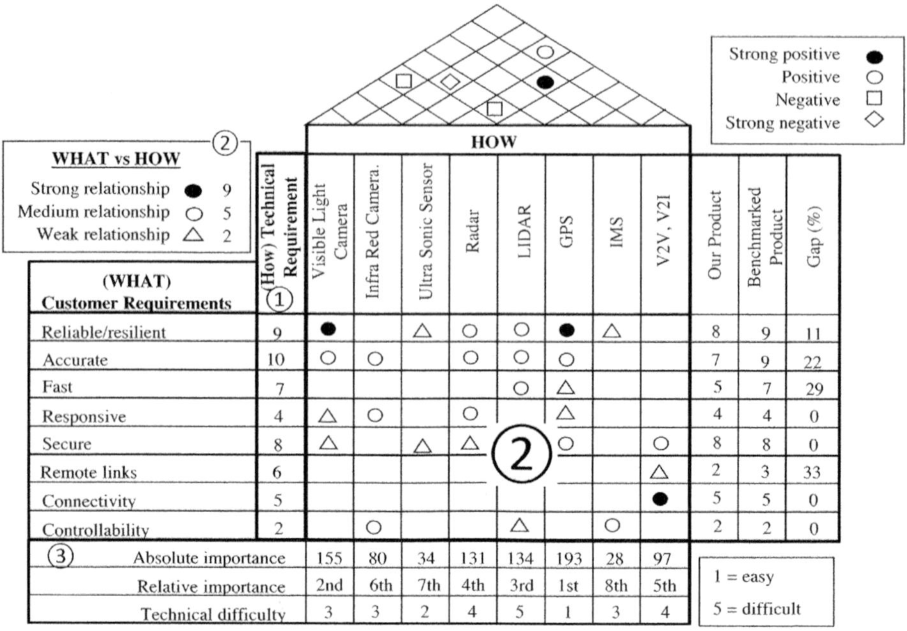

FIGURE 7.20
QFD for planning of vehicle technologies and features.

The left two columns display customer requirements and expectations for an autonomous vehicle, and the top middle columns show technical features that need to be considered, such as a video camera, LIDAR (Light Detection and Ranging), GPS, and so on.

In the middle section of the matrix we can show the relationship between the customer requirements and the technical requirements. Each customer requirement has different weight values depending on the needs. By multiplying the weight values, ① and the relationship value (the symbols in the middle, ②), we can determine the most significant technical features we have to consider for the design of an autonomous vehicle (③ = ① × ②)—refer to the priorities in the second to last row of the QFD matrix.

The example matrix (Figure 7.20) is a technical viewpoint vs. customer requirements. Therefore, a cost effect can be analyzed separately through cost-effect analysis with a similar concept. For this purpose, we can use the decision analysis table discussed in Chapter 6.

The next step is to fix the design concept. Figure 7.21 is an example of the most critical concept of a new vehicle showing the platform dimension, layout, and the white body structure. Based on this concept, detailed design activities follow.

Once the product concept has been fixed, we can determine the specific quality targets based on benchmarking activities: features, functions, performance, IQS, warranty targets, and so on. Figure 7.22 shows an image of the benchmarking activity to set up the target of vehicle performance. During

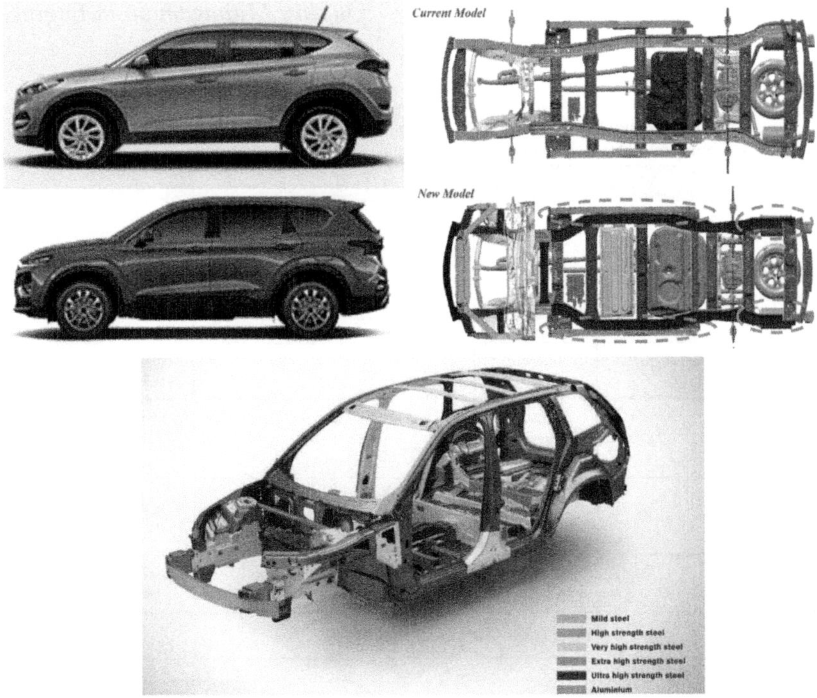

FIGURE 7.21
Fix the product concept.

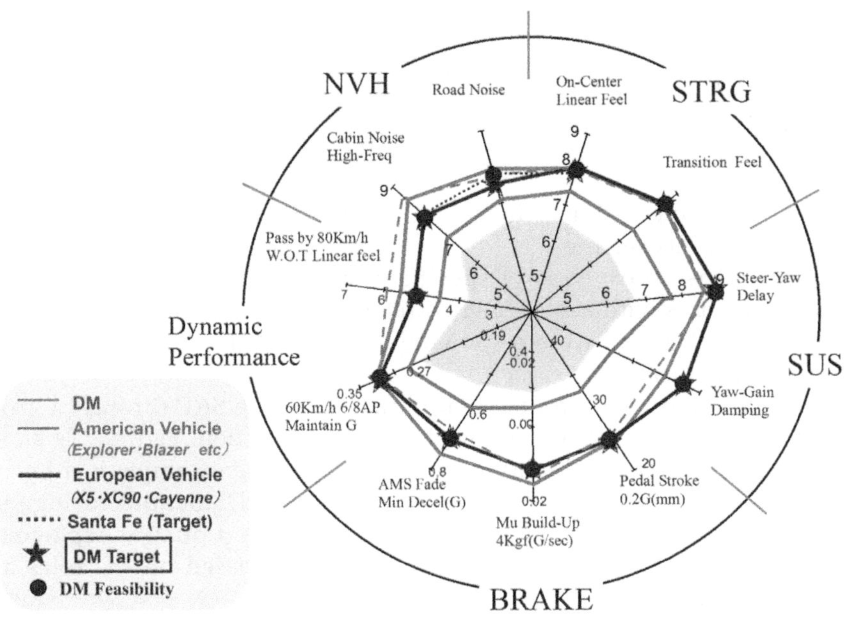

FIGURE 7.22
Benchmarking and project target set-up of major characteristics.

R&D	. Robust Design through 5-Quality Pillars: (PPH, PQ, IQS, New Technology, D-FMEA) . Concurrent Development and Engineering		
Quality Panel (Q-KPI)		**KPI Target**	**Strategy**
1. PPH **(Warranty)**	**Q-History**	• 100% CMR of Design Issues • No Repetition	• Clarify the data quality (Root Cause Analysis & CMR) • Set up the PPH DB • .PRES Process
	Claims	• 15%~ 30% Reduction	• Target breakdown to each Engineering module • Set-up system and Process
	Cost	• 15%~ 30% Reduction	
2. PQ **(Perceived Quality)**	**Internal Surveys**	• Over "7" • Top in Segment	• PQ Engineering Manual ,Training and Evaluation
3. IQS	**External Surveys by JDP**	• TOP-3 in Segment	• Analysis and Target breakdown to each Engineering module • Action Deployment by IQS Mgmt. Sheet-1 and Sheet-2
4. New Technology	**Risk Manage Mgmt.**	• 100% Deployment of FMEA • Zero Recall • IQS : < Average_PPH of IQS Items • MIS : < Average MIS_Warranty Items	• Theme-up Teams and Evaluation Process
5. D-FMEA	**S-FMEA**	• 10 D-FMEA/Module (Total > 50) • 100% Refection in DWGs. • 100% Confirmation	• System, Management and Engineering
6. Late Design Change	**DC after Die-Go**	• Late_DC /Total_DC: < 20%	• System, Management and Engineering

Mfg.	. Current Development with Engineering: (P-FMEA/Control Plan, STR, RPU, PPM) . Joint Industrialization activity with R&D and Suppliers		
Quality Panel (Q-KPI)		**KPI Target**	**Strategy**
7. P-FMEA	**D-FMEA →** **P-FMEA →** **Control Plan**	• 100% connection with D-FMEA • 100% Unique P-FEMA for New Technologies and New Process • 100% of P-FEMA for A-rank Past Problems	• Joint P-FEMA with R&D
		• 100% Deployment to Control Plan	• Deployment through Control Plan

FIGURE 7.23
Target & strategy setting.

the benchmarking activity, the performance targets need to be evaluated through feasibility testing.

Figure 7.23 is an example of specific numerical targets for internal and external qualities with achievement strategies. In this example, for the purpose of simplification, we are showing only the R&D and Manufacturing division's responsibilities.

The final step of new product planning is to create SAP through a group discussion method as discussed in the previous section. Figure 7.24 shows an image of the overall quality action schedule. This example is a Level-1 SAP covering divisional activities only, such as R&D, manufacturing, and supplier quality activities. From this Level-1 SAP and quality deployment activity, level-2 and level-3 SAP need to be made based on the roles and responsibilities of each functional group.

In the next chapter, we will have an in-depth discussion of how we can deploy the scientific quality plan that we developed here in a structured way throughout the organization.

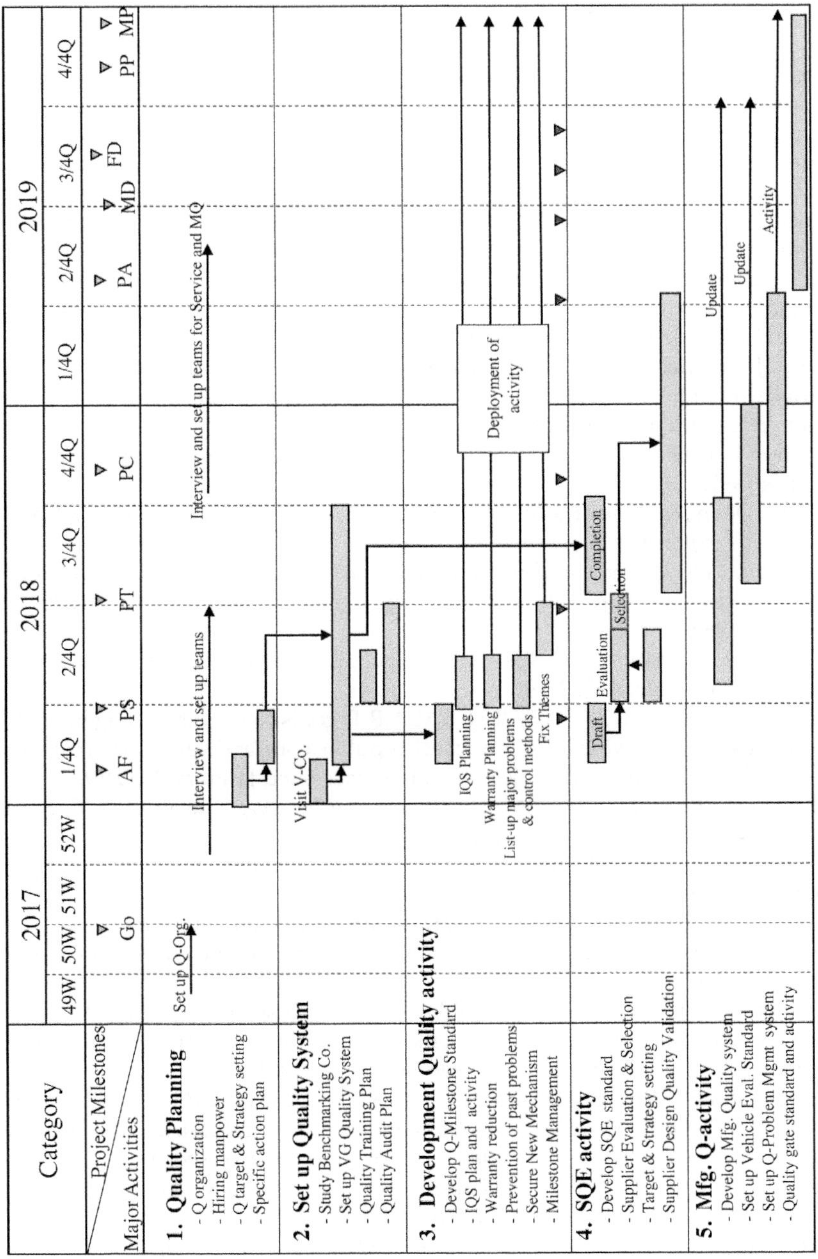

FIGURE 7.24
Project scheduling. (A simplified version of level-1 SAP.)

Notes

1. White, Sarah K., March 16, 2018, "What is CMMI? A model for optimizing development processes," www.cio.com/article/2437864/developer/process-improvement-capability-maturity-model-integration-cmmi-definition-and-solutions.html (accessed January 27, 2019)—CMMI (Capability Maturity Model Integration) is a framework for appraising the process maturity of the organization.
2. Belicove, Mikal E., September 27, 2013, "Understanding Goals, Strategy, Objectives and Tactics in the Age of Social," www.forbes.com/sites/mikalbelicove/2013/09/27/understanding-goals-strategies-objectives-and-tactics-in-the-age-of-social/#9b00ad44c796 (accessed January 27, 2019).
3. Chandler, A.D. Jr. *Strategy and Structure: Chapters in the History of the American Industrial Enterprise*. MIT Press, 1962.
4. Doerr, John. *Measure What Matters*, Penguin Random House, 2018.
5. Powell, Major J. W. "Competition as a Factor in Human Evolution" *The American Anthropologist*, 1, no. 4 (October 1888).
6. Diamond, J. (1998). *Guns, Germs and Steel* (Chapter 13). Vintage, Random House: 1997.
7. Elmuti, Dean., & Yunus Kathawala. *An overview of benchmarking process: a tool for continuous improvement and competitive advantage*, MCB UP Ltd., 1997.
8. Bateman, G.R. "Benchmarking management education: teaching and curriculum," in Camp, R. (Ed.), *Benchmarking*, Quality Resources, White Plains, NJ, p. 6., 1989.
9. Perceptual mapping, https://en.wikipedia.org/wiki/Perceptual_mapping (last modified March 15, 2019).
10. Bogue, Robert. April 25, 2005, "Use S.M.A.R.T. goals to launch management by objectives plan," *TechRepublic*, www.techrepublic.com/article/use-smart-goals-to-launch-management-by-objectives-plan/ (accessed January 28, 2019).
11. Johnson, G., K. Scholes, & R. Whittington. *Exploring Corporate Strategy: Texts and Cases, 8th Edition*, Prentice Hall International: 2008.
12. Marketing Strategy of Intel—Intel Marketing Strategy, www.marketing91.com/marketing-strategy-intel/(September 24, 2018).
13. Savov, Vlad. "Samsung is now the world's biggest chipmaker," January 31, 2018, www.theverge.com/2018/1/31/16954228/samsung-memory-chipmaker-world-biggest-2018 (September 24, 2018).
14. Jurevicius, Ovidijus. January 10, 2019, "SWOT analysis of Samsung," Strategic Management Insight, www.strategicmanagementinsight.com/swot-analyses/samsung-swot-analysis.html (accessed January 28, 2019).
15. Eassa, Ashraf. "Why Samsung's 14-Nanometer Exynos Yields Probably Aren't That Great," April 30, 2015, www.fool.com/investing/general/2015/04/30/why-samsungs-14-nanometer-exynos-yields-probably-a.aspx (accessed January 28, 2019).
16. Bhasin, Hitesh. The GE McKinsey Matrix, February 14, 2018, www.marketing91.com/ge-mckinsey-matrix/ (March 24, 2019).
17. Principles of Management, Historical Overview of Management, www.zeepedia.com/read.php?historical_overview_of_management_the_egyptian_pyramid_great_china_wall_principles_of_management&b=54&c=1 (accessed January 28, 2019).

18. Jaques, Elliott. "In Praise of Hierarchy," https://hbr.org/1990/01/in-praise-of-hierarchy (accessed January 28, 2019).
19. Forget Design Thinking and Try Hybrid Thinking, August 25, 2009, www.fastcompany.com/1338960/forget-design-thinking-and-try-hybrid-thinking (accessed January 28, 2019).
20. Concurrent Engineering, www.me.nchu.edu.tw/lab/CIM/www/courses/Computer%20Integrated%20Manufacturing/Chapter3%20-Concurrent%20Engineering.pdf (accessed January 28, 2019).
21. Rouse, Margaret. "PERT chart (Program Evaluation Review Technique)," https://searchsoftwarequality.techtarget.com/definition/PERT-chart (accessed January 29, 2019).
22. Gantt chart, https://en.wikipedia.org/wiki/Gantt_chart (last modified June 11, 2019).
23. This work has been released into the public domain by its author, Garry Booker at English Wikipedia, https://commons.wikimedia.org/wiki/File:GanttChartAnatomy.svg (last modified March 8, 2019).

8

Structured Quality Deployment

In the previous chapter we thoroughly examined how excellent quality planning can be achieved using scientific approaches. Through scientific quality planning we can define the quality goal, quality strategy, resources required, and action plans with respect to the timeline.

Deploying the plan in the actual workplace, however, is another matter altogether. Many companies suffer from a big gap between their goals and actual achievement. The less keenly we manage the deployment actions through an organizational structure, the larger the gap we discover at a later stage of the quality activity.

Proper planning requires in-depth knowledge, experience, ideas, thinking, forecasting, etc. However, *deployment is action. Action requires moving people and organizations*, whether they are hierarchical or cross-functional or matrix. Many companies with poor quality products have weaknesses in *QAD (quality action deployment)*; haphazard deployment ruins a good plan.

An authoritative approach by one leader or relying on a system alone does not work well in modern management. *QAD is the systematic management of action and leadership in a structured way*. For instance, military power does not depend solely on the power of weapons or technology but also structured power. Throughout history, the superpower nations not only had superior technology but also structured power.[1]

Structured Quality Deployment requires management activities through QAM (Quality Action Management). QAM is the process to manage QAD through intelligent management activities. The fundamental concept of intelligent management is to convince people and organizations to act in an intelligent way. People are intelligent creatures. The best way to ignite intelligence is intelligence itself. A core part of intelligence is logic and structure. *Intelligent management requires intelligent tools and methods.*

When simple and clear logic has been applied, it can move people because people are easily convinced. Good management uses IQ (intelligence quotient) appropriately alongside EQ (emotional quotient) and PQ (political quotient) as discussed in Chapter 3.

Quality actions in the organization need to be sustainable and continually improving. Management should strengthen the system, culture, and capability of the team rather than only focusing on daily management and exhausting all management energies in that area. In an efficient organization, quality activities are deployed self-sufficiently and continually grow in scope. This kind of self-powered continuous improvement activity can be called "Self-Improvement of Quality."

1. Job Tree and Target Breakdown
2. Quality Action Deployment (QAD)
3. Quality Action Management (QAM)
4. Self-Improvement of Quality
5. Leadership
6. Culture
7. Infrastructure
8. Management Principles and Practices

8.1 Job Tree and Target Breakdown

Good planning is not sufficient for success—it is only half done. A good plan has no meaning if it is not implemented in the actual workplace. During planning, we have identified all the quality goals, strategy, and actions required. We also set up a clear organizational structure as part of the planning.

Now it is time for us to cascade our quality goals and strategy appropriately into the organization and onto the actors who will perform the relevant jobs. One very useful tool is a "Job Tree" as shown in Figure 8.1. The purpose of the job tree is to approach a task like a tree structure, logically from a vital few to many, from left to right, as shown in Figure 8.1. According to Adrian Bejan's *"Constructal Theory,"* a treelike structure is prevalent in nature because it makes things flow better.[2] Similarly, in management, a job tree structure makes information flow better from point A to point B. A company's top goal and strategy are like the root and trunk of a tree, and from the trunk, a few big branches spread out, which are level-1 targets. Then from the big branches, many small branches spread out, which are level-2 control items and targets to achieve.

Through this job tree structure, each job breaks down into smaller packets, and the packets, i.e., the information, can flow from the top to lower structures more efficiently.

First, the job tree shown in Figure 8.1 clarifies the first level of targets and strategy, ①. Then it assigns next level responsibility to the next level's team defined in Figure 3.11. During the process, the scope of the responsibility of each team with respect to level-1 target is marked by a circle or a double circle in the middle section of ②. Then, the control item and target of each team is allocated as ③ with different timing of action required, ④.

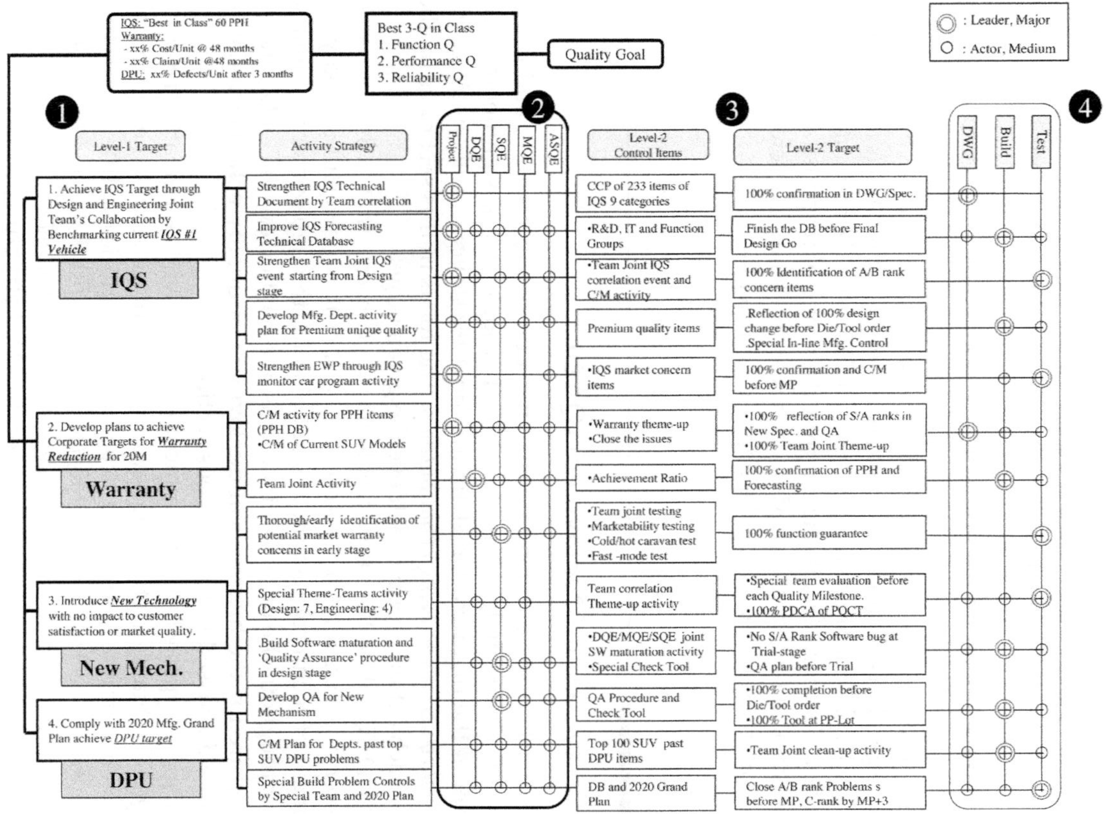

FIGURE 8.1
The breakdown of quality actions by job tree.

Once the second level of responsibility and target have been clarified and assigned to the second level of teams. The teams need to deploy the detailed plan to achieve the target assigned to them, including assigning some responsible items to the another lower level of teams.

Figure 8.2 represents how the job tree can be used by the second level and third level of the organization. For instance, the Level-2 organization takes the assigned targets from the Level-1 job tree, and then they develop their own job tree as well as logically assign the next level's responsibilities and targets to the next lower level organization or team.

A typical job tree has three levels: the first is a project leader or top management, the second is a divisional level, and the third is a department and team level. We need to note that the job tree is not only for hierarchical deployment of targets but also for cross-functional deployment. This is a potent tool for quality deployment *because the responsibilities and the relevant information flow together.* Furthermore, in the job tree, the scope of responsibility is clear with measurable targets assigned to each relevant team. Therefore, the next follow up activity, P-D-C-A, can be deployed efficiently.

8.2 Quality Action Deployment (QAD)

The previously discussed job tree is focused on strategy setting and target breakdown into hierarchical and functional groups—it is assignments to the next level of the organization structure, now *the strategies and targets should lead to quality actions to bring about outputs.* The focus in this section is on the *action* by each action group to achieve the given target.

The QFD (Quality Function Deployment) concept that was discussed in Chapter 5 is versatile in its structure so that it is powerful not only for deployment of technical parameters but also for the deployment of quality actions throughout the organization. Thus, we can borrow the QFD concept and structure for the hierarchical deployment of quality actions similar to the job tree discussed in the previous section.

This logical approach to deployment of action will be hereafter referred to as "QAD" (Quality Action Deployment). Figure 8.3 shows the concept of QAD, which is similar to the QFD structure. The significant difference is that *the customer requirements section is replaced with quality KPIs, and the technical parameters section is replaced with the department team targets.*

Based on the concept and skeleton of the QAD structure, we can develop a corporate level of QAD in a detailed format for its application purpose. Figure 8.4 shows an example of QAD applied in the auto industry, which was modified for this book. Since each company is in different businesses and has different needs, the QAD concept and the format can be adapted for varying purposes of quality action deployment.

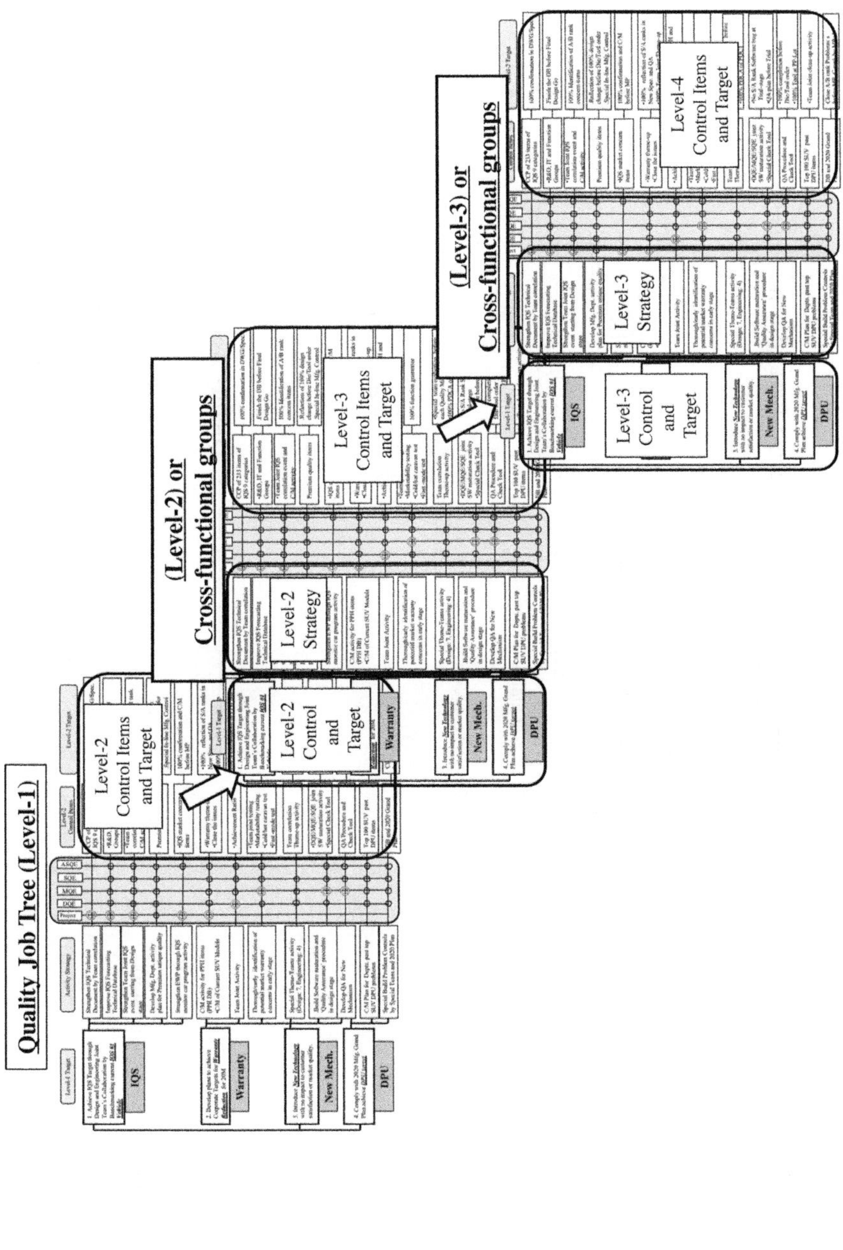

FIGURE 8.2
Quality target breakdown to lower levels.

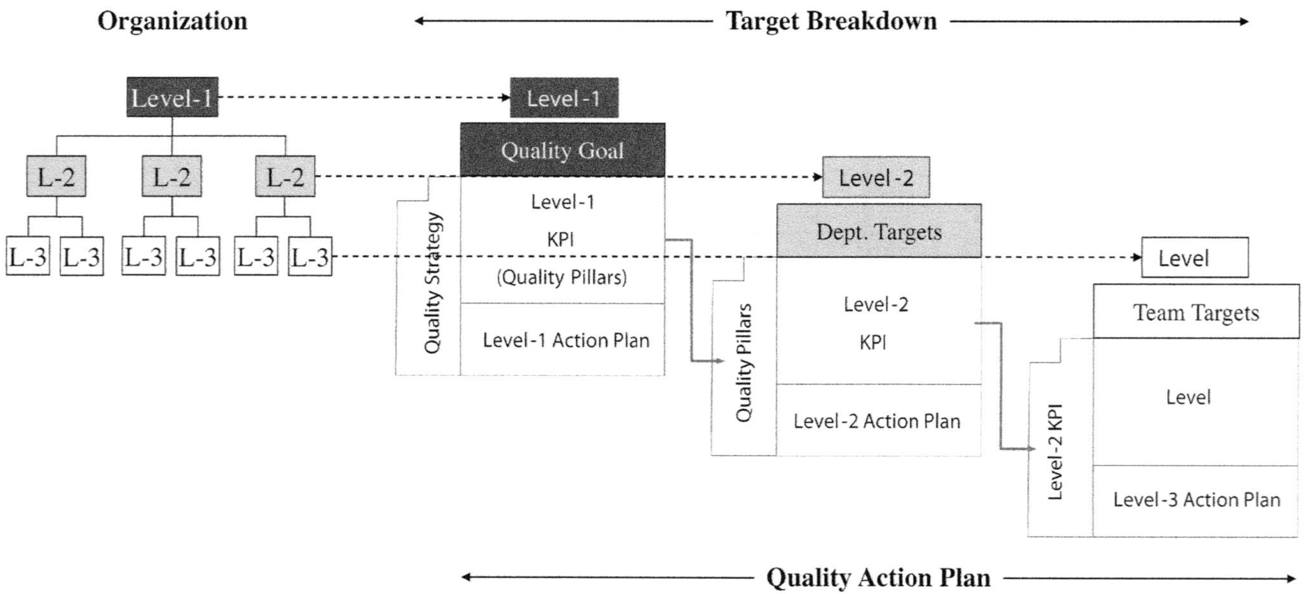

FIGURE 8.3
QAD (quality action deployment).

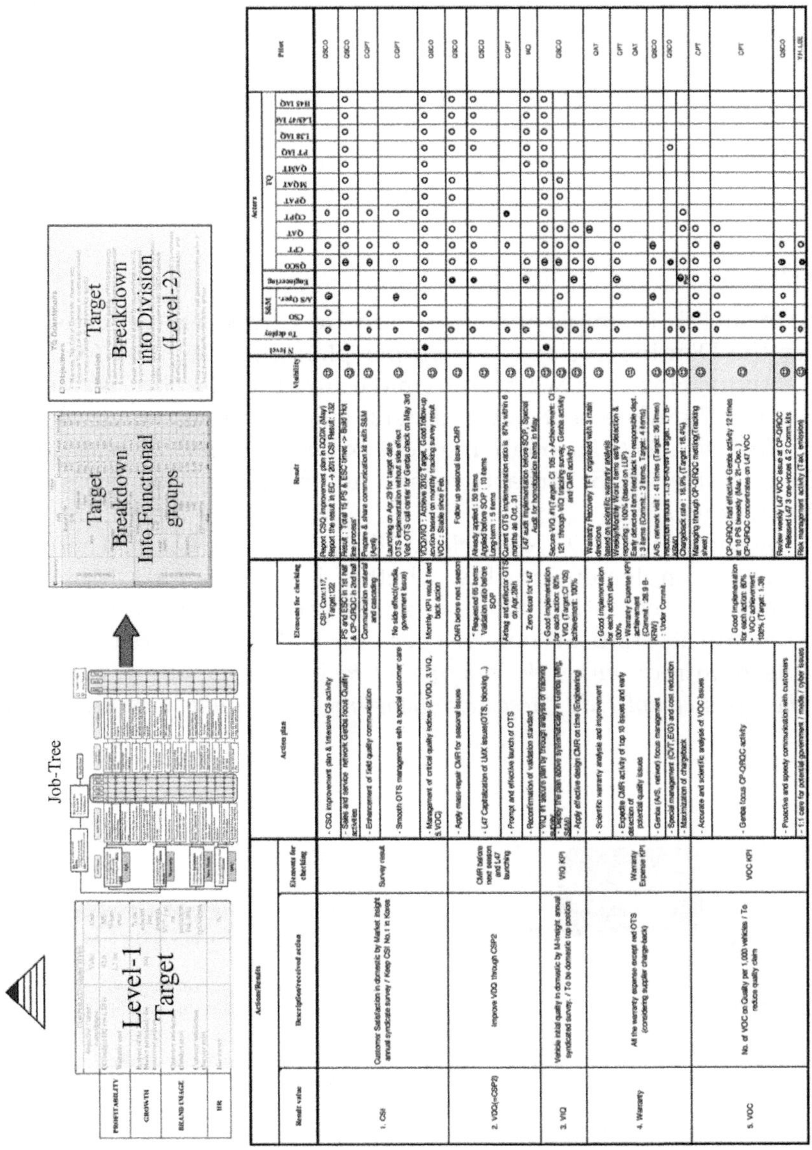

FIGURE 8.4
QAD (level-1).

8.3 Quality Action Management (QAM)

The success of scientific planning relies on *structured implementation and systematic management.* The quality actions deployed by QAD needs to be managed continually to deliver outputs to customers and the company. *QAM (Quality Action Management) is the management method to motivate, inspire, and support the organization to follow action plans made in the QAD process.* QAM requires continuous coaching and correction of the action groups by managers and team leaders. There are several critical processes required for effective QAM.

8.3.1 P-D-C-A

Deming's principle is simple and clear: Plan-Do-Check-Act (which is also known as the Deming cycle). After proper planning, we must focus on the *"do"* activity. However, we need to make sure that the "do" is done in a proper way. Therefore, a structured deployment should *include "doing" and "checking" activity.* Through the "checking" activity, we can learn about the effectiveness of our actions and any necessary adjustments or corrective actions.

Figure 8.5 shows the P-D-C-A cycle for the purpose of QAM. There are several ways of implementing this, whether it be through a formal or informal audit or evaluation. Once a quality project has been launched, we need to know its status. When we check the status, there are two major viewpoints required: *one is the progress of tasks, and the other is the contents of the tasks completed. These two items must be regularly reviewed and evaluated.*

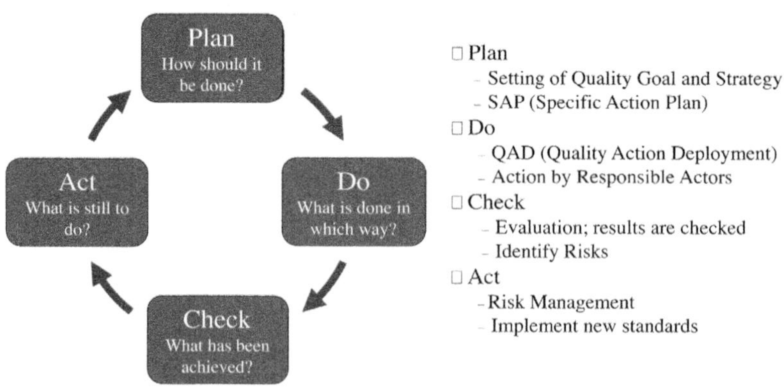

FIGURE 8.5
P-D-C-A for quality action management (QAM).

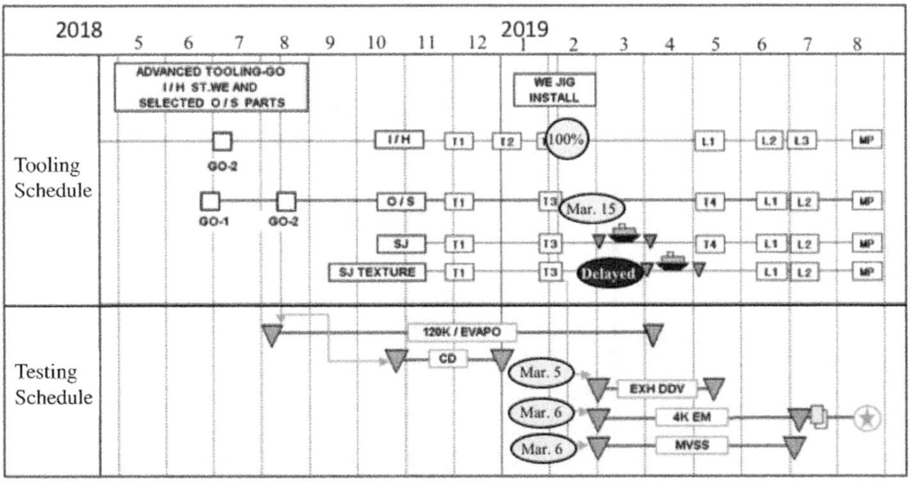

FIGURE 8.6
Progress status of specific action plan (SAP).

8.3.2 Checking the Progress Status

The first step of management is to know progress made to the plan, i.e., activity items and timeline. The broad scope of completed tasks needs to be evaluated regularly by team leaders and the project leaders to identify concerns regarding progress made to the schedule; this checking can identify any corrective or supportive actions required for the success of planned actions.

The critical activities and updated status need to be shared with the relevant teams in a controllable and manageable way, including through visual control methods, as shown in Figure 8.6.

8.3.3 Quality Milestone Evaluation

When developing a new product we need to hit milestones or events to confirm that product development has achieved specific goals while moving towards the final objectives or targets. Without achieving milestone targets it would be difficult to reach the final targets. *Milestone evaluation is not only about checking progress made to the schedule but also confirming specific activities and their results.* That is why it is essential to set up each *evaluation item and criteria, including milestone targets* during the planning stage.

For new product development, quality milestone evaluation is a critical process to apply. There are several phases of new product development. Evaluation of major milestones needs to occur during the planning phase, development phase, and the industrialization phase.

Figure 8.7 is a generic process of new product development, which is a typical type used by many automotive companies. It starts with "Scientific

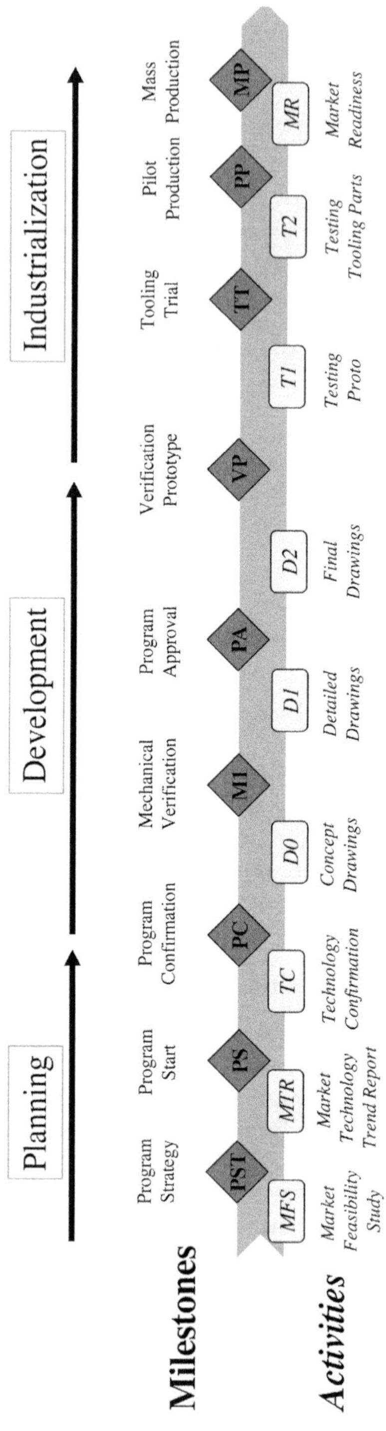

FIGURE 8.7
New product development process and quality milestones.

Planning" based on the market and technology research. At this beginning stage, whether it is a new model or a new technology, the engineers should determine what the product needs and how it should perform. Those requirements are fed into the development stage to set detailed specifications in drawings, and they go into the industrialization process to make sure quality is built through the manufacturing process.

Figure 8.8 shows the evaluation category, contents, and progress necessary at each stage of evaluation. The thorough documentation of program activities and results is critical in demonstrating the evidence of achievement at each milestone.

Each project and program has different focuses and methods. Numbers of milestones and ways of performing the evaluation depend on the complexity of a new product or a company's strategy. However, too many evaluations can drain the engineers' energy and focus.

> *In today's management styles and methods, many engineers spend much of their energy preparing presentation documents. Cynically, people call those people "PowerPoint engineers" because they have to spend more of their time making presentation materials by using PowerPoint software rather than making drawings or doing tests. Therefore, with respect to milestone management, it is essential to set up efficient milestone processes and evaluation methods rather than merely many evaluations, requiring many redundant documents.*

8.3.4 Quality Review through a Regular Process

In most quality operations, *regular quality review and management processes* are essential to ensure that outputs meet the quality targets of the company and the project. During the normal operation of a company, the company leaders should review and check overall quality progress and concerns regularly. Moreover, they need to manage them to mitigate risks through coaching, advising, and providing the necessary resources. For effective outcomes, there must be a balance between top leaders' regular review meetings and autonomous self-operations by each functional group.

The primary purpose of any quality meeting is to check the status of quality improvement and to make a decision for any corrective actions or supports required when potential risks are identified. Unlike milestone management, regular quality reviews are flexible and dynamic through communications with working teams. Through the meetings, much information can be exchanged openly; therefore, cross-functional coordination and resource arrangements are effectively possible.

Figure 8.9 shows an example of the regular quality review structure and processes inside an automotive company. Most companies already have a quality organizational structure and relevant functional teams. Therefore, instead of creating a new structure, the existing team structures should be streamlined to maximize the total output of the corporation with suggestions mentioned in this book.

Classification		No	Evaluation document	Document detail	Full-Model-Flow							Minor-Model-Flow			
					PST	PS	PC	MI	PA	VP	TT	PS	PC	PA	TT
Change/ addition, strategy		1	Model outline	Development target, major spec, variation, new mechanism/material, master schedule, etc	-	○	○Chg/add	-	-	-	-	○	○Chg/add	-	-
		2	Action to DQ eval comments	Answer to the previous evaluation comments	-	○	○	○	○	○	○	○	○	○	○
		3	DQ master plan	DQ policy (After any target change, revise policy upon presentation & approval)	-	○	-	-	-	-	-	○	-	-	-
		4	MTP inclusion in Planning-dwg	Expedite D to include plant's process/equipment requirement to Planning dwg	◎	◎	-	-	-	-	-	◎	-	-	-
		5	List of major changes & theme	Identify major changes (new mechanical/structure/material) and show reasonability of theme up items	-	○	○Chg/add	○Chg/add	-	-	-	○	○Chg/add	-	-
Plan / Actual	Changing Point	6	DQ action plan & actual	DQ master schedule (major event) & actual	-	-	○	○	○	○	○	-	○	○	○
		7	Spec maturation sheet & actual	Maturation actual of major change point (new mechanical/structure/material)			○	○	○	○			○	○	
		8	DQ theme action plan & actual	Target value of the theme. Master plan implementation-drive & actual	-	-	○	○	○	○	-	-	○	○	-
		9	Feature spec check on CBU	Check feature spec on CBU (incl non-change points)	-	-	-	-	-	○	-	-	-	○	-
		10	DQ test plan & actual	Logic behind car # setting, time duration, method, car utilization	-	-	○	○	○	○	○	-	○	○	○
		11	Test (need/no) plan/actual control sheet	Decide test need/no need, then summarize (target & actual control)	-	-	○	○	○	○	-	-	○	○	-
		12	Maru-W	Maru-W checking Schedule and Actual			○	○	○	○	◎		○	○	◎
		13	CBU 5-sense judgement criteria summary	Matrix of 5-sense criteria for each lot (set evaluation point for each category)	-	○ Set Item	○	○	○	○	○	○ Set Item	○	○	○
		14	CBU 5-sense judgement criteria for critical items	Strategies and quality level data	-	-	-	○	○	○	○	-	-	○	○
		15	Quality target / criteria	Graduated quality criteria at event, toward target full-achievement	-	○	○	○	○	○	-	-	-	-	-
	Past trouble	16	KT check list, tracking action plan & actual	For past trouble, strategy, target, and actual control.	-	-	○	○	○	○	-	-	○	○	-
		17	DFB function guarantee sheet	Cascade critical past trouble c/m using the sheet	-	-	◎	◎	◎	◎	-	-	◎	◎	-
	IQS	18	IQS improvement. spec	Critical action items to bull's-eye target point (spec action progress & actual ctrl)	-	○	○	○	○	○	○	○	○	○	○
	Claim	19	Market claim cost reduction	Critical action items (MQ to submit Verification Map)	-	○	○	○	○	○	○	○	○	○	○
	Reflect	20	CRF inclusion	Progress of event spec maturation plan. Track number of problem vs CRF approved.	-	-	○	○	○	○	○	-	○	○	○
Certification	SOFT	21	Certification progress control	Plan and progress	-	○	○	○	○	○	-	○	○	○	-
		22	Model Category, Vin#	Clearly Category and Vin#			○	○					○		
		23	To acquire Certification	Acquire date and MP Start date.	-	-	○	-	○	○	-	-	○	○	-
	TEST	24	Test of Certification	Copnfirm Test schedule and Resul	-	-	○	-	-	○	-	-	○	○	-
Issues	Issues	25	Major issues & path forward	Current issues and next action	-	-	◎	◎	◎	◎	-	-	◎	◎	-
	Out-standing	26	MP day-1 D/C & check plan	Spec change which occurs from MP VIN No.1 & the check plan for change point	-	-	-	-	-	◎	-	-	-	◎	-
		27	MP R/C items & check plan	After MP start up, check R/C and change points	-	-	-	-	-	◎	-	-	-	◎	-
Recurre-nce prevention		28	For the problem occurred after launch	Find why problem occurred after launch (especially repair) & reflect c/m to next model	-	-	-	-	-	-	○	-	-	-	○
		29	Next model reflection items	Do why-why analysis on success & failure (strategies) and reflect c/m to next model	-	-	-	-	-	-	○	-	-	-	○
		30	Achievement ratio of planning stage requirement	Calculate achievement ratio to the planning requirement, analyze gap, and apply c/m to next model	-	-	-	-	-	-	○	-	-	-	○
Budget		31	Project budget planning sheet	Budget plan (tachi 2 comments. Compare with the actual of past similar model)	•Make	•Apln	•	-	-	-	-	•Apln	•	-	-
		32	Project budget actual	Control expenses' actual and analyze gap	-	-		•	○	•	•	-		•	•

FIGURE 8.8
Milestone management of new product development (example).

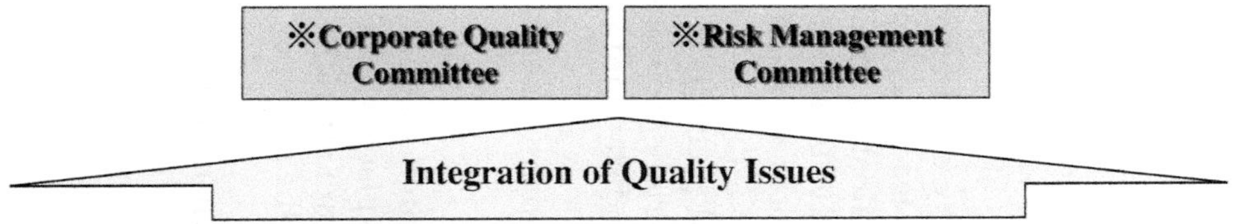

Quality Review Meeting	Purpose	Chaired	CEO	CQ	R&D	SQM	Mfg	Frequency
※Corporate Quality Committee	Confirmation company-wide overall quality progress and risks	CEO	○	◎	○	○	○	Monthly
Cross Functional Quality Meeting	Review of cross functional quality of R&D, Supplier and Mfg.	CQ Chief		◎	○	○	○	
Total Customer Satisfaction	Review of major quality concerns (Safety, Warranty, VOC)	CQ Chief		◎	○	○	○	Weekly
Customer Protection	Deploy customer protection activity on top VOC issues	CQ Chief		◎	○	○	○	
Engineering Quality	Engineering quality review meeting	R&D Chief		○	◎			
Mfg. Quality	Quick resolving of daily mfg. quality issues	Mfg. Chief		○			◎	Daily
Supplier Quality	Quick resolving of daily in-coming parts quality	SQM Chief		○		◎		
Shipping Blocking	Decision making of stopping the shipping of products	CQ Chief		◎	○	○	○	Requirement basis
※Risk Management Committee	Decision making of potential crisis issues	CEO	◎	○	○	○	○	

FIGURE 8.9
Regular quality review structure and process (example).

For instance, in Figure 8.9, each division such as R&D, Supplier Management, and Manufacturing already has its internal quality review meetings. Therefore, a key emphasis should be given to how the activities and information can be synthesized, coordinated, and linked together to deliver total optimized quality output. Otherwise, this can lead to additional wastes in quality, which we will discuss in "Lean Quality Management" in Chapter 9.

8.3.5 Evaluation Principles and Methods

Given a borderline quality issue, project members tend to be more generous in their interpretation of the severity of the problem because we as human beings have a propensity to be generous to our own work. To counteract this tendency, a milestone or a regular quality evaluation by a top leader like the CEO or Quality Chief can create positive pressure on team members to improve potential quality concerns. However, quality evaluators should not put pressure on every quality issue. Instead, they need to be realistic, understanding the situation clearly so that the project members are willing to accept advice and coaching.

When performing evaluations of technically sophisticated products, it is difficult to understand all the technical details and make the right decisions. Therefore, a proper evaluation principle, method, process, and standard are as critical as the quality reviews and evaluations themselves.

The following are specific principles and methods for company leaders to consider adopting to make the right judgments.

1. *Principle 1. Decide on the Genba:* Any type of thorough evaluation is not possible away from *Genba*, the actual workplace where things are designed, made, and tested. Today, the main strength of Japanese products' quality comes from harnessing the power of *Genba*.

   The broader meaning of *Genba* can be broken down into three things: *Genba* = actual place, *Genbutsu* = actual part, and *Genjitsu* = actual situation.

   The overall meaning is that you "go, see, hear, smell, touch, feel, and try-out" for yourself. It also means "not to solely rely on the reports," as:[3]

   - Reports are not real-time; and
   - Reports may not be objective!

   We can make better decisions from observing the actual workplace, interviewing relevant people, getting hands-on experience, and so on.

   Here is serious advice from Taiichi Ohno, the father of the Toyota production system: "Confirm with your own eyes. It is relatively easy to persuade people on the *Genba* with examples, but away from the *Genba*, there is not always a way to prove one's point, so many times each side ends up thinking their idea is a good one. Perhaps the hardest thing is for managers, senior managers, and supervisors to persuade each other."[4]

2. *Principle 2. Respect Experts:* Customers vary greatly—they are in many different circumstances, and they use products in many different ways. It is very challenging for a company to predict every customer's usage modes or to forecast a customer's perception of the product.

   Of course, companies should have robust technical standards or evaluation procedures. However, the massive amounts of data and information coming from the market or product validation tests are difficult to digest and interpret accurately. Much of a product's quality information is hidden in a soft side or under a subcomponent level. For instance, since software products are not visible and are integrated with technological products, it is more of a challenge to grasp its entire quality.

   Under such challenging circumstances, we need someone who knows things much better and deeper than the decision-makers. We can call this kind of person an "expert." Generally speaking, experts should be able to represent thousands of customers and hundreds of customers' usage modes with high technical skills, knowledge, and experience.

   We can foster experts inside the company or hire them from the outside. In any case, it costs money, time, and culture to have them and keep them. For instance, in a culture where experts are not respected or are not paid enough, it is difficult to keep them or their skills may fade away without a proper system to utilize them.

   Therefore, first, we need to create a culture and system that respects experts and utilizes them. We also should consider having a *group of experts* in each critical area. Using their expertise we can save time and money by making appropriate decisions at crucial moments. Furthermore, the judgments and evaluations made with the experts will be respected and followed by the employees on the *Genba*.

3. *Principle 3. Prepare a Backup Solution:* During the evaluation meetings previously mentioned, many grey cases may arise where it is difficult to make a definitive judgment. Customers' methods of complaints and the magnitude of each complaint can fall in a broad, varying range. In such circumstances even the experts can find it difficult to determine whether a complaint will be a definite problem or not.

   In such cases, the evaluator should ask the team to prepare a backup solution as they move to the next milestone. Preparation of a backup solution is part of the risk management method discussed in Chapter 6. The backup solution can take various forms: it can be a technical solution or a communication manual with customers, or both.

When there is electric motor noise from a steering wheel at a very moderate level, first they can refer to a technical standard. However, if the noise level is borderline, it could be difficult to make a judgment to say this is going to be okay or will be a claim in the market. Some customers who park the car in a quiet parking garage in the basement may find the noise bothers them, for example. However, it is difficult to make 100% risk-free products in the tight schedule of product development. Moreover, typically, a 100% noise-free product costs more money as well.

In such a circumstance, it is better to be safe by being critical of the potential risks. A recommended direction is that we ask the team to prepare a backup solution in case there is a claim in the market. In that case, the team can still go to the next development process while developing a backup solution as their secondary activity.

Through diligent monitoring of the product after launch, if a concern is found in the market, they can quickly apply the backup solution for after service as well as for mass production at manufacturing.

8.4 Self-Improvement of Quality

As a fact of evolution, every organism has a tendency to become better in terms of survival, growth, and reproductive ability. We, as company owners, managers, and employees, would like to see our workplace be better in performance and quality to serve customers better, to survive better, and to grow bigger. It is a natural path for evolution.

Charles Darwin observed the survival of the fittest in nature. Evolution means changing and adapting to nature and the environment. Improvement in business has the same principle of evolution—to change and adapt to the business environment and the competition.

In past decades, with the introduction of new machine and computer technology and modern management, the overall aspect of quality has evolved dramatically in all types of industries. Now the base is well established; however, at the same time, society's expectation of quality levels is increasing.

In nature, moving upward is more difficult than moving forward or downward. Improvement is more difficult than keeping the status quo. Without continuous effort to move upward, the movement will eventually stop to the dead state. That is the nature of quality as well.

Quality activity minimizes entropy increase in business operation processes. Unless we expel the internal entropy from our system, the entropy increase will disrupt our system, and it will end up in a dead state due to uncontrollable disorder. That is a simple law of physics.

In other words, if we do not clean up the office or the plant floor, what will happen? Dust will cover the desks, rust will build up on the machines, and everything will be in disorder, increasing the entropy; and eventually, everything will stop and begin to break. That is nature.

In the previous sections, we discussed how we could deploy quality in a structured way—this is just a starting point! When everybody is moving upward, how can we move upward faster than others? In this journey, do managers need to take care of every part of our quality operations tirelessly? Or do we need to police every operation? This way of management is neither practical nor sustainable.

At the maturity stage of a company, each function and team wants to take care of its operation *by itself* and continue to improve quality outputs. The author calls this process *"Self-improvement of Quality,"* which should occur *spontaneously* inside the organization. The *self-improvement of quality* is an important subject that the company leaders should think about in modern quality management.

Rather than policing or continually supervising, managers should build a system and culture of *"Self-improvement of Quality" in the organization* so that the organization and its members will grow more independently capable of being globally competitive, and as a result, the organization will become stronger, bigger, and successful.

There are critical elements for "Self-improvement of Quality" as graphically represented in Figure 8.10. The primary four elements are

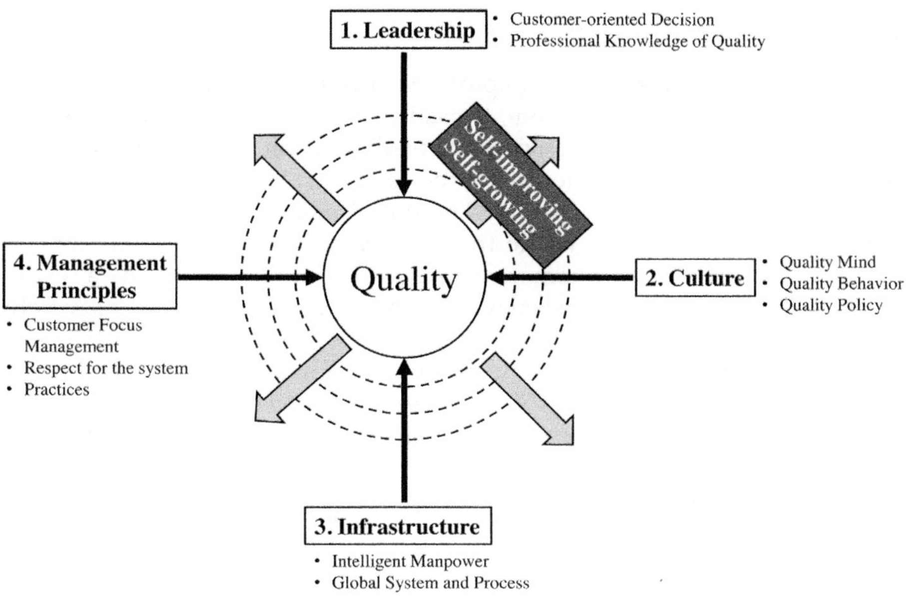

FIGURE 8.10
The key elements for self-improvement of quality.

Quality Leadership, Quality Culture, Quality Infrastructure, and Quality Management Principles.

8.5 Leadership

During the discussion of quality planning and deployment in the previous chapters, we emphasized scientific and structured methods of quality planning and deployment. The fundamental method of quality deployment is a top-down and cross-functional approach. *The success of quality deployment and management relies on people's attitudes, actions, and skills.* The performance of people has a significant variance and is difficult to manage.

People make systems and processes. *People make quality.* The operators of machines, processes, and management are people. The machines are programmable, but human beings are difficult to program. The most effective way of "programming" the employees is through inspiration, motivation, and intelligence.

People can be motivated to do things right when they have a role model. *The most effective role models are the company leaders. The employees follow in the footsteps of their bosses—not the words they say.*

There are fundamental ingredients required for quality leadership.

8.5.1 Customer-Oriented Decision

What is quality leadership? In Chapter 2, we discussed that *"Quality starts in the boardroom."*[5] In the boardroom, typically profit-oriented discussions and decisions dominate. The customer-oriented decision may have difficulty overruling the short-term interests of the company. That was the case of the big safety recalls in recent history. The board members knew the defects, risks, and potential crises. While the priority of their debates was for the interest of the company, customers were getting hurt and killed. As a result, the pride of the employees, the company, and even the country was severely damaged.

The boardroom is the highest decision-making committee of a company. *Business is not just about money; it is about creating customers.*[6] *When the top company leaders make the right decisions for customers, the employees see, learn, and copy the practices.* This is the way people are educated and programmed to do things right by themselves.

8.5.2 Professional Knowledge

When leaders make a decision, the decision should be based on rationality. Our rationality is based on knowledge. Without knowledge, there is no reference point to discern that our decision is adequate and correct. As any job position requires knowledge, quality leaders also should have a solid knowledge of quality.

The basic scope of quality knowledge is covered in this book regarding quality systems, quality management, and quality engineering. Traditionally, some MBA-type managers used to neglect engineering and technical knowledge; however, managers in the technology industry need to have minimum engineering knowledge to understand the technical part of products. As for quality managers, quality engineering and methods as discussed in Chapter 5 are essential tools for quality leadership.

8.6 Culture

Culture is the social behavior and norms found in human societies. "Culture" comes from the Latin meaning *"to cultivate."*[7] Culture is a verb, a dynamic word. Culture is built over a long period of time.

In the same way, corporate culture is not static. The corporate culture is dynamically changing, adapting, and evolving. The culture comes from within, and it's driven throughout the entire organization. The power of culture is a driving force for an organization. Culture builds power. Culture always wins.[8]

Quality culture is a quality mindset, which leads to quality behavior. The power of quality culture is that the organization is pursuing excellence and high performance, and ultimately *creates happy customers.*

The left side of Figure 8.11 explains Schein's Organizational Culture Model that American management professor Edgar Schein developed in 1980. Edgar Schein divided organizational culture into three different levels:

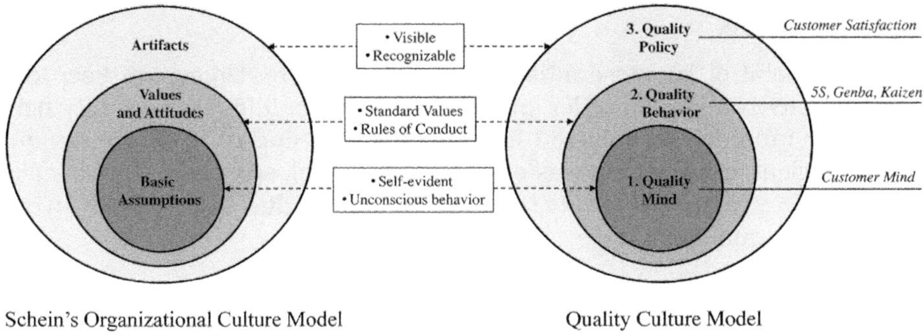

FIGURE 8.11
Quality culture model.

artifacts, values and attitude, and basic assumptions. The Schein model is like an onion model, wherein:

- artifacts are more visible things to external parties, such as logos, organization structure, etc.
- values and attitudes are goals, strategies, etc.
- basic assumptions are the beliefs, unconscious behaviors, etc.

From the Schein model, we can make an analogous model for quality culture as shown on the right side of the diagram in Figure 8.11. *The core part of the quality culture is the quality mindset that is the customer mindset, and the quality mindset leads quality behavior, and quality behavior produces customer satisfaction that is visible from the outside.*

8.6.1 Quality Mindset

The core part of the culture is invisible and unconsciously embedded in our brain in the form of mind. Our mind has a cognition filter to provide value judgments. When we have a quality mindset, we give more weighted value to quality.

When we have *the quality mindset*, our mind is trying to understand things from the standpoint of the customer. In other words, we *think as if we are in the customer's situation*. Therefore we unconsciously try to represent the customer's standpoint and protect the customer's interests.

8.6.2 Quality Behavior

Quality mindset drives good quality behavior as the outcome. As a good athlete has good routines and habits, good quality behavior comes from good working habits. According to the research, more than 40 percent of the actions people performed each day weren't actual decision, but habits.[9]

8.6.3 5S Habits

5S is the habit of Japanese industrial quality culture. Unless we keep our working environment in order and clean, it is difficult to retain quality habits in our inner being. Table 8.1 explains the meaning of 5S. 5S makes our work efficient, minimizing wastes of time, material, space, etc. *The core philosophy of 5S is the habit of quality behavior.* 5S habit produces consistency in our behavior for quality.

8.6.4 Genba

Genba represents the actual workplace. As we discussed in the previous chapters, *Genba* truly represents reality. Customers do not buy a concept, but

TABLE 8.1

5S Quality Behavior

Japanese Term	English Equivalent	Meaning from the Japanese Context
Seiri	Sort (Tidiness)	Throw away all rubbish and unrelated materials in the workplace
Seiton	Set in Order (Orderliness)	Set everything in proper place for quick retrieval and storage
Seiso	Shine (Cleanliness)	Clean the workplace; everyone should be a janitor
Seiketsu	Standardization (Keep cleanliness)	Standardize the way of maintaining cleanliness
Shitsuke	Sustain (Discipline)	Practice "Five S" daily—make it a way of life; this also means "commitment" and "habit"

things made in *Genba*. *Genba* contains problems and concerns, but it also contains the keys to solve the problems. With the customer mindset, *employees' natural behavior would be to go to Genba, to check for quality concerns, and to improve the quality*. *Genba*-focused behavior is the natural outcome of our quality mindset, and such behavior should be our routine and habit.

Genba-oriented people like to show the actual situation and demonstrate actual parts in the actual workplace rather than a nice presentation. In a reality-oriented company, Genba is the ultimate answer to questions and problems.

8.6.5 Kaizen

Kaizen represents a continuous improvement process. There is no end in company business operations, nor in quality improvement activity. *Kaizen* activity in the organization comes from the quality behavior of the organization's culture.

Many people have a misconception about the motivation for doing *Kaizen*. They think *Kaizen* is necessary because the situation is desperate. But Ohno emphasized, "The most important thing about doing *Kaizen* is to do *Kaizen* when times are good, the economy is strong, and the company is profitable."[10] When times are good, there are many things on the surface that are redundant or inefficient, etc.

Kaizen should be a quality culture habit for continual improvement activity regardless of the business situation. The basic concept comes from continuous roll-up activities of P-D-C-A.

> *During a management meeting, if someone says "No problem" because everything seems to be fine, that is the problem. From a Kaizen perspective, "No problem is the problem" because Kaizen requires people to see and find problems! For instance, the Toyota Production System is designed to see problems, make it easy to solve the problems, and make it easy to learn from mistakes. Making it easy to learn from mistakes means changing our attitude towards them for continual improvement.*[11]

TABLE 8.2

Continuous Improvement vs. Traditional Approach (examples)[12]

Traditional Approach to Quality	Kaizen Approach for Quality
Quality should be improved until it conforms and adheres to design requirements and specifications.	Meeting the specifications is not enough. The variance still needs to decrease by increasing the capability of the process.
Achieve the quality up to the point that ordinary customers can perceive the value.	Continue to improve the quality to meet demanding customers' levels and professional levels.
Reduce defect ratio to the point of the design limit.	Improve the design, improve the parts, and improve the process. Make a continuous cycle for zero defects.
Quality functions are deployed by quality relevant groups	Quality is everybody's responsibility. Quality should be deployed by every functional group.
Quality improvement activity costs money.	Deploy zero-cost quality improvement activity. The output should be bigger than the input cost.
If there are no particular issues in the product, keep the current design and process.	Find opportunities to improve design and process to increase quality while reducing cost.
Always respect and practice the current QMS (Quality Management System).	Practice the current QMS, and improve the QMS to enhance its efficiency and effectiveness.

Table 8.2 provides some examples of *Kaizen,* continuous improvement. Continuous improvement has different viewpoints and standards for momentum to improve quality continually compared with the traditional approach to quality.

8.6.6 Quality Policy and Communication

The quality policy is a manifesto of how the company operates regarding quality value to customers. This is a formal statement by the organization. This is an official declaration in front of customers.

Most companies say "Customer first" as illustrated in Table 8.3. The policy is a mission statement that the company is striving to achieve. Saying is, of course, always easier than doing. Since the employees and customers try to connect what the company says to their experience of the products or services, an ambitious yet substanceless quality policy has no meaning.

The policy should be the representation of the company's quality culture. It should be a mirror image of what the company has done, is doing, and is going to do. It should have supporting quality pillars like leadership, culture, infrastructure, and management principles.

The most important thing is that the policy should be practiced through daily business practices as a constitution. Otherwise, the policy becomes just a slogan. *The quality policy also should be effectively communicated with employees directly and indirectly.*

TABLE 8.3

Quality Policy Statement (Examples[13])

TOYOTA	HONDA	FORD	HYUNDAI
Toyota grows by adding values to our *customers*. "Customer First" and "Quality First"	Achieve customer satisfaction by producing products that meet or exceed our *customer's* expectations at a reasonable price.	Improve our products and services continually to meet our *customer's* needs, and provide a reasonable return to our stockholders.	Adhere to the highest quality as the most important promise to our *customers*.
IBM	**SAMSUNG**	**LG**	**SONY**
IBM is committed to the goal of achieving the highest standards of quality delivering value to our *clients* each time they interact with IBM	It is our Quality Policy, based on an effective management system, that ensures our delivery of the best products and services that exceed our *customer* requirements and expectations.	Provide *consumers* with the high-safety products by promoting enterprise quality activities and quality awareness.	Sony is committed to improving product and service quality from its *customers'* viewpoints and works hard to maintain and enhance *customer* satisfaction, confidence, and trust.

The following example is the standard practice and communication of quality policy by H-Company:

The quality policy and objectives shall:

- *be appropriate to the purpose of the organization*
- *include a commitment to comply with requirements and continually improve the effectiveness of the quality management system*
- *provide a framework for establishing and reviewing quality objectives*
- *be communicated and understood within the company*
- *be posted in meeting rooms (where appropriate) and department offices*
- *be reviewed for continuing suitability*

When the policy is consistent with what the organization is striving to do, the declaration naturally demands attention and actions from the employees. *That is the power of quality policy and its communication.*

8.7 Infrastructure

The efficiency of human activities or technology development depends on the supporting infrastructure. The infrastructure is not only physical systems but also invisible structures like constitution, laws, IT, and the financial system. A good foundation makes things flow better; for instance, information, business, and management flow efficiently with the help of a well set up infrastructure.

A good quality infrastructure in a company has the power to self-improve quality, starting from the inner organization. There are two major quality infrastructures: *intellectual infrastructure* and *system infrastructure.*

8.7.1 Intellectual Infrastructure

Intellectual infrastructure is a group of talented workforce and the organization's knowledge, including patents. This is the biggest asset of the company's daily operation and growth. The quality infrastructure needs a good pool of employees having quality mindsets and technical skills as well as capable management skills.

This intellectual infrastructure is not necessarily proportionate to the size of the company, because it is not given but needs to be built. Many globally competitive companies have a very systematic process to build intellectual infrastructure starting with hiring a talented workforce to fostering growth through internal and external education and training programs.

> *S-Company has a special program to hire talented people globally. They also have special internal and external education and training programs. The company requires its employees and managers to regularly attend specific and general technical, quality, and management courses. They pull out the employees from their workplaces during the core working hours to provide them with the most updated knowledge and expertise whether they are directly or indirectly related to their work.*

Figure 8.12 shows the contents of the standard quality program that S-Company uses to build quality intellectual infrastructure. This quality program was designed and taught by the author. Without proper quality knowledge and skills of the employees, it is difficult to become globally competitive.

Building robust quality intellectual infrastructure starts with fostering the workforce to have competent quality engineering and management skills.

8.7.2 System Infrastructure

Every country, society, and organization has its own systems and processes that the members are expected to respect and follow. They can take the form of laws, regulations, rules, signs, etiquette, and so on. The more internationalized the systems and standards, the easier they are for people (including visitors) to observe and follow.

The same principle applies to company operations and quality management systems. "Quality System" means the construction of organizations, responsibilities, procedures, processes, and resources (personnel, material, funds, information) for performing quality management.

Course I : Quality Fundamentals	
I-A. Introduction to Quality	**I-B. Fundamentals of Quality Management**
Why Quality?	Quality Planning
Meaning of Quality	QMS (Quality Management System)
No Safety, no Quality	5S Principles and Productivity
Productivity vs. Quality	Quality Management 5-Principles
Cost of Poor Quality	Quality House with 5-Pillars
Award of Premium Quality	Scientific Quality Management
How to deploy quality?	Total customer Satisfaction
How to measure quality?	

Course II : Quality Engineering	
II-A. Quality Engineering System & Process	**II-B. Quality Engineering Tools**
New Product Development Process	Cause and Effect Analysis
Quality Engineering	QFD
Design Quality Engineering	FMEA
Supplier Quality Engineering	FTA
Manufacturing Quality Engineering	8-D report, QC Story, 5P
Market Quality Engineering	SPC
Quality System Engineering	- Statistics in Quality Engineering
Engineering Principles for DFQ	- Descriptive and Inferential Statistics
Software Engineering Principles	- ANOVA
Software Quality Management	DOE

Course III : Advanced Quality Management	
III-A. Quality Management System	**III-B. Risk Management Process**
QC-Circles	New Product Evaluation
Kaizen	New Product Launch Management
Monotsukuri	Management of Media and Cyber
ISO 9000	Customer Relationship Management
ISO/TS 16949	Quality Audit
TMS (Toyota Management System)	Recall Prevention Management
TQM (Total Quality Management)	Crisis Management
6-Sigma	

FIGURE 8.12
Quality training program of S-company. (Designed by Dr. Lim.)

With global business operations, one company's policies and systems do not stay only in one city or country. Instead, they go out into the world. Therefore, international-minded systems and processes are essential for better communication with the outside world. By maintaining international viewpoints of quality operations, we can supply products and services for customers worldwide in the same language. For instance, there are many kinds of global standards in the area of quality, engineering, and business management, such as ISO 9001, IATF 16949 (International Automotive Task Force), 6-Sigma, ISO 26262, SPICE (Software Process Improvement and Capability Determination), CMMI (Capability Maturity Model Integration), and so on.

We need to continue to improve our system infrastructure, choosing the right standards for the organization and using them in the daily operations of the company to actually improve quality and bring about business outcomes.

8.8 Management Principles and Practices

When problems exist in the system, inadequate outputs are inevitable. Who designs the quality system? It is management. One of the vital elements for self-improvement of quality is management power.

Management's mindset, policy, leadership, and skills are critical elements for self-improvement of quality. The power of management makes continuous improvement possible in the organization and moves employees.

The following are the exemplary cases we can consider as management policy or principles.

Lim's 5-Management Principles discussed in Chapter 4:

1. Start from the customer
2. Cultivate a quality culture
3. Make a closed loop: P-D-C-A
4. Focused and detailed management
5. Transparent management

GE's 4E evaluation criteria for managers set up by Jack Welch:

- Energy—this is about high personal energy.
- Energizers—power and energy to be able to energize others to carry out the organization's vision, mission, and policy.
- Edge—excellent professional knowledge, management skills, decision capability, etc.
- Execute—deliver outputs through right actions.

Examples of management policy that are universally applicable:

- Always proceed with a "customer first" mindset
- Respect sound theory and develop fresh ideas for innovation of technology and quality breakthroughs
- Make the most effective use of time to finish the job during the core working hours
- Strive continually to improve the efficiency and quality of work
- Respect the quality management system and practice it during the daily work
- Be ever mindful of the value of research and endeavor

The message here is that a company should have certain management principles matching the company's vision and policy. The company should maintain and promote the principles by evaluating the managers using the applicable criteria. In this way a company keeps the momentum and power for self-improvement of quality coming from inside the organization itself.

Notes

1. Diamond. Jared, *Guns, Germs, and Steel*, W. W. Norton, 1997.
2. Bejan, Adrian. *Design in Nature: How the Constructal Law Governs Evolution in Biology, Physics, Technology, and Social Organization*, Doubleday, New York, 2012.
3. Mogarei, Hormoz, February 2, 2016, "Genchi Genbutsu Genjitsu (3Gs), (Go to the Genba & see for yourself!)," www.linkedin.com/pulse/genchi-genbutsu-genjitsu-3gs-go-genba-see-yourself-mogarei-mba/ (accessed February 1, 2019).
4. Ohno, Taiichi. *Taiichi Ohno's Workplace Management: Special 100th Birthday Edition*, McGraw-Hill Education, 2013.
5. Nilsson, Joyce, *The Essential Deming*, McGraw-Hill, 2013.
6. Drucker, Peter, *The Essential Drucker*, HarperCollins Publishers, Inc., 2001.
7. Harper, Douglas. Online Etymology Dictionary, 2001.
8. Vanderbloemen, William. *Culture Wins*, Post Hill Press, 2018.
9. Duhigg, Charles, *The Power of Habit: Why We Do What We Do in Life and Business*, Random House, 2014.
10. Ohno, Taiichi. *Taiichi Ohno's Workplace Management: Special 100th Birthday Edition*, McGraw-Hill Education, 2013.
11. Shook, John. "How to Change a Culture: Lessons from NUMMI," January 1, 2010, https://sloanreview.mit.edu/article/how-to-change-a-culture-lessons-from-nummi/ (accessed February 2, 2019).
12. Warwick Business School: Continuous Improvement module, https://tools.patientsafetyinstitute.ca/Communities/qim/Shared%20Documents/General/Comparing%20Traditional%20and%20Continuous%20Improvement%20Approaches.pdf (accessed August 10, 2018).
13. TOYOTA: www.toyota-global.com/sustainability/society/quality/ (accessed August 12, 2018); HONDA: https://world.honda.com/profile/philosophy/ (accessed August 12, 2018); FORD: Liker, Jeffrey K. *The Toyota Way*, McGraw Hill Professional, 2003; HYUNDAI: http://hyundaipng.com/quality.html (accessed August 12, 2018); IBM: www.ibm.com/ibm/responsibility/policy10.shtml (accessed August 12, 2018); SAMSUNG: www.samsung.com/semiconductor/support/quality-management/ (accessed August 12, 2018); LG: www.lg.com/global/sustainability/customer/product-quality-and-safety (accessed August 12, 2018); SONY: www.sony.net/SonyInfo/csr_report/quality/policy/ (accessed August 12, 2018).

9

Lean Quality Management

Since the Second Industrial Revolution, mass production became the norm in the manufacturing industry. People enjoyed abundant industrial products at affordable prices. Ford's mass production system in the form of the moving assembly line introduced in 1924 made T-model cars affordable at a $300 price range, a drastic decrease from the $850 price in 1908. Thanks to the affordable price, the everyday person could afford to buy a car—realizing Ford's goal to "democratize the automobile."

Since then, we have focused on increasing quantity with more plants, more people, more equipment, and more materials. Industrialists thought, the more they make, the more they could sell. They thought of their growth as an unstoppable, successful journey as long as they continued to pedal the bicycle. Economies of scale pushed manufacturers to churn out standardized, mass-produced goods to meet the high demand of consumers. With only one black color with the same design features, Ford was able to sell as many as 15 million T-model cars between 1908 and 1927 at an affordable price.[1]

This approach worked well at the dawn of the mass production era. Once we made something at a reasonable price, people were willing to pay for it. However, through several economic difficulties in modern history like the Great Depression in 1929 and the oil crisis in the 1970s, industrialists realized simple mass production was not the answer for future industrialization because the demand-driven economy could not be sustained. Customers began to change and make more careful purchasing decisions based on price, value, preferences, performance, function, and quality. In this environment, without continued strong demand, mass production capability was no longer a potent tool. The industry began to realize they needed to change from the manufacturer viewpoint to the customer viewpoint, i.e., customer's perception of the value of a product.

In the 1950s in Japan, the traditional American-style mass production system was not suitable for Japanese conditions where a small market required manufacturing in small quantities with many varieties. To make manufacturing more efficient in a mass production environment, Toyota developed *JIT (Just In Time)* and *Jidoka* that became the backbone of Toyota's Production System[2] in the 1960s.

Nowadays, the demand for personalized products has become the norm in the market. The industry responded to this demand by small batch production of various products with numerous options. To support this direction a new type of quality management must improve quality *holistically: moving from Little q to Big Q and defining* quality as the *result of work effort* versus *input cost*.

Many managers reflexively ask for more workforce and more investment in equipment, machines, and robots. They provide convincing numbers such as ROI (return on investment) to justify the investment. This approach completely deviates from the lean management concept.

Modern management should consider all aspects of business activities with a lean perspective, as waste is inherently embedded in every part of business systems and processes. From this perspective, we need *to improve quality with lean management*. In the manufacturing industry, quality with lean management would enhance the competitiveness of engineering, manufacturing, and sales and service throughout the business value chains. *Lean Quality Management* comprises quality operations in a lean way: reduction of defects and variances as well as the elimination of waste in quality systems, management, and engineering.

1. History of Lean Manufacturing
2. Value vs. Waste
3. The Elements of the Lean Production System (Toyota)
4. Lean Six Sigma
5. Quality Management with Lean Thinking

9.1 History of Lean Manufacturing

Efficiency is the backbone of any business operation. Any animate or inanimate thing seeks efficiency to flow better, to grow better, and to reproduce better. This is a natural path of evolution—without efficiency, there is no evolution.

The history of civilization reflects the search for efficiency from the age of hunter-gathers to that of industrialists. For instance, during the agricultural age, humans developed efficient water management systems to fully utilize rain to be fed into the soil. However, some negative mindsets carried over from the agricultural industry into the manufacturing industry. The agriculture industry depends heavily on natural conditions; depending on the weather, the harvest can be bountiful, or it can be scarce. Therefore, agricultural people like to build an inventory of crop in case of a bad crop season. As we went through the industrial revolution, we carried over some agricultural thoughts and habits. When the machines are running well, we think, "let's all work hard today and make as much as we can because you never know when machines might break down or when worker absenteeism might get worse."[3]

There are some significant differences in efficiency management between agricultural and manufacturing industries. In the manufacturing industry, things can be reasonably well-controlled under good management. We can forecast material demand, parts cost, labor cost, market demand, technology trend, etc. Moreover, we can also manage potential risks.

> *When 9/11 occurred, there was a significant shortage of special parts that were being imported from Japan to North American plants. Since there were strict inspection procedures implemented at the U.S. ports for security reasons, many shipped parts from Japan were held at U.S. customs.*
>
> *H-Company smartly managed the situation. At the time of 9/11, there were many H-Company employees who were on a business trip to Japan from the North America plants. H-purchasing division asked those employees to carry urgent parts to the North American plants to minimize the plant shutdown possibility due to the shortage of the particular components. Even in such a crisis, the situation was well-managed with few harms to global manufacturing activities.*

Since the industrial revolution, our thinking for efficiency management has been forever changed. The major impact is described in Adrian Bejan's book, *The Physics of Life*: "The big change in the history of human access to power was the development of heat engines, which consume fuel, not food. The heat engines triggered two revolutions, the industrialization and electrification of the globe."[4] The core purpose of running a heat engine is to deliver work efficiently. Efficiency is determined by the second law of thermodynamics that concerns irreversibility. In the new era of the manufacturing industry, we have to be sensitive to irreversibility in total business processes. *Waste in the manufacturing process is irreversible in terms of economy and business, for not all human activities can be turned into value or profit.*

In modern manufacturing after the industrial revolution, there were pioneers who thought of work efficiency seriously. Their approaches were logical, rational, and empirical based on realities in the workplace, the factory floor.

In the 1890s, Frederick Taylor, the father of scientific management, studied productivity by scientific methods while working for the manufacturing industry. When he was working at the Midvale Steel Company from 1867 until 1876, he was always constantly mindful of work efficiency. His time and motion study as part of the scientific method was to Figure out standard human motions to achieve maximum task output in a given time. For example, his research on pig-iron handling was very scientific and determined how a first-class pig-iron handler can load 47 tons of pig iron into a railcar per day without being exhausted. Working methods, resting time for recovery, and the right size of shovel, etc., were carefully analyzed throughout his time and motion study. The study of human motion efficiency was summarized later in his book published in 1911, *The Principles of Scientific Management*: "The enormous saving of time and therefore increase in the output which it is possible to effect through eliminating unnecessary motions… which results from a thorough motion and time study, made by a competent man."[5]

His innovative ideas, like the division of labor, management methods, and standardized work ways, made a significant contribution to the early manufacturing industry.[6]

Henry Ford emphasized value-driven manufacturing when he started his company in 1903. He warned against "waste" many times in his book, *My Life and Work*: "For anyone to be required to use more force than is absolutely necessary for the job in hand is waste.… Waste is due largely to not understanding what one does, or being careless in doing of it." He also emphasized the importance of quality and engineering, saying, "The design has to balance. Men die because a part gives out. Machines wreck themselves because some parts are weaker than others. Therefore, a part of the problem in designing a universal car was to have as nearly as possible all parts of equal strength considering their purpose."[7]

During World War II, the Willow Run airplane manufacturing plant located between Ypsilanti and Belleville, Michigan, was constructed by the Ford Motor Company. The Ford system was so efficient that the plant could produce one B-24 Liberator heavy bomber every hour. Willow Run was the physical embodiment of the Ford Production system which was later transformed by Toyota into the concept of "Just In Time" and Lean Manufacturing.[8]

In the late 1940s, "Made in Japan" meant cheap, poorly crafted items.[9] After World War II, with the influence of American quality control, Japanese goods changed radically in quality. "The Japanese welcomed input from foreign companies and lecturers, including two American quality experts: Deming and Juran."[10] Such a drastic improvement was not coincidence but was possible with the Japanese culture of learning and diligence.

Taichii Ohno graduated from Nagoya Technical High School in 1932. His first job was at Toyoda Boshoku, a company making cotton threads. His company was using machines from Toyota Automatic Loom Works to weave cloth. He was impressed by Sakichi Toyoda's invention of a special device that automatically stopped the machine when a thread broke or ran out. The machine stops making defective parts until a human comes to fix the problem by connecting the thread. This concept later became the foundation of Toyota's Jidoka system—"automation with a human element."[11]

After the war, Taichii Ohno began to work at Toyota Motors as a section manager. He was deeply troubled by a significant productivity difference between the United States and Japan that he had heard about; the gap was about eight to one. At that point, he began to focus on improving Toyota's productivity ten-fold to catch up to American companies. He thought he could not achieve this target unless Toyota Motors changed its thinking. That was the spark that led to the Ohno System. His ingenious idea, JIT (Just In Time), came by working with Kiichiro Toyoda, Toyota Motor's founder.

Later, Ohno scientifically studied and developed many lean systems, processes, and tools to improve the productivity of Toyota. Ohno's "Seven Wastes" are the critical steps in the lean process. The details of the Toyota ways were later revealed to the world throughout his book, *Toyota Production System*, in 1988.[12]

9.2 Value vs. Waste

Any activity of human or machine contains inefficiencies. This is the law of entropy generation. When we move an object from point A to B, we always generate inefficiencies. The amount of inefficiencies depends on many parameters: mass (m), gravity (g), height (h), distance(s) between A and B, friction (μ), and air resistance (c_d). However, in an actual situation, it comprises more than those physical parameters. The amount of work required will also depend on other parameters: how to lift the object, how to walk or carry it from A to B, and how to unload and load it.

Taylor was famous for studying human motions and time scientifically. He tried to determine the most efficient human motion for a particular work. And Ford made a conveyor belt assembly line to make the parts move to people rather than having people move to the parts. Ford said, "Save ten steps a day for each of twelve thousand employees, and you will have saved fifty miles of wasted motion and misspent energy."[13]

Their concepts and purposes were the same—how to minimize waste and increase the value. Here, the definition of *Value* is *anything that the customer wants and is willing to pay for.* And *anything that the customer does not want and won't pay for* is *Waste.*

Business activities are any activity that is engaged in for the primary purpose of making a profit, including operations, marketing, production, and administration.[14] Even a small business requires complex operations: setting up the organization and systems, managing suppliers, sales and marketing, etc. *"Value" and "waste" inherently coexist in these business activities.*

Figure 9.1 shows how value-added work can be distinguishable from non-value-added work.

We can define the business efficiency by the ratio of the value-added activities (a) to the total business activities, (a)+(b), as follows:

$$\text{Business Efficiency} = \frac{(a)}{(a)+(b1)+(b2)} \tag{9.1}$$

According to Armand Feigenbaum, an American quality expert in Total Quality Management, the average waste embedded in American products is about 15% to 40%,[15] which means the business efficiency ratio creating real value can vary between 60% and 85%.

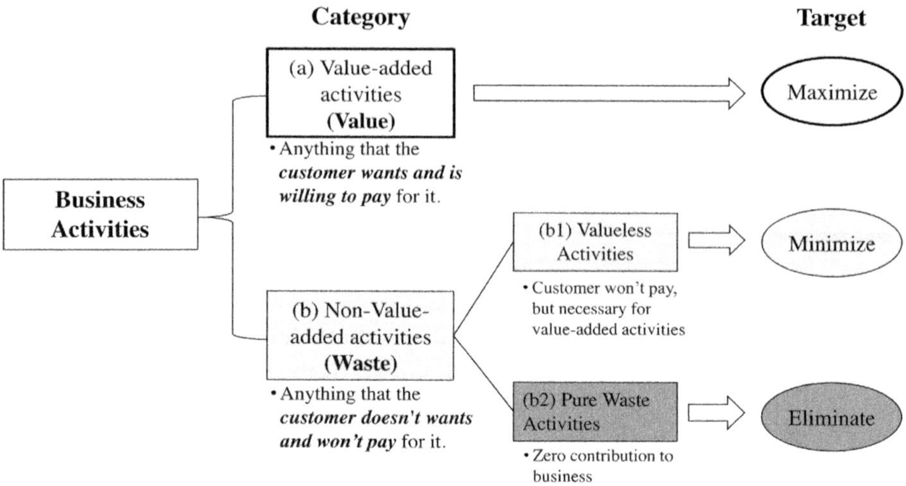

FIGURE 9.1
Value-added vs. non-value-added business activities.

9.2.1 Ohno's Seven Forms of Waste

Ohno, the father of the Toyota Production System, said, "The most important part of the Toyota Production System has been to increase the production efficiency by constantly and thoroughly eliminating waste."[16] The "Business Efficiency" defined in Figure 9.1 is comparable with the production efficiency defined by Ohno. He also defined the waste in production as "anything other than the minimum amount of equipment, materials, parts, and working time absolutely essential to production."

Ohno carefully studied waste during the manufacturing process in Toyota in the 1960s and 1970s. He identified the following seven forms of waste and published them in his book in Japan in 1978[17] and the English version later in 1988: *Toyota Production System: Beyond Large-Scale Production*.[18]

1. *Transportation:* Every time a product is touched or moved unnecessarily there is a risk that it could be damaged, lost, delayed, etc. It increases the cost for no added value. Transportation does not add value to the product, i.e., it is not a transformation for which the consumer is willing to pay.
2. *Inventory:* Whether in the form of raw materials, work-in-progress (WIP), or finished goods, *inventory* represents a capital outlay that cannot yet produce an income. The longer a product sits in one of these states, the more it contributes to waste. The smooth, continuous flow of work through each process ensures excess amounts of inventory are minimized.

3. *Motion:* In contrast to transportation, which refers to damage and transaction costs associated with moving the product, motion refers to the damage and costs inflicted on what creates the product. This can include wear and tear for equipment, repetitive strain injuries for workers, or unnecessary downtime.
4. *Waiting:* Whenever the product is not in transportation or being processed, it is waiting (typically in a queue). In traditional processes, a large part of an individual product's life is spent waiting to be worked on.
5. *Over-production:* Making more of a product than is required results in several forms of waste, typically caused by production in large batches. The customer's needs often change over the time it takes to produce a larger batch. Over-production is the worst kind of waste.[19]
6. *Over-processing:* Doing more to a product than is required by the end-customer results in it taking longer and costing more to produce. This also includes using components that are more precise, complex, expensive, or higher quality than absolutely required.
7. *Defects:* "Defects are not free. Somebody makes them, and gets paid for making them."[20] Having to discard or rework a product due to earlier defective work or components results in additional cost and delays. The company pays good money to make defects as well as to correct them.

9.3 The Elements of the Lean Production System (Toyota)

It is essential to build the right foundation and key elements for a lean production system to maximize production efficiency by consistently and thoroughly eliminating waste. In the case of the Toyota Lean Production System, there are two major pillars: JIT and Jidoka. Figure 9.2 represents the two pillars and vital elements of Toyota's lean production system. There is some uniqueness in Toyota, but some elements are common in Japanese industries, which have been deeply rooted in Japanese history and culture.

9.3.1 JIT (Just In Time)

The core concept of JIT is a *pulling system* instead of a pushing system. Just In Time was first initiated by Kiichiro Toyoda, who was the first president of

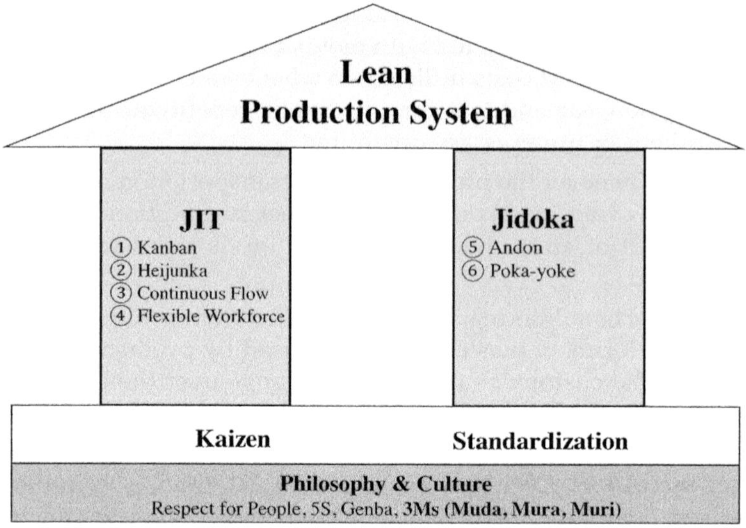

FIGURE 9.2
Key elements of Toyota's lean production system.

Toyota Motors. Kiichiro Toyoda was the son of Sakichi Toyoda, the founder of Toyoda Automatic Loom Works. In the 1950s, Toyota Motors was suffering from many parts delivery problems. Kiichiro emphasized that *the most efficient way to assemble parts in an assembly plant is when each part arrives Just In Time.*[21] His initiative motivated Ohno to deeply study the pulling system like the U.S. supermarket system of replacing products on the shelves just as customers purchased them. That is why, at the beginning, the system was called a "supermarket system."

JIT is a system in which the upstream process needs to produce only what is needed when it is needed and in the amount needed as the downstream process takes it away. Later, Ohno developed many lean elements to make JIT run efficiently.

9.3.1.1 Kanban: Card-Kanban and e-Kanban

In the TPS (Toyota Production System), a unique production control method called the "Kanban system" plays a critical role. Kanban is not a magic word—the meaning of Kanban is "a signboard." Kanban is a small sign card playing a vital communication role between downstream and upstream processes.

The Kanban method is simple and effective in delivering real-time requirements of downstream processes to upstream processes: where parts are needed; which parts are needed; when parts are needed; and how many parts are needed.

JIT philosophy drove out the method—Kanban is one of the tools to support JIT, not the other way.

9.3.1.1.1 Card-Kanban System

Figure 9.3 shows the original concept of Toyota's Kanban system. *Card-Kanban system represents parts flow and Kanban flow* that is information flow. In this flow, two kinds of Kanban (the production instruction Kanban and the parts retrieval Kanban) are used for managing parts.[22] The *production instruction (PI) Kanban* physically travels back to the parts manufacturer for production instruction: part number, part name, quantity required, and delivery location. And *parts retrieval (PR) Kanban* travels inside the assembly process with the assembly.

9.3.1.1.2 e-Kanban System

Through continuous technological improvements, the Kanban system has evolved into the *"e-Kanban,"* which is managed using IT methodologies, as represented in Figure 9.4, and has increased productivity even further.

First, a worker removes the PR Kanban from the parts container and puts it into a Kanban collection box before taking the first part out of the container. Then those Kanbans are scanned by a Kanban reader, and purchase orders are electronically issued to parts manufacturers through the internet. The supplier then immediately prints off the production Kanban, produces the quantity of components indicated on the card and ships both the parts and Kanban to the client.[23]

The advantage of this e-Kanban system is that the supplier can get the production information from the clients *immediately* as the parts are being used, and it can also eliminate the concern of losing Kanbans during the process.

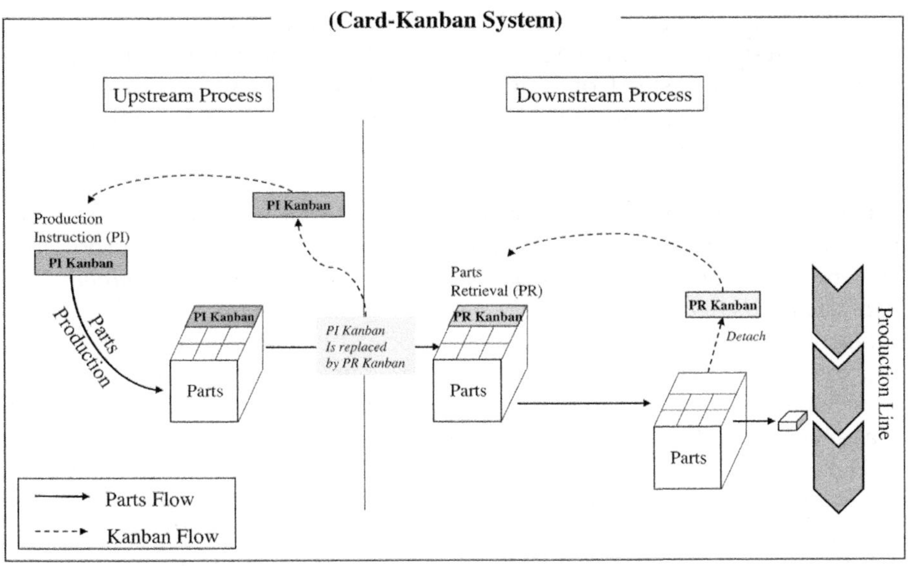

FIGURE 9.3
Card-kanban system (parts and information flow).

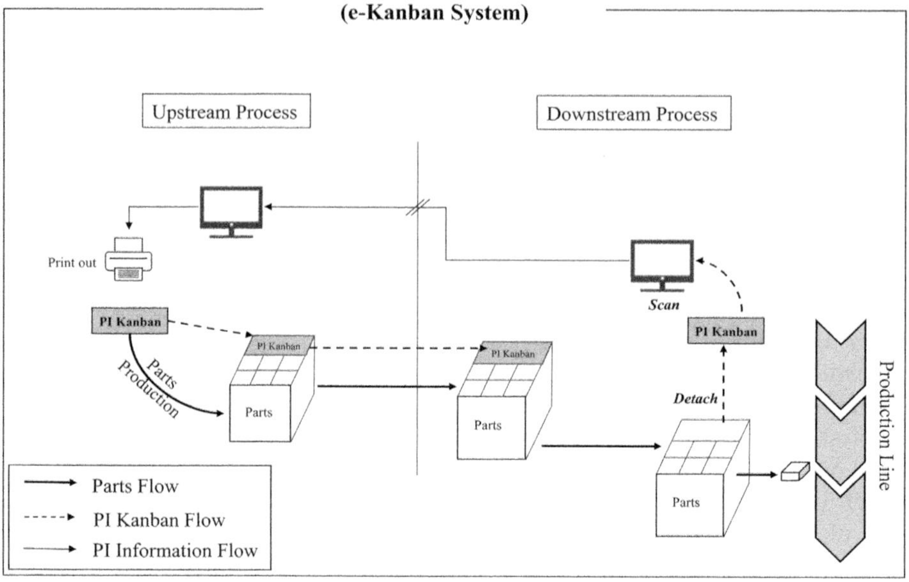

FIGURE 9.4
e-Kanban system (parts and information flow).

9.3.1.2 Heijunka: Leveled Production

The primary motivation of JIT is to respect the market demands. Customers make a purchasing order, which is the initiation of the pulling system. Production should follow the pulling by producing the products in sequence. In other words, instead of producing batches of the same model, mixed models are produced on the same production line according to the market demand of the customer. We can increase customer satisfaction by demand-based sequence production.

Figure 9.5 represents the difference between an unleveled production and a leveled production method. The key advantage of the leveled production system is that each day's *production workload has lower variance, improving productivity and better meeting market demands.*

During production, the production workload is changed not only by production volume but also by the different models due to the contents of products. Some products and models have higher option contents than others. If we make higher grade models only on certain days, as the case of the first column in Figure 9.5, we might need to hire additional human resources and increase investment on more machines to meet the high workloads on Monday and Friday. However, when we make lower grade models, we are still bound to pay idle workers or idle machines, such as on Tuesday and Wednesday.

With leveled production, we can effectively manage our cost by balancing our production workload uniformly. The leveled production of mixed

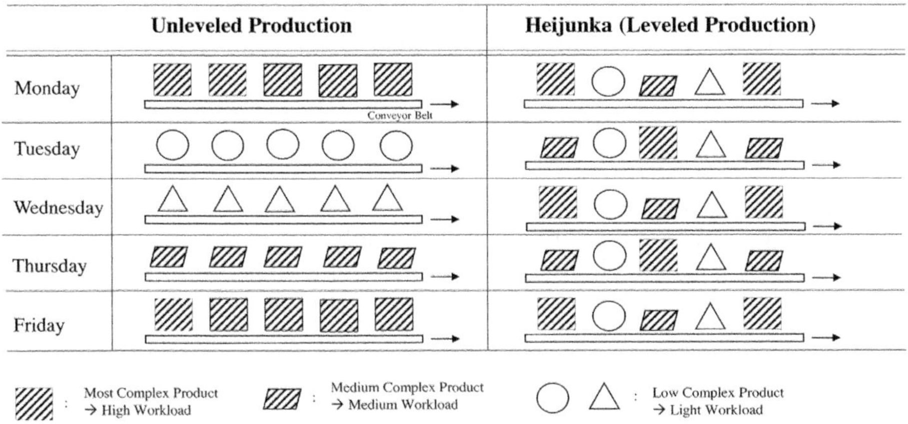

FIGURE 9.5
Heijunka (leveled production).

models can also help suppliers to equally manage their production workload. The supplier can run their production more uniformly to be able to meet JIT production.

When each day's workload is significantly different, it results in high variance in the processing line performance and the parts consumption speed—this kind of variance results in degradation of the productivity in keeping JIT.

J-Company had only one assembly line, which had a maximum production capacity of 300,000 units/year. The company's 4-door D-segment sedan was so popular, with sales of more than 220,000 units/year such that they could make a profit from selling just one model for five years. In 2005, market research showed that there was a strong demand for SUV models in the local market. They seriously considered building another production line with substantial investments. The labor union also pushed for the building of another assembly line, explaining the efficiency increase in making the SUV platform model. However, the company leaders' opinions were divided.

They performed a scientific situation analysis and a conservative market forecast. The sales forecasting of a new SUV model was over 150,000 units/year. It was very tempting to build another production line. However, they decided not to build another assembly line, but to produce mixed models on the one assembly line. They developed a new SUV model to be launched in 2007. In 2008, the financial crisis in the United States hit the local economy, and the market demand for a new SUV was not as high as expected, and the sales of 4-door D-segment sedan dampened as well so that they had to introduce other new models to fully utilize one assembly line: a compact 4-door sedan as well as a compact SUV model.

Today, the company is using the same line to produce 4-door D-segment sedans, 4-door C-segment sedans, full-size SUVs, and compact-size SUV models. Their daily production is a mix of four different models on three different platforms. Their yearly production is nearly 250,000 units/year, but have never exceeded more than 300,000 units/year.

Leveled production of mixed models has the flexibility to be able to meet the dynamic changes of the business environment—*this is the secret to how such a small-volume company can still survive and prosper in the market.*

9.3.1.3 Continuous Flow Production

Traditional batch production has many issues. The number one issue is that a lot of WIP (Work in Process) inventory inevitably happens, which increases a hidden cost. It is called a hidden cost because from an accounting viewpoint WIP is recognized as a company asset. Figure 9.6 shows the clear advantage of continuous flow production. It does not have any WIP inventory. We can ship the first lot within 12 minutes instead of waiting 30 minutes as in the 10-batch production method.

The WIP inventory behaves like buffer or safety stock, which make the JIT system run inefficiently. Without continuous flow production, it is difficult to build a lean manufacturing system.

From a quality viewpoint, batch production has a lag effect on the disclosure of quality issues and corrective actions. Figure 9.6 shows that the final quality checkpoint is in "Process-3." When we find a concern in "Process-3" in the batch production, many WIP inventory parts have already been processed in the pipeline.

In contrast, continuous production has faster reaction capability to quality issues. When a quality issue is found in "Process-3," we can immediately react to stop the production and correct the problem in the upstream. Since there are only a few parts in the production stream (in this example case), we can manage the impact much more efficiently through repairing or scrapping the defective parts that are in the process.

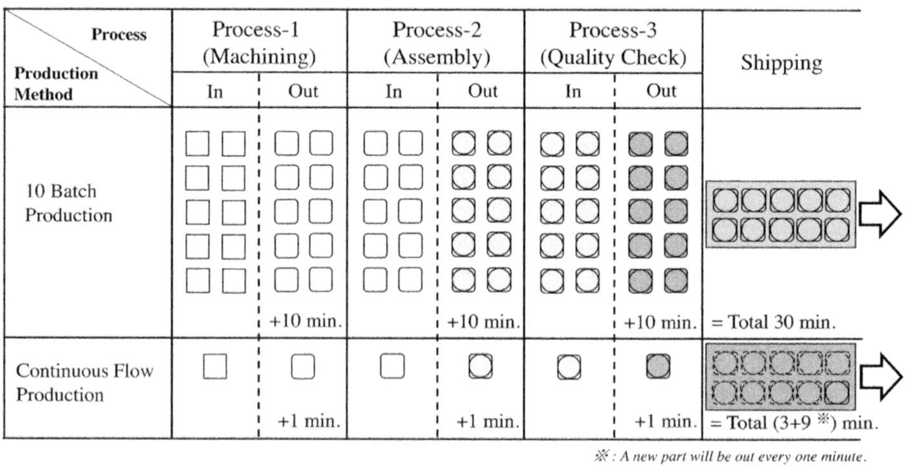

FIGURE 9.6
Continuous flow production.

9.3.1.4 Flexible Workforce Production

The manufacturing process requires four essential elements: man, material, machine, and method, which are called the 4Ms. A company builds products by using the 4Ms and makes profits by adding value to the 4Ms. At the same time, the 4Ms become the source of the cost, which should be controlled and managed.

As we discussed earlier, the most efficient form of lean production is to meet market demand. Since the market demands changes, it is difficult to follow the exact trajectory of the demand line. The market is dynamic and flexible—the production system should be as well.

Amongst the 4Ms, the most flexible, manageable element is "man," the human element. Other elements like machines can be more flexible when human elements are involved. For example, with machine plus man, we can build a semi-automated process, where the workforce takes care of particular processes that are expensive or inefficient to produce using full automation.

How we use machines and human resources together is a crucial factor in managing to increase value while decreasing the waste cost. To run such a semi-automated manufacturing line efficiently, human and machine should be integrated smartly, especially as many machine processes require a human operator. There are typically three motions required: loading the part, setting and operating the machine, and unloading the part. Usually, *the loading and unloading operations are costly to automate.* While the machine is running, the operator has nothing to do except watch it. However, watching the machine does not improve the job quality, so it is a complete waste.

Figure 9.7 represents a *cellular production* environment. The concept is transformed from "one operator, one machine" to "one operator, many machines in different processes."[24] In such an arrangement, *one operator can flexibly handle various processes by operating multiple machines.* The operator can operate five machines with different settings depending on the models. One (or more) workers walk around the line, add parts to the processes, and then start the process. While the machine process works on the part automatically, the worker adds the next part to the following process, and so on.

The flexible workforce production method is a necessary *intermediary step* for smart manufacturing to increase productivity efficiently by designing a production system with a combination of humans and machines in an intelligent mode of operation.

9.3.2 Jidoka

Jidoka is smart automation—automation with a human touch.[25] The concept was originally conceived by Sakichi Toyoda back in 1926. When Sakichi Toyoda founded Toyoda Automatic Loom Works, he made an ingenious device in his automatic machine that automatically stopped the process when the thread broke or ran out.[26]

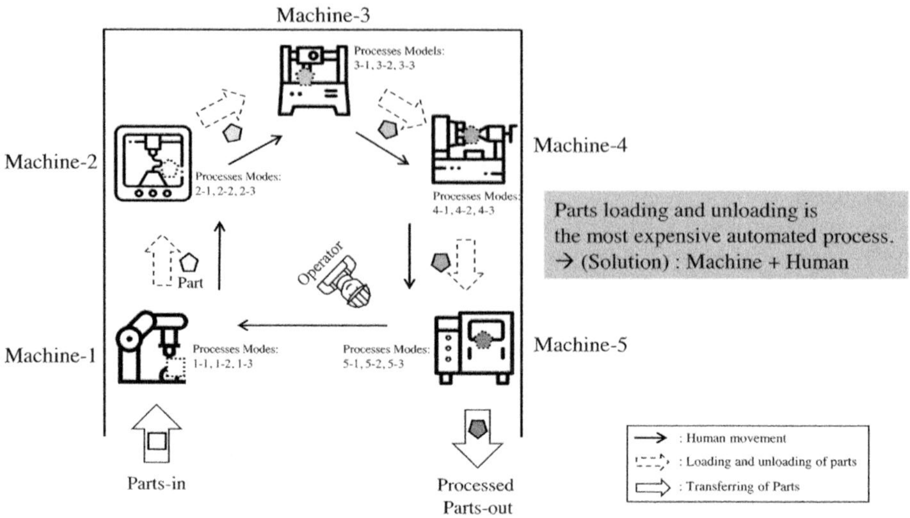

FIGURE 9.7
Flexible workforce production.

Toyota's *Jidoka* system became the pillar of the TPS, and reflects Toyota's philosophy and spirit. With this pillar, Toyota has demonstrated many good examples in the industry of how to improve quality while reducing waste. Whenever we invest in a machine or equipment, we need to make sure it does a proper quality job. If the machine makes defective parts, it decreases productivity instead of improving it.

There are other benefits of *Jidoka* as well, as represented in Figure 9.8. When a machine stops, production will stop, and people will be eager to resolve the

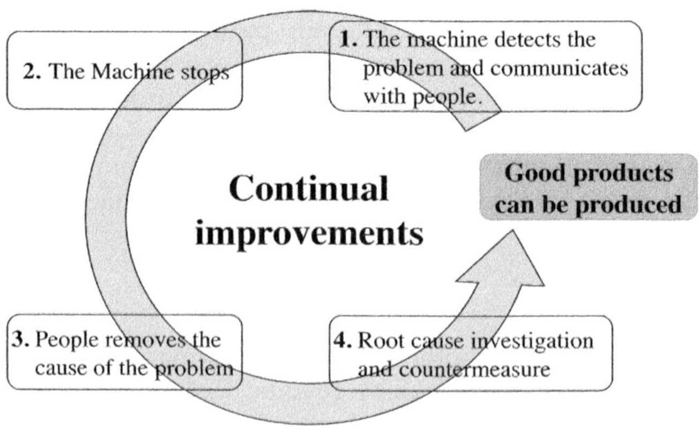

FIGURE 9.8
Four steps of Toyota's Jidoka system.

problem as soon as possible. At the same time, they want to make sure it won't happen again.

The underlying philosophy of Toyota's Jidoka system is smart automation plus the human element. If a machine finds a problem or does not do a proper job, it should stop processing and communicate with people. Similarly, if a human detects a problem, whether it is from machines or human operators, we also should be able to stop our work and communicate with the machines or people.

9.3.2.1 Andon

Toyota's *Andon*[27] is a component of Toyota's *Jidoka* system, as shown in Figure 9.2. Almost all equipment made for Toyota has the *Andon* system, and it was first installed in the Kamigo Plant in 1966.[28] When the machine is in trouble, it stops production and sends an alarm signal with a visual light warning. When workers in the assembly line also find a problem or make a mistake in the job, they stop the production line by pulling the Andon cord installed beside the line.

Stopping the machine when there is trouble forces awareness on everyone. The concept of Andon is to improve automation with human interference. Figure 9.9 shows several examples of the Andon system: the Andon lamp displaying the simple information of warning, stop, or going-well, and the Andon board showing the detailed information of the production situation.

9.3.2.2 Poka-yoke

Poka-yoke (Foolproof)[29] plays a significant role in Toyota's *Jidoka* manufacturing. Poka-yoke can be linked with the Andon system as shown in the Andon lamp of Type-B of Figure 9.9. For instance, for foolproof torque processing in the assembly line, Andon is connected with an impact torque wrench. The light can show whether or not the right torque is being applied by the wrench while tightening the bolt. Blue light goes on when a worker starts tightening a bolt. When the worker finishes the tightening process successfully, the blue light changes to green. If there is a concern on the torque range close to the upper or lower spec, the yellow light comes on to draw attention to the processing or the tool. If a bolt has been tightened out of specification, e.g., too loose or too tight, a buzzer sounds, and a red light goes on to advise the worker of the concern. If the worker does not resolve the problem, *the production line stops.*[30]

> *The most effective way to implement Poka-yoke in production is through a robust product design. During the design review, the production department should bring concerns to the design engineering team to build a foolproof concept into the product design. Some Poka-yoke ideas are simple in their application, like two different*

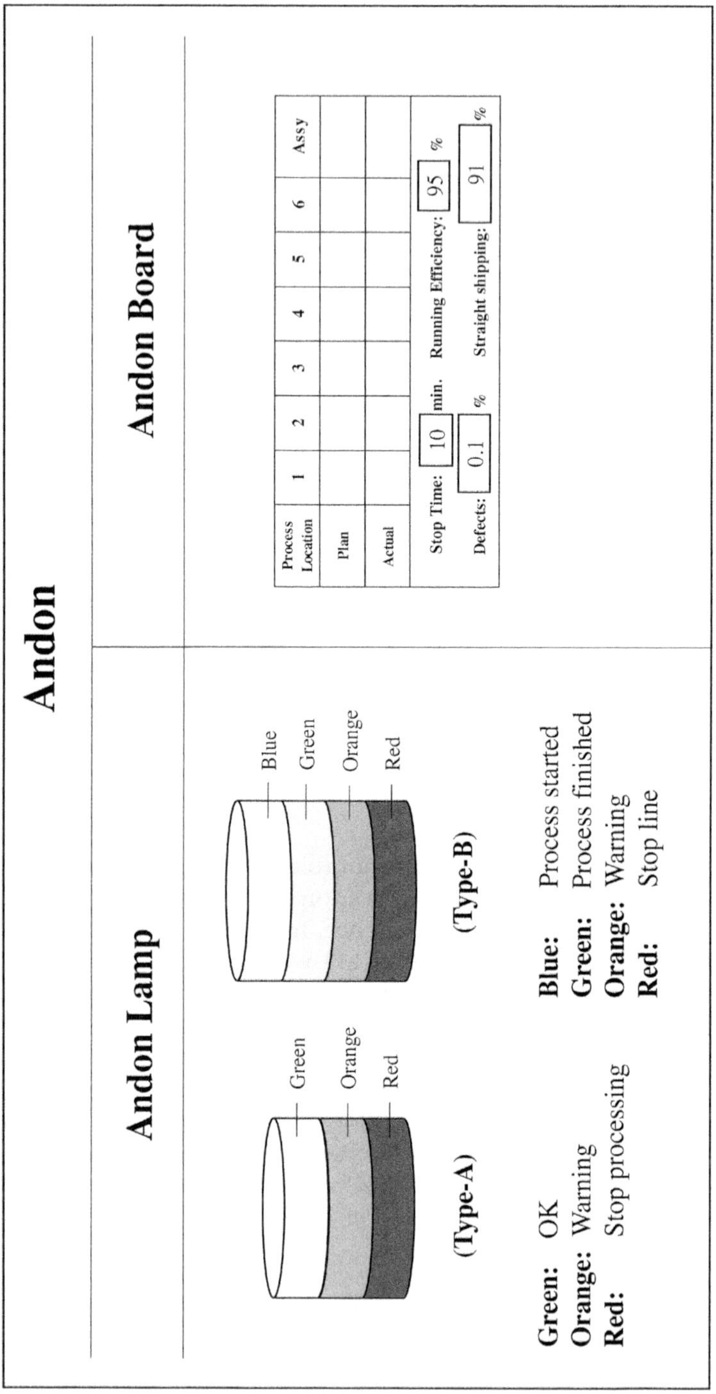

FIGURE 9.9
Toyota's Andon system.

sizes of holes in the left and right side of a part to prevent potential mis-directional assembly. However, there are complex and difficult types as well. In such a case, several Poka-yoke ideas should be combined to eliminate any possible mistakes.

For example, a coupler assembly process is a labor intensive job for production workers to manage. Most of the couplers have unique colors to match the correct male and female parts. However, this is not an assurance method that will guarantee the right quality job—sometimes, the workers can still misassemble the wrong male and female parts due to the common shapes of the couplers.

Therefore, first, they should consider designing unique shape couplers in proximity of the wire harness bundles as a design Poka-yoke.

Second, if the unique shape of the coupler design is difficult because it is virtually impossible to design every coupler to be unique in the circuit design, designing different length of wires is another Poka-yoke idea so that the same shape of the couplers cannot reach wrong circuitry couplers in their neighboring areas.

9.3.3 3Ms (Muri, Muda, Mura)

There is a famous saying by Eiji Toyota, the fifth president of Toyota Motors: "You can squeeze water from a dry towel." This expression is the epitome of Toyota's philosophy and culture toward lean manufacturing. Eiji Toyoda talks about the elimination of waste, which is the fundamental foundation of Lean Manufacturing.[31] To better understand the philosophy and practical approaches, it is necessary to understand the viewpoints in terms of 3Ms (Muri, Muda, Mura).

Muri in Japanese means *"you do more than your capability."* It is about an overload condition. Muri is pushing a machine or person beyond capable limits *resulting in abnormal issues.* Figure 9.10 shows that during the manufacturing

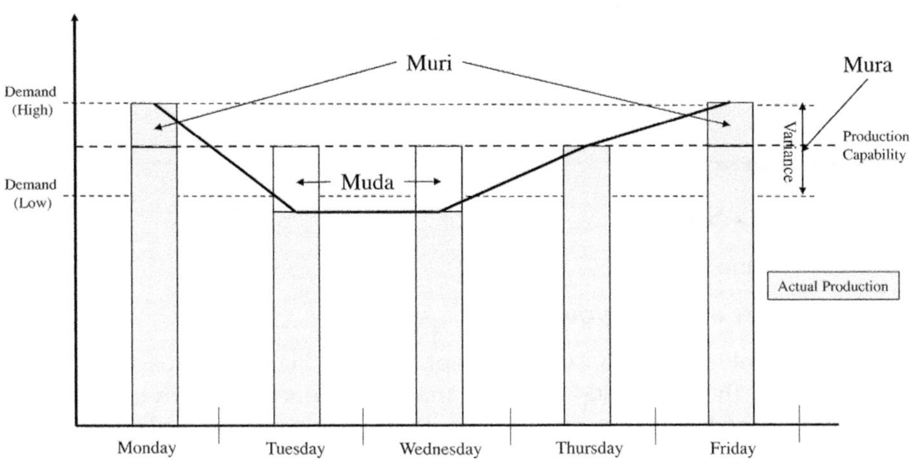

FIGURE 9.10
Toyota's 3Ms (Muri-Muda-Mura).

on Monday and Friday, the production department is requested to produce more than its capacity. As a result, the employees need to run the equipment at a level greater than its capacity and work overtime. However, the actual output is not proportional to what we input. The machines may not have enough maintenance opportunity, which results in breakdown. The company may need to pay overtime to the employees, and fatigue will build up, resulting in unusual safety or quality defects. We may also build up high inventory. We can avoid the Muri situation by better production planning and management.

Muda in Japanese means *"you do something that is wasteful."* Previously we discussed Ohno's seven wastes—they do not create any value that the customers will pay for. When we do not use all our capability, that is Muda wasting our capability. Figure 9.10 shows that the actual workload on Tuesday and Wednesday is well below the production capability. Poor production planning, poor management, abnormal issues, etc. are the typical causes of this problem. Since we still need to pay costs for the workforce, machines, electricity, and utility, this becomes *"Muda,"* a complete waste that we could have saved.

Mura in Japanese means unevenness. *Our production capability should be the average of the market demand.* Some fluctuations in the market demand can be managed by flexible manufacturing methods as discussed previously. However, if we have a variation in the actual workload due to management, technical, or quality issues, it could result in a significant gap compared to the market demand. Figure 9.10 represents a *Mura* situation: variance and fluctuation in the actual workload each day. With the *Mura* situation, it would be difficult to make a production plan to the *JIT*. As a ripple effect, the suppliers will suffer, as well as impacting the *JIT*. The *Mura* results in ineffectiveness and waste. Our target is to minimize the variance. To achieve this purpose, we can mobilize tools like Heijunka, a leveled production method by mixing the high and low workload models.

9.4 Lean Six Sigma

9.4.1 Six Sigma

9.4.1.1 History and Background

Six Sigma evolved from a TQM concept. The Six Sigma approach was first developed in the late 1980s within a mass manufacturing environment at Motorola by Mikel Harry and Bill Smith to meet demanding quality targets on complex manufactured products.[32] With the Six Sigma methodology, Motorola saved billions of dollars by optimizing many processes throughout the company.[33]

Later, in the 1990s, Six Sigma became widely known when GE adopted it. Jack Welch was an advocate of Six Sigma, and under his leadership, Six Sigma became fully rooted in the GE organization.[34] In bringing Six Sigma quality to GE, Jack Welch recognized that the program was not only for engineering and technical projects but also for overall business improvement activities. He noted the many application areas of Six Sigma in GE:[35]

- Plant managers can use Six Sigma to reduce waste, improve product consistency, solve equipment problems, or create capacity.
- Human resources managers can use it to reduce the cycle time for hiring employees.
- Regional sales managers can use it to improve forecast reliability, pricing strategies, or pricing variation.

9.4.1.2 The Approach of Six Sigma

Six Sigma is a data-driven statistical methodology to reduce variation and defects in business processes. Six Sigma is applied to specially selected projects that need significant improvements in cost and quality.

Six Sigma is a business-driven, multi-dimensionally structured quality management methods for goals such as:

- Reducing defects
- Reducing variance in manufacturing and business processes
- Reducing costs
- Increasing profits
- Increasing customer satisfaction

9.4.1.3 Average vs. Variance

You cannot survive in a harsh business environment with mediocre quality targets and standard quality assurance approaches. Ninety-nine percent of the quality targets we usually think are okay are actually very bad in most industries.

Six Sigma's quality target is six standard deviations between the mean and the nearest specification limit in any process. In the case of a normal distribution in a long-term process (1.5 sigma *shift*), the expected defects are less than 3.4 ppm (problem per million).[36] Some people see this target as a tough challenge. However, the actual quality in many industries is far better than Six Sigma's target. For example, in the airplane industry, global accidents per million flights decreased to 2.3 ppm in 2016 from 2.8 ppm in 2015.[37] The defect ratio of many electronic components and products is below 3.4 ppm. Six Sigma's quality target is realistically achievable through proper management methods.

> *For a customer who experiences even a single quality issue in a product, the average quality of the product is recognized as a 100% defect to that person. Customers do not count the average, but they count their own experience. In this perspective, lowering the potential defect rate is critical to improving the brand image of a product to customers.*

The "average" approach in business, manufacturing, and quality performance does not tell the whole story. Understanding the variance is a way of forecasting our business or process performance. For example, if the average delivery time of a certain product is only two days, but the range is 1 to 7 days, such a high variance means that the process is not stable.

Variance is the average amount of deviation between the statistical mean (μ) and all data points (x_i) that are used to calculate the mean, and it can be represented by the standard deviation (σ):

$$\sigma = \sqrt{\frac{1}{N}\sum_{i=1}^{N}(x_i - \mu)^2} \tag{9.1}$$

With an understanding of the variance, we can estimate the distribution of the population such that we can forecast the performance of a business process. During a Six Sigma project, knowing, calculating, and reducing the statistical term (σ) is as important as the average (μ).

9.4.1.4 Six Sigma Methodology

Six Sigma has a distinctive standard method to attain a reduction in variation through process improvement. DMAIC is applied as a generic problem-solving and improvement method as represented in Figure 9.11. The DMAIC

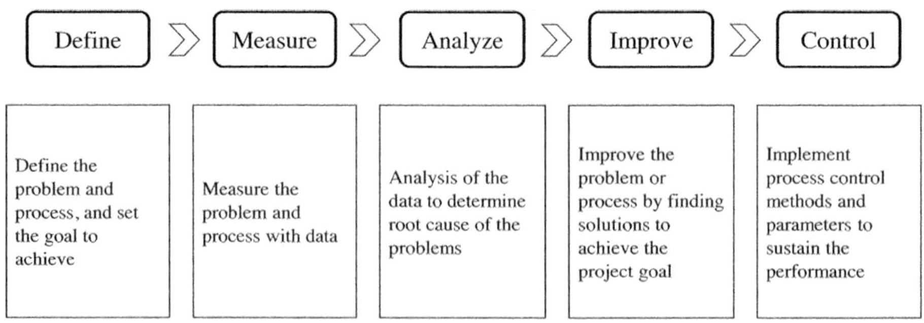

FIGURE 9.11
DMAIC process of Six Sigma.

approach is similar to Plan-Do-Check-Act developed by Deming. DMAIC applies to an existing product or process that needs improvement in customer satisfaction and business performance.[38]

First, we need to identify the most critical issues in customer's satisfaction or the business performance of a company, and then select projects to improve them. The general DMAIC process is as follows:

9.4.1.5 Define

Write down what we currently know. Seek to clarify facts, set objectives, and form the project team. Define the following:

- The problem and process
- The customer(s)
- Voice of the customer (VOC)
- Critical to Quality—what are the critical process outputs?

9.4.1.6 Measure

This step clarifies the current situation, the problem, and the process, with data. The principal object of this process is to collect the current baseline data to compare with future improvement. The team decides on what should be measured and how to measure it.[39]

9.4.1.7 Analyze

Analysis of the problem is the most crucial step for problem-solving. The primary goal is to identify the root cause of the problem. We can use similar principles and processes that we discussed in "Scientific Problem Management" in Chapter 6. Moreover, we also can utilize the QC-tools and QE (Quality Engineering) methods explained in Chapter 5.

9.4.1.8 Improve

After we have identified the root causes of the problem and process, we need to devise a method to improve the problem and process by eliminating the major root causes. At the same time, we have to determine the right parameters to improve. For this purpose, quality engineering tools like FMEA (Failure Mode and Effects Analysis), FTA (Fault Tree Analysis), DOE (Design of Experiment), and SQC (Statistical Quality Control) can be used. As for improvement options and methods, we can use the scientific problem management methodology described in Chapter 6.

9.4.1.9 Control

Even though we have implemented a solution to the problems and the processes, it is not guaranteed that the solution will remain effective throughout our continual business flow.

Any organization has a habit of going back to the original status unless there is a mechanism to prevent it. Therefore, we should implement a new system to prevent the problems and the processes from recurring. Also, a regular audit and review process should be followed.

9.4.1.10 Organizational Structure of Six Sigma

One of the strengths of Six Sigma is a robust organizational structure with a clarification of the roles and responsibilities of project team members. Depending on the training level of Six Sigma, each hierarchical person can take the role and responsibility of the project as summarized in Figure 9.12.

> *V-Company had an ambition of launching a Six Sigma project under the influence of F-Company in 2001. Their hasty approach to Six Sigma failed. Most Six Sigma projects that were launched did not achieve the projects' targets. After analysis of the reasons, they identified critical causes of the failures: the project members were not trained well enough to deploy the projects effectively, and the commitment of the employees and the support from the top management was weak. Since then, the company has launched a systematic Six Sigma training program as well as an improved project management system. Currently, the company has more than 250 Black Belts and has closed more than 450 projects.*

9.4.2 Lean Six Sigma

9.4.2.1 History and Background

Six Sigma has played an important role in improving quality in manufacturing and service industry since GE's success story in the 1990s. Since Six Sigma was developed at Motorola in the United States, it was initially quickly adopted by many American firms. The tool has been applied not only in the manufacturing sector but also in the service industry to improve overall business performance.

In spite of its enormous success, Six Sigma had some difficulties and challenges. For example, the Six Sigma program that Robert Nardelli brought over from GE in 2000 failed at Home Depot.

A Six Sigma project costs money and time to see a substantial output. It also needs to build cultural infrastructure to make the project output effective. It takes money to save money using the Six Sigma quality methodology. As we discussed the case of V-Company in the previous section, we can't expect to significantly reduce costs and increase sales using Six Sigma without investing in training, organizational infrastructure, and cultural evolution.[40]

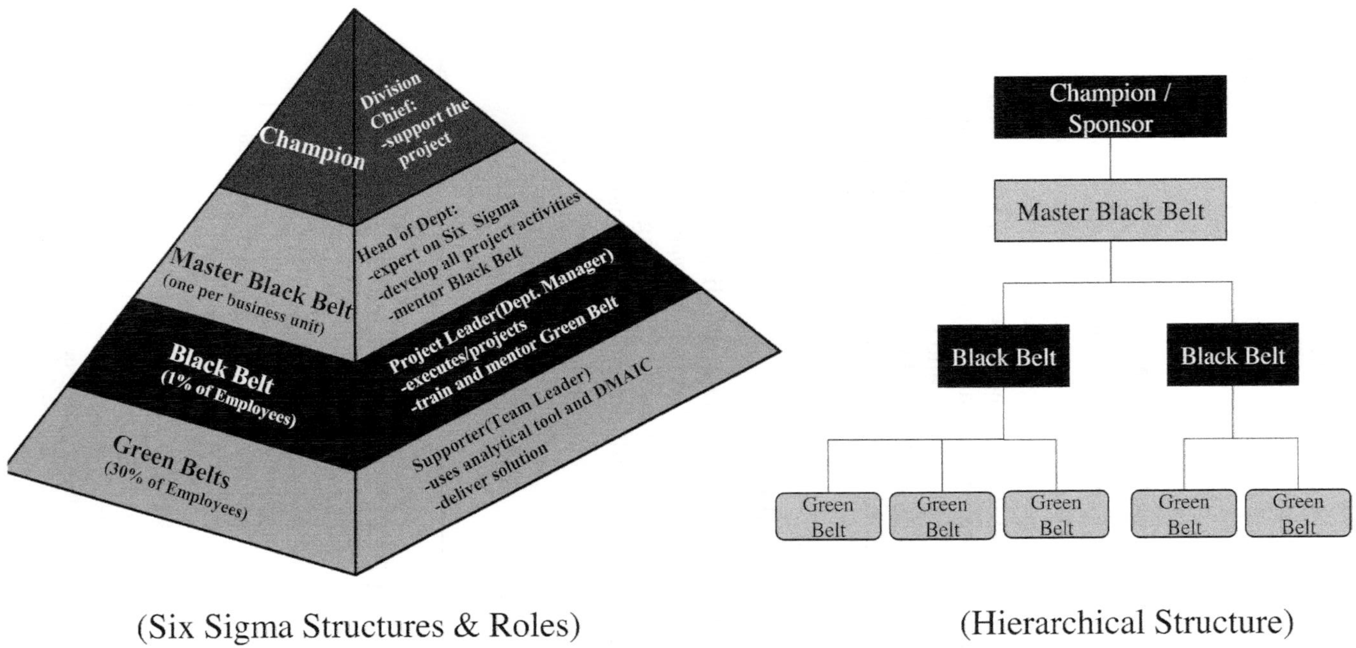

FIGURE 9.12
Roles and responsibilities: Organizational hierarchy of Six Sigma.

Lean manufacturing emphasizes efficiency by eliminating waste, leading to reduced delay and cost. In this perspective, *lean manufacturing philosophy and methodology is an excellent complementary solution to the concerns of Six Sigma. The solution is Lean plus Six Sigma.* That is why many companies are now turning to Lean Six Sigma to satisfy their need *promptly* for a structured continuous improvement program in terms of cost and speed.

9.4.2.2 Approaches of Lean Six Sigma

Six Sigma is already a powerful tool, comprising of fundamental elements such as defect and variance reduction, cost reduction, and profit increase as shown in Figure 9.13. Lean Six Sigma starts from "Cultural Infrastructure" such as quality mindset, quality attitude, and quality leadership. Then it adds the "Lean" elements of lean manufacturing such as waste elimination, reduction of non-value-added steps, and speed increase as the core elements, as represented in the center pillar of Figure 9.13. With the lean activities on top of the Six Sigma approach we can significantly reduce cost and further increase profits.

9.4.2.3 The Elimination of Waste

Six Sigma emphasizes effectiveness by eliminating defects and reducing variance whereas *the "Lean" system emphasizes efficiency by eliminating waste.* As we discussed previously regarding "value" vs. "waste," any activities that the customer will not pay for are wastes, and they should be

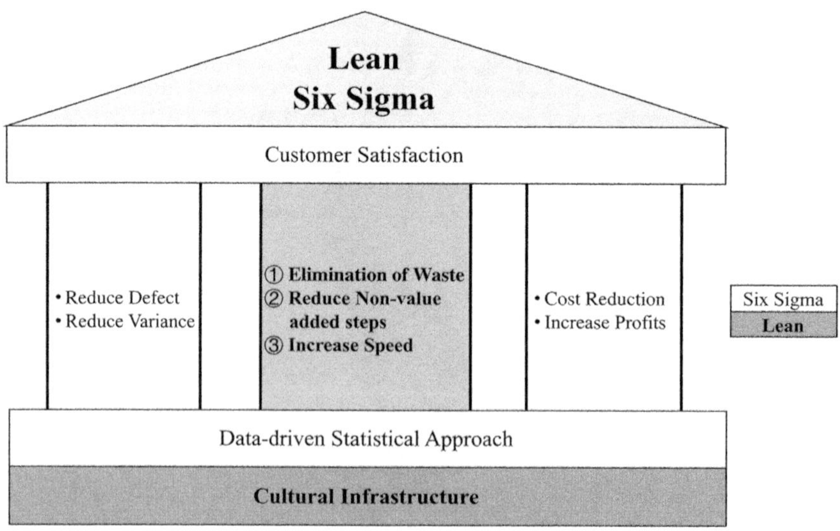

FIGURE 9.13
The structure of Lean Six Sigma.

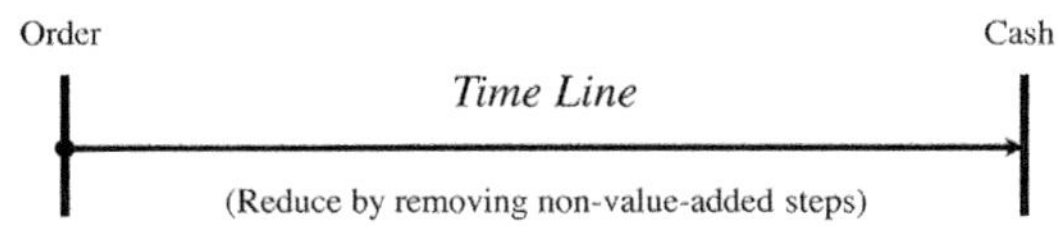

FIGURE 9.14
Business steps and timeline.[41]

eliminated or minimized—the elimination of waste is the first step in the Lean Six Sigma process.

In lean manufacturing, the seven wastes were defined: transportation, inventory, motion, waiting, over-production, over-processing, and defects. In addition to the seven wastes in manufacturing, there are many other types of waste existing in overall business activities. First, we need to define the types of wastes by using a tool like VSM (Value Stream Mapping), as discussed below and represented in Figure 9.15.

9.4.2.4 Reduce Non-Value-Added Steps

Business activities are composed of multiple steps of processes; the more business steps, the more possibilities of inefficiencies and waste. As illustrated in Figure 9.14, "Time Line" is an important indicator representing the efficiency of the whole business process. Non-value-added steps delay the time from the placement of the order to receiving the cash.

When we reduce non-value-added steps, the process becomes streamlined from production to delivery and receiving cash. To be lean in business, it is critical to identify non-value-added steps.

To identify the waste and non-value-added steps, we can use a tool called "VSM (Value Stream Mapping)." VSM is a visual tool that helps us document all the activities required to remove waste and reduce non-value-added steps to make the business flow better in creating value while meeting customer requirements.[42] Figure 9.15 shows how VSM can be used to represent *multiple business steps in a macroscopic view*. VSM visually displays all the steps, tasks, or activities in a process and documents their sequence from start to finish.

From the current state of VSM, as in Figure 9.15, we can create an ideal future state VSM:

$$\text{Current State of VSM} + \text{Lean Basics} = \text{Future State of VSM} \tag{9.2}$$

We need to analyze and compare them to identify the seven wastes and non-value-added steps. Then we can increase future VSM by reducing wastes and non-value-added steps.

9.4.2.5 Increase Speed

Business speed is system and culture. Speed is a big part of the Lean Six Sigma approach. Six Sigma is a data-driven statistical approach, which

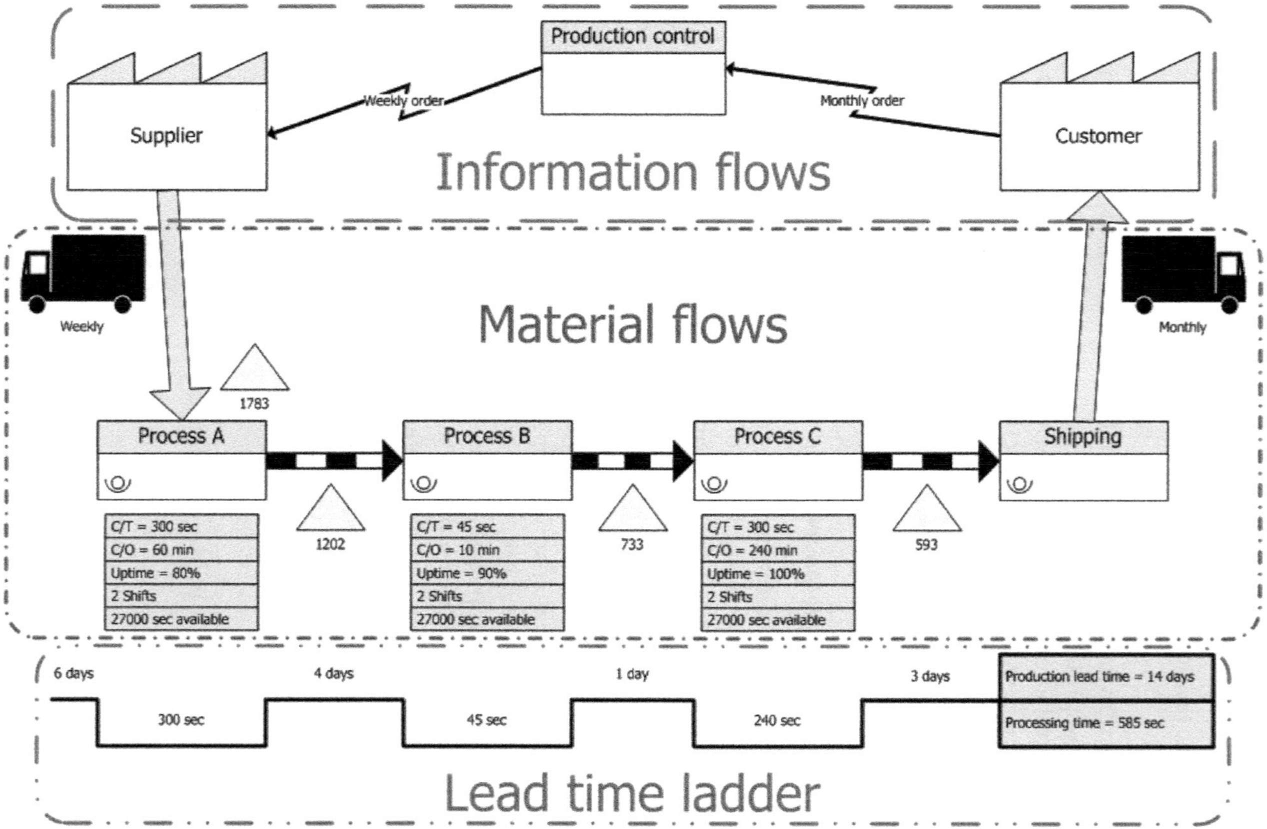

FIGURE 9.15
Value stream mapping. (Image credit by Daniel Penfield, CC by-SA 3.0.)[43]

is effective in improving the accuracy of business processes. However, good data won't guarantee a good decision.

Moreover, data collection takes time—*good-enough data now is better than perfect data later.*[44] The employees should have training in the data collection process and analysis methods, and the managers should be trained to make speedy decisions by interpreting the data scientifically and systematically in a timely manner. Most importantly, a *speed culture should be implemented through the quality system, management, and the continuous technical training of the employees.*

> *Korea's Samsung Electronics Co. is famous for "ppalli" speed—a faster speed of decision making and faster action. Samsung Electronics Co. is a leading company in speed, innovation, and a new style of management in Korea. Samsung's "speed culture" is the representation of Korea's "ppalli ppalli culture"—many Korean companies have this speed culture to remain competitive in technology and markets.*
>
> *When an organization is left to grow naturally, it becomes fatter, and the decision-making process is slower. Every year, Samsung Electronics Co. makes large-scale organizational and procedural changes to be slimmer and faster in its management, decision making, and driving innovations. It reduces the number of levels in its management hierarchy and encourages a faster decision process by holding more frequent online discussions between business division heads and employees. The primary purpose of these changes is to instill "ppalli ppalli" speed culture inside the company so that the employees are motivated to move faster in the planning, engineering, and business processes. "We aim to reform our internal culture, execute as quickly as a startup company and push towards open communication and continuously innovate"—this is the motto declared by Samsung's executives.*[45]

9.4.2.6 Reduce Lead Time

There are two aspects of lead time to be considered. *One is product lead time, to bring new products to the market faster, and the other is business process lead time.* The approach of reducing lead times for both cases is the same—shortening the time it takes to complete the tasks within a process.

Every organization and business process has bottleneck points. The Lean Six Sigma team needs to identify and remove them. These reductions enable the process to become more responsive and flexible to customers or other processes.[46]

9.4.2.7 Reduce Changeover Time

To follow a lean production strategy we need to make many different variations of products in the same production line. To make this possible, first, we need to make lot sizes smaller. When the lot sizes are smaller, more frequent changeovers are necessary. Therefore, reducing the changeover time of different dies, tools, machines, etc. is critical for the small-lot production system.

9.4.2.8 Effect of Lean Six Sigma

The more complex a process is, the greater the possibility of waste. The more non-value-added steps there are in the business steps, the more difficult it will be to achieve the Six Sigma effect—the actual defects will be higher than the theoretical value due to noise and unstable processes.

The chart in Figure 9.16 reveals that the *overall yield value* at various sigma levels significantly improves as we reduce the number of non-value-added steps by lean activity. Lean eliminates non-value-added steps or waste from the process. Six Sigma improves the quality of the process by reducing variance and defects. Lean Six Sigma improves quality, speed, and cost simultaneously.[47]

Therefore, for the Lean Six Sigma approach, we have to take the Lean and Six Sigma approaches together. First, the non-value-added steps in each process should be removed or reduced. At the same time, the process capability of each process needs to be improved through the Six Sigma activity.

The effects that Lean and Six Sigma bring are:

- Improving productivity by removing waste
- Reducing non-value-added steps
- Faster processes
- Cost reduction

Cost of Quality	25–40% of sales	15–25% of sales	5–15% of sales	<1% of sales
	Overall Yield vs. Sigma (Distribution Shifted ±1.5σ)			
No. of Steps	±3σ	±4σ	±5σ	±6σ
1	93.32%	99.379%	99.9767%	99.99966%
7	61.63%		98.839%	99.9976%
10	50.08%		99.768%	99.9966%
20	25.08%	88.29%	99.536%	99.9932%
40	6.29%	77.94%	99.074%	99.9864%

Lean Six Sigma improves quality, speed, and cost simultaneously

Lean reduces non-value-added steps

Six Sigma improves quality of value-added steps

FIGURE 9.16
Synergistic effect of Lean and Six Sigma. (Yield data: Motorola[48] and cost data: Harry and Schroeder.[49])

- Higher quality: Improvement of process stability and process capability
- Increasing customer satisfaction
- Sales increase

9.5 Quality Management with Lean Thinking

Some people in functional departments complain that quality management methods by their quality department are a nuisance to their work. They create complex evaluation procedures, issue over-quality requirements, change documentation formats, call frequent meetings, release great amounts of data, and so on.

They also believe that quality activities led by the quality department interfere with their core job functions: the quality activities increase their workload or lead time, create more tasks than benefits, and create more cost than savings. Moreover, some people think the quality department polices them rather than helps or supports their difficulties and struggles.

This is one of the reasons why the quality department is generally not welcomed in the engineering and manufacturing operations in many companies.

Previously, we demonstrated that the lean approach is effective in bringing benefits to the company in terms of quality, cost, and profits. For example, when we improve quality through the Lean and Six Sigma approaches, as explained in Figure 9.16, we reduce not only defect and variance but also improve quality cost. When we have fewer than 3.4 defects per one million opportunities, the quality cost drops to less than 1% of sales[50] (Figure 9.16).

Lean **Quality Management** *is the management method and practice that develops a system and process to improve quality by using lean methodology while minimizing input cost including preventive quality cost.* A lean approach in quality management can bring about opportunities in effective quality management regarding contribution to quality as well as contribution to the overall business. When we take lean approaches in quality management, first, we need to *eliminate waste and reduce non-value-added steps while increasing speed in overall quality management.*

The first step is to identify wastes and eliminate them in our quality management process.

According to the author's experience, there are **Seven Wastes in Quality Management** *when we do not manage with discretion and scientific methods:*

9.5.1 Wastes in Inspection

Management people always excuse themselves for poor quality or defective outputs. Typical types of excuses are: (a) defects occur because they do not have enough inspectors to check products in the inspection line; and (b) a supplier's parts quality is not stable because of their poor quality control. These two examples sound contrasting. However, both (a) and (b) cases are relevant to the same root cause, which is *poor management of process control.*

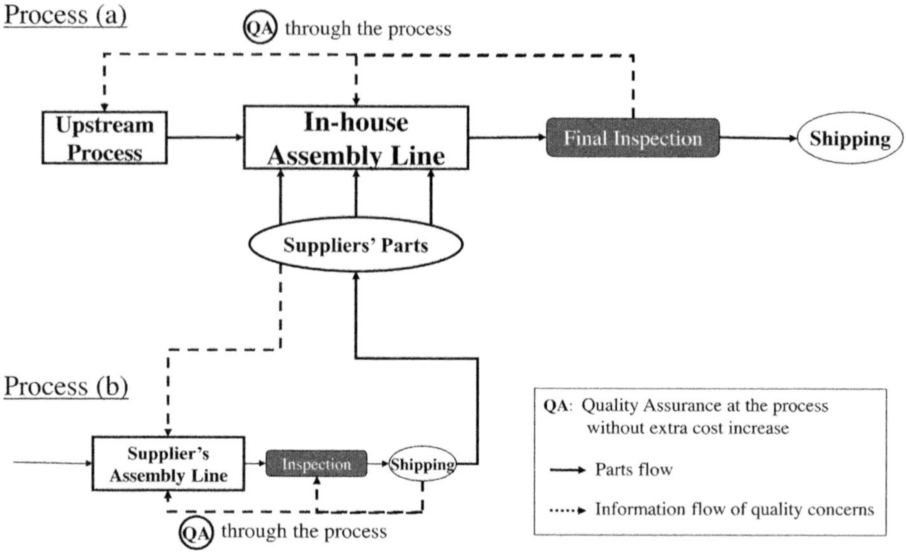

FIGURE 9.17
Improvement of process control with lean quality management.

Let's look at Processes (a) and (b) in Figure 9.17—both have the same mirror image. Production people tend to believe that they come to have poor quality because of not enough inspection. However, inspection cost is the waste—if we had done the job right the first time, there is no need for inspection. Of course, we know we cannot completely eliminate the inspection process. However, *quality management relying on inspection process creates waste.*

Moreover, there are so many processes that we cannot even check for through inspection. For instance, it is difficult to check the inner assembly condition of a vehicle after so many parts have been assembled. Therefore, any quality issues relevant to processes or parts issues should be *assured through processes rather than inspection,* and the issues found in the downstream should be fed to the upstream as illustrated in Figure 9.17.

As for critical quality checkpoints, we should devise a *Poka-yoke,* foolproof method or *QA (quality assurance) equipment.* As for other necessary quality checkpoints, we need to *check the quality status at each process or just at the next process before the final inspection line.* We can deploy minimum checking activities at the local process areas. *The primary purpose of this type of checking is to investigate the cause of problems at the spot and preempt the risks rather than for corrective action.*

The final inspection items and procedures should be minimized. Making it zero is ideal. We keep minimum inspection items that are only possible after complete assembly like operation noise or emissions check of a vehicle, As the overall process becomes stable and reliable, *we can reduce more inspection activities that are waste. That is the way of lean quality management—refer to Honda Engine Plant in Section 2.3.1.*

9.5.2 Wastes in Repair or Scrap

This is a type of waste already defined by Ohno, which was discussed in Section 9.2. However, quality issues inevitably occur, and as a result, repair or scrapping of the final products at times seem to be inevitable if they are physical products.

Most of the time, repair or scrapping decisions come only after the final assembly of the products. This type of management is a waste that we have to eliminate.

First, the decision to repair or scrap needs to be made before the process or right after the process instead of later in the process. We need to discard bad parts or stop the production line very promptly. Otherwise, we will continue to build more processes into the problem product, which will add more waste because we will eventually have to move the defected part out of the final product.

Preventing it from happening again is as necessary as the effective management of repairing or scrapping is. Therefore, proper analysis and record-keeping of the problem should be completed, with the calculation of the total wasted cost, which can become valuable lessons for the employees and the company.

We can use the Pareto principle to prioritize the issues and solve the problems scientifically, as we discussed in Chapter 5. Then appropriate preventive actions should be implemented, as discussed in Chapter 6.

9.5.3 Wastes in Data Collection

It is common to collect quality data during the manufacturing process or in the service process. Most of the time, we collect data manually either by sampling or a 100% check. Then in special cases, we use the data to draw x-bar and R-chart for statistical process control.

Unfortunately, there is a problem with the management side when people routinely collect data without further analysis to improve the process. In that case, the measurement data are not used to reveal the causes of poor quality or productivity. Routinely collecting data may provide peace of mind regarding quality control, or may show the evidence of quality control to internal or external auditors. In such a case, the data collection activity becomes a waste.

Therefore, if we decide to collect any data, the purpose should be clear. Collecting data for just quality assurance purposes is a waste unless a follow-up activity is made. Moreover, if processes become stable and there is no quality or legal risk in not collecting the data, we can reduce the sampling frequency or eliminate the data collection process altogether.

Torque measurement is a critical quality assurance method in the automotive assembly process. However, there are so many critical processes that must be measured. The more complex models or technologies that are applied, the more points that need to be measured. The problem is that the number of torque measurement locations always increase with the passage of time. Unless we reduce some previous, unnecessary measurements, the number of measurement activities will continue to grow. Those uncontrolled measurements and data collection induce a waste.

Any type of data collection costs time and money. The purpose of data collection at each process should be regularly reviewed and dynamically updated—if no concerns exist in the market, we can consider to eliminate the data collection or minimize the sample size.

We need to ask the questions, *"Is this measurement absolutely necessary for customer value, what are we going to do with the data, what is the impact if we stop collecting the data?"*

9.5.4 Wastes in Documentation

Some people have the misconception that quality is all about data and documentation. They perform more documentation to experience the vicarious satisfaction of performing real quality activities. In this respect, those managers ask for a document, frequently change the format, check the document, and then close the activity. *They do not go any further.* Value-added quality management produces business values, but weak management creates workloads—see Section 8.3.3.

Today's engineers spend more time making presentation documents than making drawings. Documentation and a nice presentation do not improve quality and productivity. It is wasting employees' valuable time and the resources of the company.

The engineering department engages in FMEA activity during new product development. Since this is a required process for engineering, engineers follow regular routines. The problem is that FMEA becomes excellent paperwork—however, there is no teamwork activity from the beginning, no update from past and current issues from the manufacturing processes or the market, no follow-up activity, and no path-forward activity. It just becomes superficial routine paperwork to prove they are doing the engineering processes required. This kind of FMEA activity has no value and is a waste.

This kind of practice is relevant to the culture of a company. As we discussed in the previous chapter, unless we improve the quality culture fundamentally, this kind of practice is difficult to abolish. When managers do not understand *Genba,* they are inclined to ask for more and more documents, exhausting the employees in an actual workplace—this becomes waste. *Customers do not pay for excellent paperwork; instead, they buy a product from the Genba.*

Moreover, there are numerous cases in the industry demonstrating that the Genba and the documents do not match each other. *Truth exists only in Genba, and working with the dynamic changing reality leads to lean quality management.*

9.5.5 Wastes in Quality Meetings

Many quality leaders want to set up regular quality meetings like morning and after-work meetings and weekly and monthly meetings. Some managers think having more quality meetings is the best way of improving quality—this is another misconception. More meetings mean people are busier attending the meetings rather than solving and improving quality issues.

The best location to have a quality meeting is in actual workplace, accompanied by actual parts, actual customers' voices, and other supporting evidence. Having meetings in a nice cozy office or a meeting room *requires more paperwork and more presentations, which results in waste. Checking the reality visually and by simple testing can substitute for ten thousand words and a hundred pages of documentation, not to mention the time savings.*

The purpose of a quality meeting is to make the right decision based on sound data, accurate information, and scientific analysis. *Any quality meeting without making a decision or improvement is a waste.*

Every quality meeting should have a clear agenda and strict time management. Daily routine quality meetings like a morning meeting should be very short, less than 5 to 10 minutes. *Lengthy meetings are waste.* Any quality meeting should have a P-D-C-A routine, starting from the remaining issues of the previous sessions to checking their progress.

9.5.6 Wastes in Quality Survey

Typically a quality department has a budget to perform quality surveys in the market. People think an external, third-party survey is accurate and trustworthy. The primary purpose of the quality survey is to understand the weaknesses and problems of the company's products. Sometimes, the external, third-party survey may have vague information that is difficult to track and follow up with actual complaining customers. Without concrete information, the quality department has difficulty convincing the relevant departments like R&D or manufacturing to engage in an in-depth investigation of the problem.

Spending money for a quality survey without taking prompt and concrete actions is waste. The best way to do a quality survey for action is by an internal department of a company in *the guise of a third party.* Then, when critical issues are discovered, we can do in-depth analysis with the complaining customers in a direct way. This type of management can also lead to communication with customers concerning misunderstanding of the product.

Any quality survey that does not lead to quality improvement actions is waste!

9.5.7 Wastes in Certification

Many industry leaders like to seek quality certificates with fervor. The number one reason is to take advantage of business opportunities. For example, many customers require ISO 9001 certification, IATF 16949, CMMI, SPICE, etc. from their suppliers. For some enterprises, achieving these certifications is all about attracting new clients, not necessarily improving the internal processes and quality capabilities.

If we seek certification for marketing purposes only, there is no benefit in the business compared to all the efforts and money invested to become certified—*this is waste.* Some companies claim to have excellent quality processes just because they have certificates on the walls. Since quality is not built into their culture and practices, they would continue with the same traditions and fire-fighting styles they had before. During the certification process they spend a great amount of energy and money for initial certification, surveillance audit, and recertification, etc. However, this kind of certification actually does not pay off in improving the quality performance in process and product.

> *C-company touts many certificates and awards displayed in its lobby. Visitors can see the certificates and award plaques on the wall, and beautiful trophies inside glass display cases. Their quality manager makes an impressive presentation about their quality system and certificates that they received. However, the reality is quite different. The calibration of measurement equipment in their test lab has been outdated, an in-line body accuracy check machine by six vision cameras has been left broken, and employees do not practice their job to the defined quality procedures, and so on. How could such a company be certified and recertified for ISO 9001?*
>
> *The company's internal information revealed that the primary mission of quality department during the compliance or recertification audit was just to pass the audit without negative comments. Therefore, during the audit, the quality leader spends more time treating auditors: long hours of luxury lunch and dinners, etc.*

Juran adamantly stated that, "In my view, adherence or certification to ISO 9000 doesn't ensure that a company will become a quality leader.... From my viewpoint, if somebody adheres to ISO 9000 and doesn't go any further, it almost assures that they will not be quality leaders, ..."[51] For instance, Toyota Motors does not have ISO certification; however, Toyota is a quality leader.

If we stop our quality improvement after a certification, no added value is created—*it becomes waste.*

Notes

1. Ford Motor Company unveils the Model T, www.history.com/this-day-in-history/ford-motor-company-unveils-the-model-t (accessed March 15, 2019).
2. "JIT Just-in-Time manufacturing," www.ifm.eng.cam.ac.uk/research/dstools/jit-just-in-time-manufacturing/ (accessed February 3, 2019).
3. Ohno, Taiichi. *Taiichi Ohno's Workplace Management: Special 100th Birthday Edition,* McGraw-Hill Education, 2013, p. 43.
4. Bejan, Adrian. *The Physics of Life: The Evolution of Everything,* St. Martin's Press, 2016, p. 34.
5. Taylor, Frederick. *The Principles of Scientific Management,* Harper & Brothers Publishers, 1911.
6. Frederick W. Taylor: Master of Scientific Management, www.skymark.com/resources/leaders/taylor.asp (accessed February 3, 2019).
7. Ford, Henry. *My Life and Work,* Doubleday, 1923.

8. "A Bomber An Hour, Lean Manufacturing circa 1944," www.strategosinc.com/willow_run.htm (accessed February 3, 2019).
9. Reif, Rita. Oct. 9, 1994, www.nytimes.com/1994/10/09/arts/made-in-japan-without-the-inferiority-complex.html (accessed February 3, 2019).
10. Total Quality, http://asq.org/learn-about-quality/history-of-quality/overview/total-quality.html (accessed February 3, 2019).
11. Ohno, Taiichi. *Taiichi Ohno's Workplace Management: Special 100th Birthday Edition*, McGraw-Hill Education, 2013.
12. Ohno, Taiichi. *Toyota Production System: Beyond Large Scale Production*, Productivity Press, Portland, Oregon, 1988.
13. Ford, Henry. *My Life and Work*, Doubleday, 1923.
14. Kenton, Will. June 12, 2018, "Business Activities," www.investopedia.com/terms/b/business-activities.asp (accessed February 3, 2019).
15. Feigenbaum, Armand V., "Quality and Productivity," *Quality Progress* 10, no. 11: QICID: 6646 November 1977, pp. 18–21.
16. Ohno, Taiichi. *Toyota Production System: Beyond Large-Scale Production*, Original Japanese Edition, Toyota seisan hoshiki, published by Diamond Inc., Japan, 1978, English Translation, Productivity Inc., 1988.
17. Taiichi Ohno, *Toyota Seisan Hoshiki* (Toyota Production Method) Diamond Inc., Japan, 1978.
18. Ohno, Taiichi. *Toyota Production System: Beyond Large-Scale Production*, Productivity Inc., 1988.
19. Perrin, Xavier. January 2015. "Why is Overproduction the Worst Muda?" *Le BLOG*. XP Consulting. www.consulting-xp.com/blog/?p=546 (accessed February 4, 2019).
20. Deming, W. Edwards, *Out of Crisis*, The MIT Press, 1982.
21. Ohno, Taiichi. *Taiichi Ohno's Workplace Management: Special 100th Birthday Edition*, McGraw-Hill Education, 2013.
22. "Just-in-Time—Philosophy of complete elimination of waste," www.toyota-global.com/company/vision_philosophy/toyota_production_system/just-in-time.html (accessed February 5, 2019).
23. "Toyota speeds parts delivery with e-kanban," www.autonews.com/article/20020923/SUB/209230839/toyota-speeds-parts-delivery-with-e-kanban (accessed May 7, 2019).
24. Ohno, Taiichi. *Toyota Production System: Beyond Large-Scale Production. Cambridge,* MA: Productivity Press, 1988.
25. Ohno, Taiichi. *Toyota Production System: Beyond Large-Scale Production. Cambridge,* MA: Productivity Press, 1988.
26. Ohno, Taiichi. *Taiichi Ohno's Workplace Management: Special 100th Birthday Edition*, McGraw-Hill Education, 2013.
27. *Andon* fixed paper-enclosed lantern; paper glued over a wooden or metal frame housing an (oil) lamp. Originally portable, later mainly with a wooden stand. The term *Andon* comes from Japanese paper lanterns used for lighting since the Edo period. www.allaboutlean.com/andon/ (accessed February 7, 2019).
28. 75 Years of Toyota, www.toyota-global.com/company/history_of_toyota/75years/text/entering_the_automotive_business/chapter1/section4/item4.html (accessed February 7, 2019).
29. It's a four syllable word, there's no silent E. "Poka" is a sort of slangy word for "mistake" or "error," and "yokeru" means "to avoid." Basically it means "foolproofing," www.quora.com/What-is-the-etymology-of-the-word-poka-yoke (accessed February 7, 2019).

30. Toyota Andon concept MotoMachi factory, Japan, www.youtube.com/watch?v=r_-Pw49ecEU.
31. Shmula. June 14, 2011, "Squeeze Water From a Dry Towel: Eiji Toyota Lean Manufacturing," www.shmula.com/squeeze-water-from-a-dry-towel/8768/ (accessed February 7, 2019).
32. Harry, Mikel. *Six Sigma: A breakthrough strategy for profitability.* Quality Progress, 1988
33. Alan J. Stolzer, Six Sigma Applied to Flight Operations Quality Assurance: An Exemplar Case Study, Embry-Riddle Aeronautical University, Spring, 2006.
34. Folaron, J., Morgan, J.P. "The evolution of Six Sigma," *ASQ Six Sigma Forum Magazine,* 2(4), pp. 38–45, 2003.
35. Welch, Jack, *Straight from the Gut,* Warner Books, 2003.
36. Stolzer, Alan J. "Six Sigma Applied to Flight Operations Quality Assurance: An Exemplar Case Study," Embry-Riddle Aeronautical University, Spring 2006.
37. ICAO (international civil aviation organization), www.icao.int/annual-report-2016/Pages/the-world-of-air-transport-in-2016.aspx (accessed February 7, 2019).
38. De Mast, Jeroen, and Joran Lokkerbol, "Analysis of Six Sigma DMAIC method from the perspective of problem solving," *International Journal of Production Economics* 139, 2012, pp. 604–614.
39. DMAIC, https://en.wikipedia.org/wiki/DMAIC (last modified May 4, 2019).
40. Cyger, Michael. "Six Sigma Costs and Savings," www.isixsigma.com/implementation/financial-analysis/six-sigma-costs-and-savings/ (accessed February 8, 2019).
41. Ohno, Taiichi. *Toyota Production System: Beyond Large-Scale Production.* Cambridge, MA: Productivity Press, 1988.
42. Murman, Earll. [et al.]. *Lean Enterprise Value: Insights from MIT's Lean Aerospace Initiative,* New York: Palgrave, 2002.
43. *Image credit: By Daniel Penfield, CC BY-SA 3.0, https://commons.wikimedia.org/w/index.php?curid=2855399.*
44. Shvetank, Shah, Andrew Horne, Jaime Capellá, "Good Data Won't Guarantee Good Decisions," *Harvard Business Review,* April 2012, https://hbr.org/2012/04/good-data-wont-guarantee-good-decisions (accessed March 16, 2019).
45. Lee, Se Young. March 24, 2016, "World's biggest startup? Samsung Electronics to reform corporate culture," www.reuters.com/article/us-samsung-elec-culture-idUSKCN0WQ0CP (accessed February 7, 2019).
46. O'Rourke, Peter M., "A Multiple-Case Analysis of Lean Six Sigma Deployment and Implementation Strategies," *Air Force Institute of Technology,* 2005.
47. Lean Six Sigma Institute, www.sixsigmainstitute.com/leansigma/index_leansigma.shtml (accessed February 8, 2019).
48. Original Data: Motorola Six Sigma Research Institute, Motorola University, Motorola Inc.
49. Harry, Mikel J. and Schroeder Richard R., *Six Sigma: The Breakthrough Management Strategy Revolutionizing the World's Top Corporations,* 2000.
50. Harry, Mikel J. and Schroeder Richard R., *Six Sigma: The Breakthrough Management Strategy Revolutionizing the World's Top Corporations,* 2000.
51. "A Century of Quality, An Interview with Quality Legend, Joseph M. Juran by Scott M. Paton," www.qualitydigest.com/feb99/html/body_juran.html (accessed February 9, 2019).

10

Software Quality Management

Our industrialized society has become much more productive through the utilization of machines and equipment in manufacturing plants. Machines and equipment took tedious, dangerous, and repetitive jobs from the human workers in the plants. Increased efficiency of production and improved uniformity of products were made possible by this type of automation.

However, real productivity and quality increases did not occur until those machines became connected with computers. The computer initiated the third industrial revolution. At the beginning of the computer era, hardware performance was the primary focus, and companies worked to increase processing speeds, build more versatile computer chips, and create larger memory drives to store more data. As the hardware of the computer and communication technologies became more powerful, adopters increasingly desired smarter uses for the new technology. *The advanced development of software brought about the real information age.*

A new IT age began with the creation of versatile computer chips together with advanced development of software. Information technologies like internet communication, big data processing, intelligent machine controls, artificial intelligence, etc., all became possible with the power of software. Today, software is ubiquitous, from the fields of engineering and manufacturing to smart consumer devices.

Software quality is critical to provide benefits from future-oriented products such as autonomous vehicles or smart robots directed by artificial intelligence (AI). Software glitches in those products can lead to life-threatening dangers. The scope of software quality is not only about error-free coding but also about robust design and algorithms.

Software quality management (SQM) is a management process that aims to develop and manage the quality of software in such a way as to ensure the product meets the quality standards expected by the customer while also meeting any necessary regulatory and developer requirements.[1]

As we discussed in Chapter 3, the overall concept of quality management is defined as: QM = QP + QC + QA + QI, which means quality management is the combined activity of quality planning (QP), quality control (QC), quality assurance (QA), and quality improvement (QI). A software product is not manufactured, but designed. Therefore, QM activity should focus on the development process and related activities, which must be integrated in QP, QC, QA, and QI.

Our focus on software quality management in this chapter is about how we can manage software quality from *the viewpoint of an embedded system manufacturer*, whose main business is to design and manufacture industrial or consumer products, i.e., smart hardware products that contain an embedded software system. Thus, if the embedded software does not work correctly, the hardware product will not function as designed. Therefore, manufacturers of smart products should be concerned about and involved in total software quality management, whether the software is insourced or outsourced.

1. Embedded System and Software
2. Supplier Selection for Software Development
3. Software Quality Assurance
4. Software Quality Confirmation
5. Software Quality Engineering

10.1 Embedded System and Software

Today's smart products have many complex features and sophisticated functions that would be too complicated or inefficient to control manually by using old methods such as mechanical tools or timer relays. Old control methods using conventional control mechanisms are not productive in modern, advanced-technology products. For example, in contrast to the early refrigerators of the mid-1900s, today's refrigerator has smart functions and operation logic to keep the food fresh while managing energy efficiency. For example, it optimizes the running time of the compressor, periodically defrosts ice build-up on the evaporator coil, and varies the fan speed depending on the needs of the users and the working environment, taking into account such factors as humidity and temperature.

In such a system, we incorporate a computer within a hardware product to perform the control functions. This then becomes an *embedded system*, which is defined as "a system whose principal function is not computational, but which is controlled by a computer embedded within it."[2]

The embedded system is also known as an integrated system because it combines hardware and software. It is built-in in most modern smart products or automatically controlled devices such as car engine control systems, anti-lock braking systems, navigation systems, machines, smartphones, home appliances, toys, LED lamps, etc. Table 10.1 lists various examples of smart products that are likely to have embedded systems, categorized by their applications.[3] Since many of these products have severe limitations in respect to space, weight, cost, and power consumption, they need to integrate both digital and analog systems, special purpose sensors and actuators, software, and mechanical items altogether in one system, which becomes an embedded system.

TABLE 10.1

Examples of Embedded Systems

Home / Office Products	Consumer / Industrial Products	Vehicle (Driver / Driverless)
• TV • Refrigerator • Laundry Machine • Microwave • Air Conditioner • Burglar Alarm System • Automatic Cooker • LED Lamp • Robot Vacuum Cleaner • Smart Home Control	• Smartphone • Drone • Hobby Remote Control Car • Game Machine • Smart Toys • GPS Watch • Smart Watch	• Engine Control • Transmission Control • Navigation System • Infotainment System • ABS Brakes • Air Bag • Climate Control • Immobilization System • Door Lock Control • Vehicle Stability Control • BUS Communication • Lighting Control System • Tire Pressure Monitoring System • Seat Belt Control Module
• Photocopier • Printer • Scanner • Cash Register • Self-checkout Machine	• Robots (Painting, Welding, Assembly) • Automatic Torque Control Machine • AGV (Autonomous Guided Vehicle) • Autonomous Logistic Control Robot	• Proximity Warning System • Adaptive Cruise Control • Crash Avoidance System • Lane Keeping System • Vision Control System • RADAR / LIDAR System • High Accuracy Mapping

10.1.1 Firmware (Embedded Software)

The definition of firmware by ISO 24765 [ISO 17a] is "combination of a hardware and computer instructions or computer data that reside as read-only software on the hardware device." In other words, *firmware is a computer program, i.e., embedded software in an embedded system* that provides smart functions for hardware products. The firmware does not need an operating system like a computer. However, it is able to tell a microcontroller how to work. For example, firmware in the microcontroller of a refrigerator can control complex functions without a computer operating system.

Firmware is held in non-volatile memory devices such as ROM, EPROM, EEPROM, or flash memory and integrated into an embedded system. Changing the firmware of a device may rarely or never be done during its lifetime depending on the types of applications and ROM types used. Common reasons for updating firmware are to fix bugs or add features to the device. This may require ROM-integrated circuits to be physically replaced, or flash memory to be reprogrammed through a particular procedure.[4]

There are thousands of different types of microcontrollers used in industry today, made by many different manufacturers.[5] However, in advanced modern products, the true smartness comes from the power of software.

10.1.2 Special Characteristics of Software Products

The common feature of the smart products mentioned in Table 10.1 is software. There exist unique quality challenges for software products that are different from those of physical hardware products. Since software is not visible or touchable, it is impossible for the software developer to declare that the software is free of bugs or defects.

Typically, hardware products without software have limited possibilities of operations due to the limitation of their physical design, while meeting the requirements of clients or users. However, a software product, i.e., a sellable software package, allows millions of possible operations depending on the complexity of the input data types and contents.[6] Table 10.2 summarizes the major differences between hardware and software products.

The most significant quality challenges of a software product are due to the invisibility, complexity, and numerous possibilities of the operations.

TABLE 10.2

Major Differences between Hardware Products and Software Products

	Hardware Product (w/o Software)	Software Product (Software)
Characteristics	• The product is visible and touchable. • Most of its defects or the absence of a part can be detected during the manufacturing process.	• The program codes are neither visible nor touchable. • *Difficult to detect problems during the manufacturing stage.*
Functions	The product allows only less than a few thousand modes of operation with different combinations of its machine settings or usage conditions.	Software packages allow millions of operational possibilities with different input data, producing different output commands.
Development	• Product Development: Customer's requirements and usage conditions are generally well defined. Physical prototypes are made for design verification. • Product Production Planning: Production equipment, tools, process, and training need to be prepared. There are many opportunities to check for defects. • Manufacturing: Materials and parts supply, quality maturity levels, and productivity are important management parameters.	• Product Development: *Customer's requirements and usage conditions are difficult to define succinctly.* • Product Production Planning: Production preparation tools or capacity need to be prepared to embed programs into different types of ROM. *Potential defects of programs are difficult to confirm.* • Manufacturing: Embedding compiled machine code into different types of ROM needs to be done—*chances for detecting defects is very limited.*
Maintenance	Typical repair or maintenance service requires a partial or full disassembly of the product.	*Wire, wireless, or remote update or upgrade* of the software is possible.
Continual Improvement	Upgrading of a product requires new tools or new parts.	Updating or upgrading can be done *flexibly* without new tool development.

Software is not manufactured in a production line—it is designed. Thus, software developers should be concerned about a broader scope of quality management from *scientific quality planning, an appropriate software development process, competent product and process standards, structured deployment of planned activities, and quality assurance process.* There is no second chance to catch problems after the completion of the development.

10.1.3 The Power of Software

The program memory in a computer or an embedded system contains specific encoded instructions that are pre-programmed by the suppliers or users of a product. *Such a collection of intangible special instructions or programs that tell the hardware what to do is called software.*

The software tells a computer or a microcontroller how to work. Even with the same type of microcontroller or computer, the variability of a hardware product's operations, functions, and performance can be millions of different combinations depending on the software's instructions. The primary force propelling today's IoT (internet of things) tech and smart products is software. When devices can communicate and interact with others over the Internet, or controllable and adaptable to human's demanding needs or environment, they become IoT and smart products. The power of the software increases the versatility and value of a product, while not necessarily increasing its cost.

The power of software lies in its flexibility and versatility. Programs can be updated flexibly, adapting to the working conditions or the needs of a user, and updates can be done remotely without even the knowledge of its users. In the case of AI software, programs can learn by themselves to improve the control performance of products, machines, and equipment through machine learning. In 2017, AlphaGo beat the best human professionals in the game of Go. *This power did not come from hardware but software.*

Today's modern vehicle has more than 40 electronic control units (ECU) embedded in the various car parts, all of which have embedded software (see Figure 10.1). The role of software is becoming ubiquitous in the car industry (see Figure 10.2).[7] The autonomous vehicle is revolutionizing the car industry, shifting cars from a conventional definition to a digital definition. The scope of automotive engineering is shifting from "upper body and lower body" concept to "hardware and software concept"—software creates future mobility value.

It is evident that the power of the future comes from the power of software.

10.1.4 Hardware and Software in Embedded Systems

Previously we discussed the general characteristic differences between hardware products and software products. When hardware products have a brain, a control system, they become smart products, where sensors and actuators are interacting with controllers.

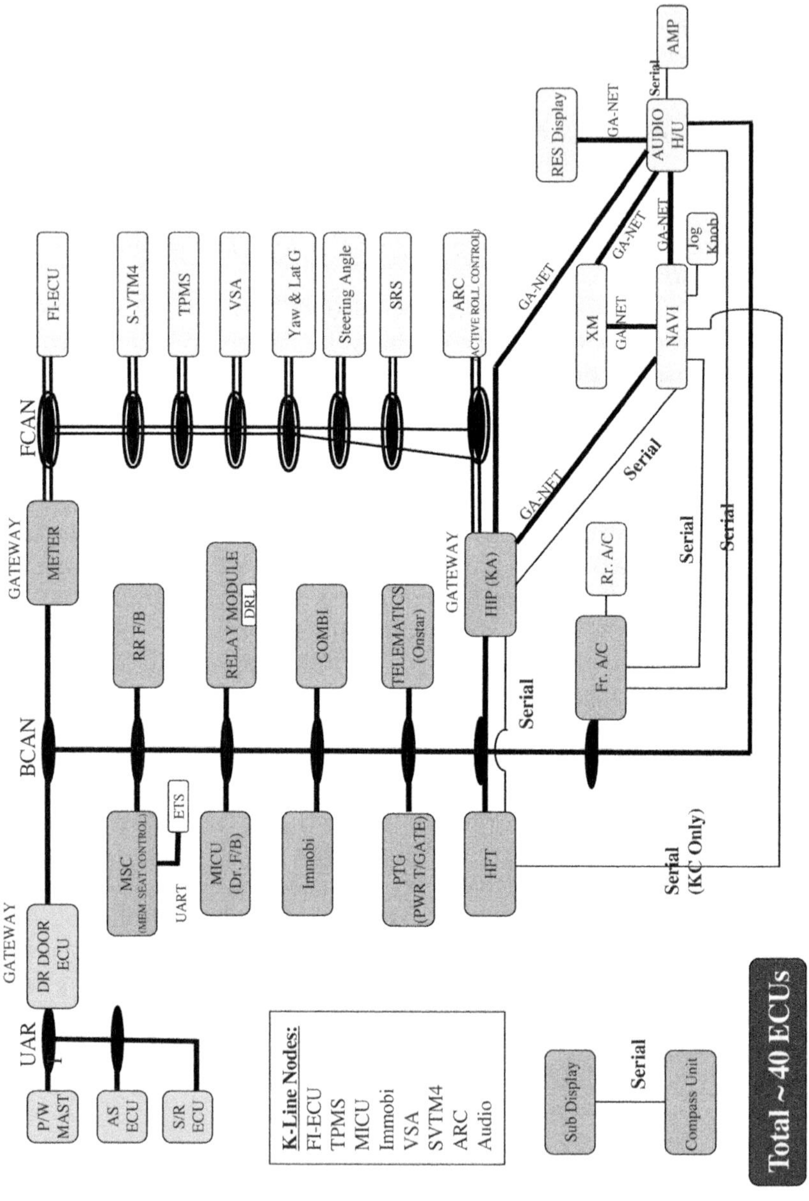

FIGURE 10.1
ECU (electronic control units) in a modern automobile.

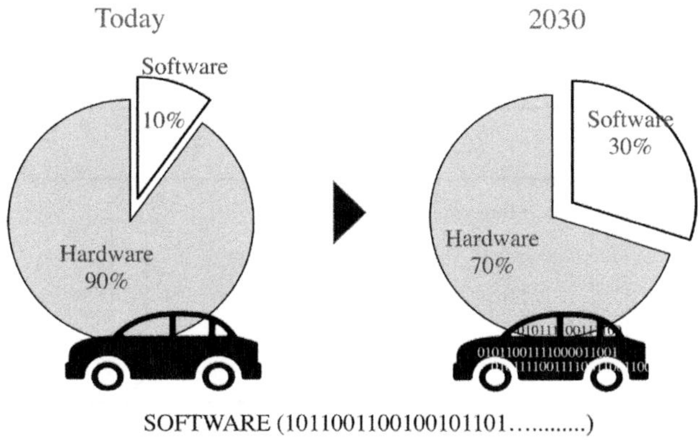

FIGURE 10.2
The value of hardware vs. software in future mobility.

As shown in Figure 10.3, an end-user system is composed of sensors, actuators, and controllers. The embedded control system processes input signals and sends output signals to actuators or other control systems. Typically, *such an embedded system requires real-time control*; therefore, specific hardware and software designs and speed performance are critical factors to warrant total system quality. For example, smart consumer products, such as laundry machines, microwaves, and digital cameras, perform real-time functions supporting users' expectations.

Figure 10.4 illustrates the overall architecture of an embedded system showing *how the end-user requirements are related to the hardware and software.* The selection of an execution platform (electronic hardware) is the first step to make sure there is sufficient capability to support customer requirements, maintaining economic balance at the same time. In addition, the embedded software under the environment of the execution platform needs to provide

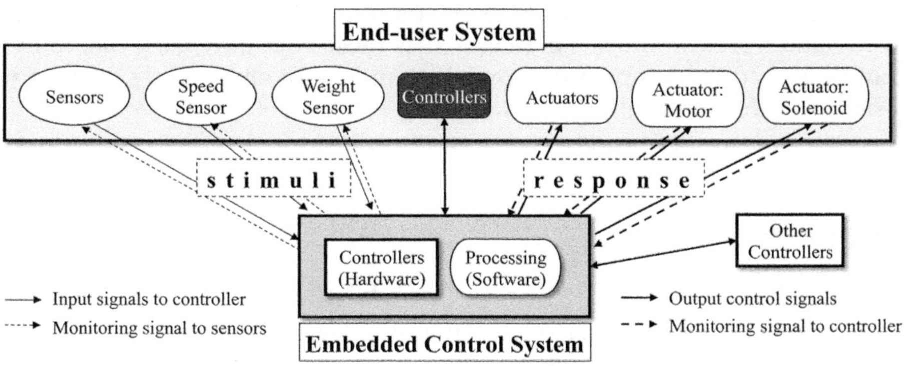

FIGURE 10.3
End-user product and embedded system.

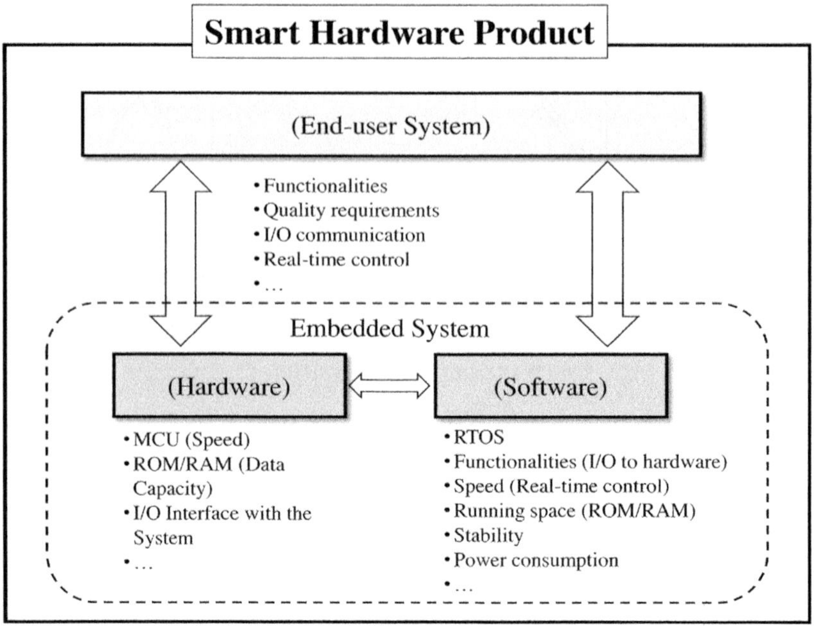

FIGURE 10.4
Architecture of an embedded system.

critical attributes to customers, such as real-time control speed, minimal space, stability, minimal power consumption, and so on.

A computer requires an operating system, which interfaces between hardware and software. However, a regular computer operating system, such as Linux, Windows, or Mac OS, takes up a significant amount of memory space and slows down the operation of programs. Therefore, in an embedded system, the industry uses a very compact real-time operating system (RTOS) such as VxWorks, QNX, eCos, RTLinux, or Window CE. With respect to computer programming languages, C is most popularly used due to the compactness of its compiler design, which makes an embedded system run faster and more efficiently.[8]

10.2 Supplier Selection for Software Development

The primary focus of this chapter is not on general computer engineering, but on advanced industrial and consumer products that need intelligent control via software. Thus, our focus in this section is on embedded systems and the integrated firmware to support those advanced-technology products.

As discussed in the previous sections, firmware is defined as software embedded in an embedded system, and the purpose of firmware is to instruct an embedded system to run predefined special instructions to control an end-user hardware product. In many cases in industry, *firmware is outsourced to software developers and embedded into hardware products by hardware manufacturers.*

10.2.1 Sourcing Model of Software Development

Many advanced smart products need intelligent control. Manufacturers of such products recognize that installing an embedded system in their hardware products is an economical and practical option. When hardware product manufacturers develop their products that need computer control, they have several options, as illustrated in Figure 10.5.

The first option, Model-A, is to develop everything in-house including electronic hardware and software for an embedded control system. This model requires the manufacturer to have significant knowledge of computer hardware and software engineering as well as highly skilled engineers in both areas of hardware and software.

In the case of Model-B and Model-C, the manufacturer can focus on the end-products and outsource the software (Model-B) or outsource both electronic hardware and software (Model-C). In this case, they rely on the expertise of suppliers either for embedded system development or only firmware development. In such a purchasing model, *they do not have to maintain many skilled engineers for both the hardware and software development of an electronic system.*

Many manufacturers are continually looking for outsourcing partners that can meet their system's needs with embedded programming solutions in

	Sourcing Model of Embedded System and Firmware		
Domain	All Insourcing (Model-A)	Insourcing + outsourcing (Model-B)	Complete outsourcing (Model-C)
Hardware Product (examples) - Automobiles - Robots - Drones - or Subsystems	Hardware Product + Electronic Hardware + Software	Hardware Product + Electronic Hardware	Hardware Product
Tier-1 Supplier		Software	Electronic Hardware (and Software)
Tier-2 Supplier			Software

FIGURE 10.5
Sourcing model of embedded system and firmware.

the best way possible. Worldwide, many renowned embedded software outsourcing companies are available depending on the application. The internet is one source of leads, for example for outsourcing companies for automotive embedded software development.[9] The scope of their services is broad from PCB (Printed Circuit Board) design to total embedded system and software development.

The more complex the product is, the greater the need to outsource hardware and software using Model-B or Model-C. A typical product of such a sourcing model (Model-C) is an automobile. Since there are so many electronic parts to control, the automobile manufacturer or the ECU suppliers rely on lower-tier suppliers for the development of electronic hardware and software. In this model, typically both hardware and software are developed by Tier-1, but either can be outsourced to Tier-2 by a Tier-1 supplier.

Such outsourcing strategies provide many benefits to hardware product manufacturers, such as shortening a product's development time, reaping the benefits of the supplier's rich expertise, and cost savings in respect of workforce and engineering expenses. The most significant advantage is that the manufacturer can focus on its own core business—the development and production of hardware products. However, a critical concern still remains to make sure the outsourced software meets the product's requirements. A key starting point is the right selection of capable suppliers for both electronic hardware and software development.

10.2.2 Selection of Software Developer

The success of the outsourcing strategy for software development depends on a supplier's software engineering and quality management capabilities. The selection process for a supplier for software development is different from the selection process for a supplier for hardware parts, as software is not visible or touchable. Figure 10.6 illustrates a selection process for a software developer by a hardware product manufacturer.

1. *Pre-Selection Process:* First, the hardware product company should clarify its requirements, such as the inputs and outputs of its product and the detailed specifications of its hardware product design. Then it can ask candidate suppliers to send them a project and technical proposal with a quote for the expected development cost.

   During the pre-selection process, the project manager needs to evaluate each software developer's project proposals concerning cost and development plan. The R&D manager's function is to judge the suppliers' technology compatibility with the required specifications, and the quality manager's function is to check the supplier's quality history.

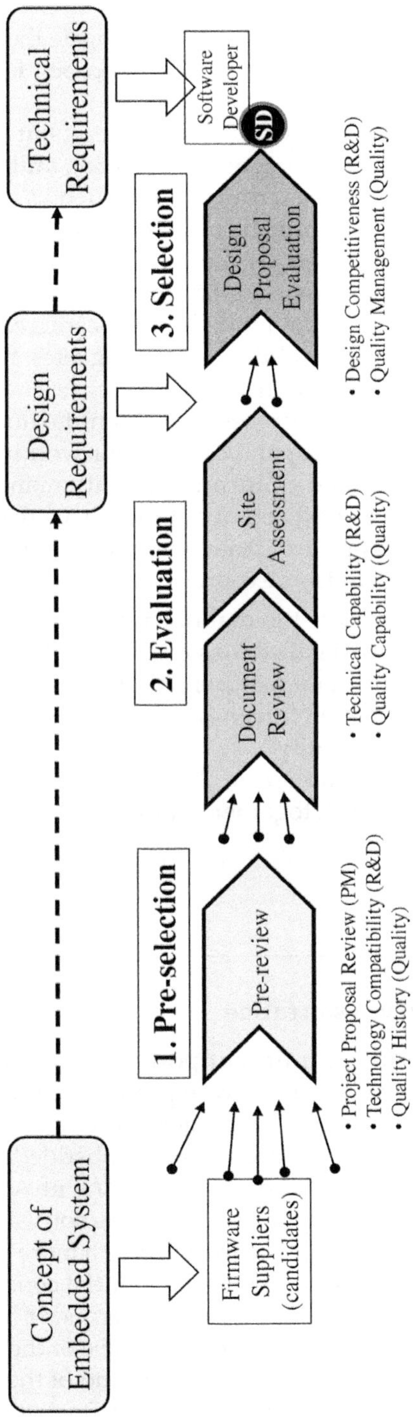

FIGURE 10.6
Selection process for a software developer.

2. *Evaluation Process:* Once the manufacturer has selected two or three competitive candidate suppliers, it must investigate the suppliers' capabilities through *document review and audit processes,* focusing on their R&D and quality capabilities.

   For document review, they first need to send out a questionnaire with questions in different categories, including multiple-choice and open response questions, regarding topics such as software project experience, quality management systems (ISO, CMMI, or others), quality assurance processes, workforce of software engineers and their skill levels, and software quality maintenance methods. While reviewing the document, the manufacturer can create a *checklist* to use during a visit to the candidates' working sites, mainly to check their R&D and quality functions.

   Next, through the audit process, the manufacturer can check a software developer's true capabilities at their working site by interviewing the workforce and confirming quality management activities. The output of this audit activity at the software developer's site is summarized in a scorecard, from which they can select two final candidates to proceed to the next process.

3. *Selection:* At the final stage, the manufacturer can ask the candidate developers to create a detailed design proposal. By checking the detailed proposals, the manufacturer should make a final judgment on each developer's competitiveness on technology, performance, delivery schedule, and quality.

   Finally, the manufacturer can choose one software developer, with whom they will work together to develop the software for their end-product.

10.3 Software Quality Assurance

Software is becoming a more significant portion of the sophisticated functions and performance of today's hardware products. Software makes hardware products and machines more intelligent and versatile for consumers. Most of today's intelligent devices are already embedded with special logic programs like *fuzzy* or *chaotic* logic, or *AI integrated* with smart sensors.

Hence, there are also growing concerns regarding software quality. As software programs become more complex and become a more significant portion of today's products, there are corresponding potential concerns and risks.

Software quality is not only about coding errors or mistakes but also *design,* including algorithm mistakes like in the case of the Boeing 737 MAX. The plane model was ground globally after a second of the aircraft, operated by Ethiopian Airlines, crashed on March 12, 2019. Boeing's cockpit software algorithm relied on one single sensor input to prevent a possible engine stall when the sensor determined that the upward pitch of the airplane was too steep. However, erroneous data from the single sensor, which measures the angle of the plane's nose, caused the system to misfire. Then, a series of events put the aircraft into a dangerous dive.[10]

The fundamentals of quality management are similar to both hardware and software products. However, there are some unique approaches required for software quality management due to the distinctive characteristics of a software product as explained in Table 10.2.

Since a software product is not manufactured but developed, *software quality management should focus on the software development process and quality maturation activities.*

10.3.1 Software Development Process

The quality of a hardware product starts from its product planning and development to its manufacturing. In the case of a software product, the completion of the development means the finishing of major quality activities as discussed in Table 10.2.

Software quality is determined through the development process, and proper processes are more likely to lead to a good-quality software product. Primary common steps of software development are the analysis of the client's requirements, high-level and low-level design, implementation by coding, verification and validation, and maintenance as shown in Figure 10.7.

- *Software specification:* Customer requirements are analyzed and transformed into specifications that are communicated to the system developers.
- *Software design and implementation:* Software architecture is decided and followed by programming to meet the specifications.
- *Software validation:* The software must be validated to ensure it meets the customers' and end-users' requirements in the market.
- *Software evolution:* The software is modified or updated to reflect changing customers' and end-users' requirements in the market.

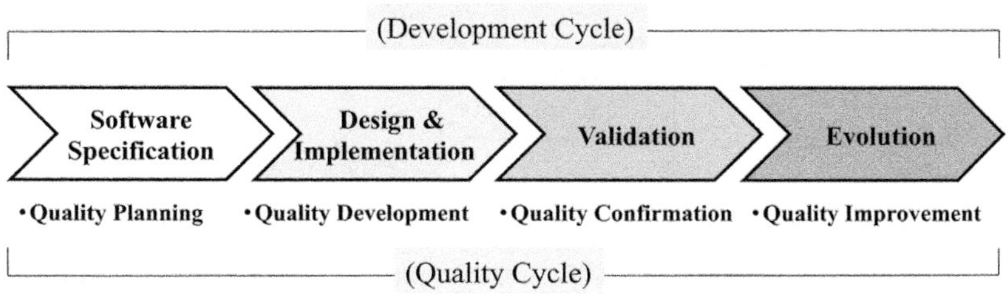

FIGURE 10.7
Software development process.

A software development process is a set of related activities that leads to specific engineering and quality activities in each development stage. Software development processes can be categorized largely into two types, depending on the development methods: *plan-driven* or *agile process.*

In the *plan-driven development process, all development activities are planned in advance, and the progress is measured against the plan. In the agile development process, planning is incremental,* and it is easier to change the process to reflect changing requirements from the client. There are pros and cons for each process. One or a combination of these processes can be used, depending on the purpose and scale of a project.

1. *Waterfall Model:* The waterfall model is a *plan-driven sequential development model,* which is appropriate for a project when requirements are clearly defined and not likely to change during development.

   The advantage is that this process is simple and clear, and every action plan and schedule can be arranged in a linear order before the start of a project (see Figure 10.8). However, the weak point of this model is the lack of flexibility to accommodate modification or changes at a later stage.
2. *V-Model:* V-Model means a Verification and Validation model, which is an extension of a waterfall model, emphasizing the confirmation process of software quality. *Software testing is planned in parallel with a corresponding phase of the development* as shown in Figure 10.9. Just like the waterfall model, the V-Shaped life cycle is a *sequential path of execution of processes.* Each phase must be completed before the next phase begins.

   Instead of moving down in a linear way, the process steps are bent upwards after the coding phase, to form a V shape. *The V-Model demonstrates relationships between each phase of the development life cycle*

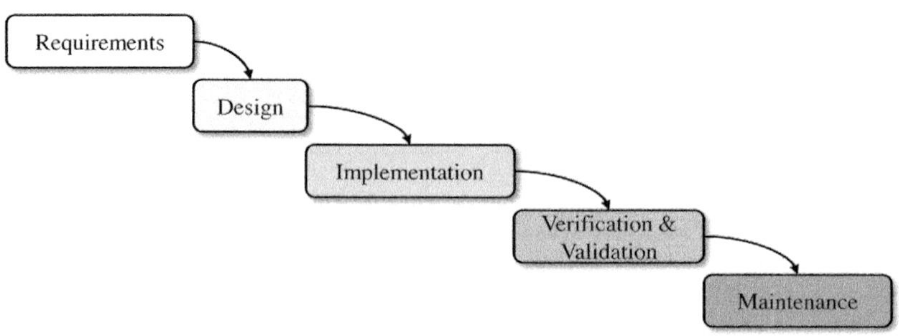

FIGURE 10.8
Software development process (waterfall model).

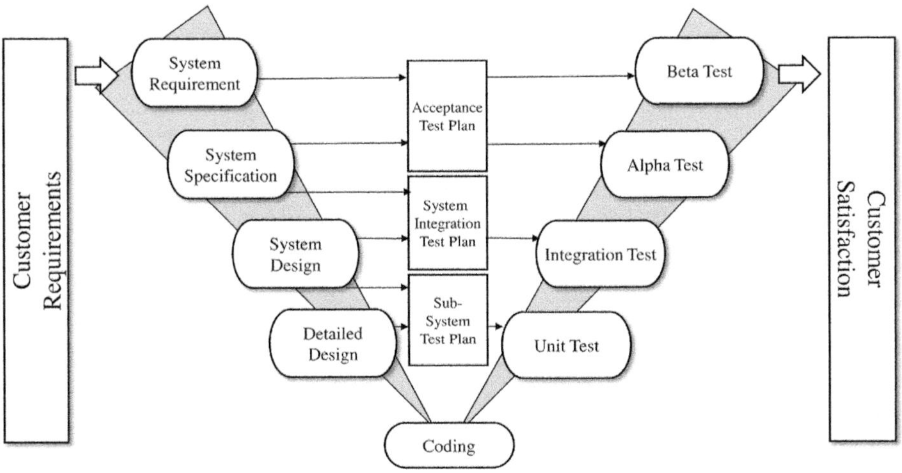

FIGURE 10.9
Software development process (V-model).

and its associated phase of testing. The horizontal and vertical axes represent time or project completeness (left-to-right).[11]

For instance, the system design phase focuses on system architecture and design. A system *integration test plan* is created in this phase as well as a sub-system test plan in order to test the software's ability to work together.

The detailed design phase is where the actual software components are designed. It defines the logic and algorithm for each component of the system. In this stage, the *unit test plan* is created.

Coding process is the implementation phase. Once the coding is complete, the path of execution continues up the right side of the V where the test plans developed earlier are now put to use. Normally, unit testing is performed by the developers on the code written by them. Other remaining validation testing, such as integration testing, alpha, and beta testing continue until the final delivery of the software product to the client.

3. *Incremental Model:* A plan-driven approach is effective when client requirements are clear at the beginning of a project. However, the requirements can change depending on obstacles and priorities. For example, a new technology product confronts many obstacles and unknown factors during development. The incremental model is an *agile process* to accommodate such a dynamically changing environment.

   The incremental model is "multiple development waterfall" cycles—the cycles are divided up into smaller, more manageable modules. In this model, each module passes through requirements,

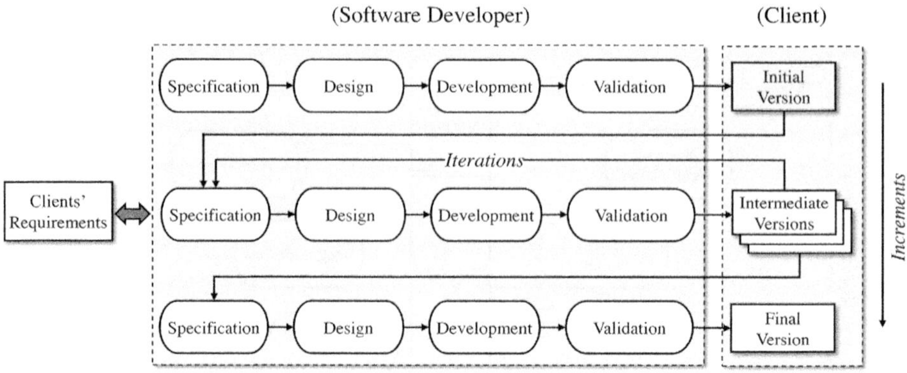

FIGURE 10.10
Software development process (incremental model).[13]

design, implementation, and testing phases. A working version of a software is produced at each stage, which helps the client to see the feasibility and intermediate progress of the software in the embedded system environment. Each subsequent release of the module adds more functions to the previous release. The process continues until the complete system is achieved (see Figure 10.10).[12]

The most significant advantage of this model is early risk detection and management by working together with the client during each iteration process.

4. *Prototyping Model:* The prototyping model is a fast development model, *a blend of plan-driven and agile processes.* It requires a simpler architecture and process, and programs can be developed quickly to check the customer's requirements and feasibility of a proposed design structure, algorithms, and response time.

   During the development process, the client can be involved to understand the concerns or limitations of the software design and dynamically update the requirements. Once the concept has been tested to work to meet the client's requirements, the same concept and structure are implemented to develop full operational programs (see Figure 10.11). While some prototypes are discarded after a feasibility study, it can evolve from prototype to working systems.

10.3.2 Software Quality Criteria

Quality software refers to a software product that is reasonably *fault (defect) and failure-free, meeting the customer requirements.* In other words, they are represented by two aspects of quality similar to hardware product quality: Little q and Big Q. *Little q represents defect and failure-free, and Big Q represents better features and performing quality* (*refer to* Table 5.1).

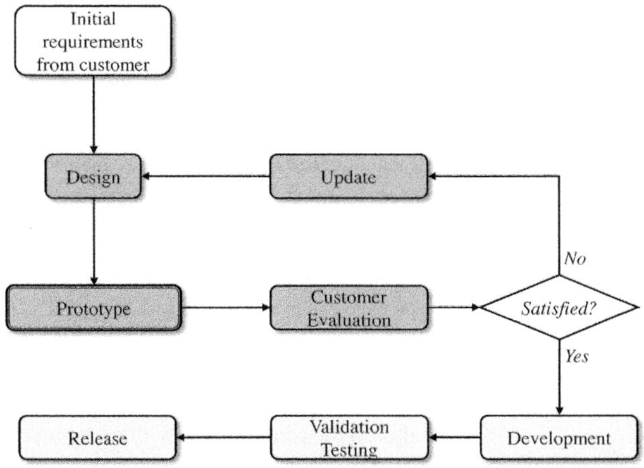

FIGURE 10.11
Software development process (prototyping model).

The "defects" and "failures" can be more specifically explained in the following ways:[14]

- Internal Quality: refers to aspects that are code-dependent, and that are not visible to the end user—internal quality problems are *errors or faults (defects)* which can lead to external quality problems.
- External Quality: is what a user experiences when running the software in its operation mode—problems experienced by the end user are *failures*.

From the perspective of Big Q, software quality is determined by the end-user experience—the major attributes of this experience are summarized in Table 10.3. Like the perceived quality of hardware products, the judgment of software attribute quality is subjective. For example, some attributes are more significant in a particular embedded system than in other applications.

> *In an automobile electronic control system, ECU development is the biggest challenge both for the suppliers and the automobile manufacturers. As Figure 10.1 shows, many ECUs are interconnected over serial or CAN (controller area network) bus. They expect other inputs from other ECUs and send their own signals to the relevant ECUs and actuators flawlessly. The software that is working in a lab cannot guarantee the same quality functions in a field, not to mention to the right performance. The system response can be slow, resulting in the delay of critical control functions, errors, or unstable conditions in the vehicle. Hence, real-time communication and control speed of the system, minimum usage of memory space and power, and system stability of running the instructions are the critical quality standards for both ECU suppliers and the automobile manufacturers.*

TABLE 10.3

Software Quality Criteria by Major Attributes (Big Q)[15]

Attribute	Quality Criteria
Features	Functional outputs by the software to control an embedded system—e.g., output signals to other sensors, actuators, speakers, or displays.
Speed	How quickly does the program process the data or information to make output decisions—e.g., efficiency of the algorithms, logic, and conciseness of the program codes?
Space	How much ROM or RAM does the software require to run the instructions of the program?
Stability	How stably and robustly can the program run all the features under maximum workload conditions?
Security	How safely can the program run without being compromised by intruders—e.g., EMI or hackers?
Scalability	How easily and flexibly can the software be upgraded, or expanded in the quality attributes and features?
Power Consumption	How much power does the program require to run instructions?
Determinism	Does the program produce the same result given the same input?
Testability	The ability to be able to track the program running conditions line by line during and after the development of the program.

Therefore, during quality planning, clear quality standards, targets, and specifications of ECUs in both hardware and software should be set up and communicated with ECU developers by the automobile manufacturer. Vague requirements lead to the delay of ECU development, resulting in vehicle testing delay followed by the delay of a new product launch as a chain reaction.

10.3.3 Software Problems and Cost Impact

Usually a hardware product allows less than a few thousand modes of operation using the limited combinations of its machine settings. However, a software package (product) allows millions of operational possibilities. Since a hardware product is tangible, most of its defects can be detected during the manufacturing process. Also, the absence of a part in a hardware product can be easily detected. However, defects in a software product which is stored on ROM are invisible.

Such unique characteristics of a software product produce unique challenges and problems during development and after its launch into the market. Software problems can come from any of the instructions, program components, or system, with numerous possible failure causes and modes.

Software problems can stem from many different stages of project development activities: requirement analysis, design, coding, integration, testing, installation, and operation in the market. Such software problems can occur in a specific flow as in Figure 10.12; problems occur through three stages—errors, faults, and failures.[16]

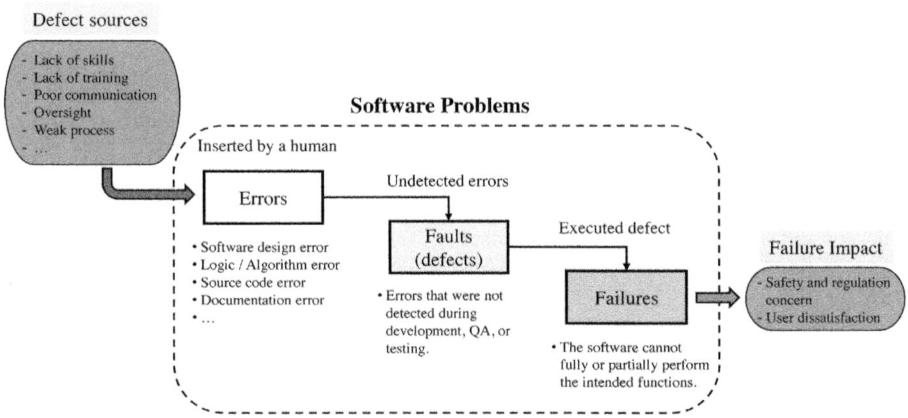

FIGURE 10.12
Software problems in development flow.

An **error** is a software developer's mistake, misconception, or misunderstanding of a project or a program—software project managers, engineers, programmers, and testers are in the category of the developer. For example, a developer may misunderstand a design notation, start with a wrong algorithm and/or mistype a variable name, which can all lead to errors in the software.

A *fault* (defect) is introduced into software as the result of an error that was not detected during software development, quality assurance, or testing. A fault, if not corrected, causes an application to either fail or to produce incorrect results (ISO 24765). It is an anomaly in the software that may cause it to behave incorrectly, not according to its specifications. The commonly used term *"bugs" also belongs to the category of faults or defects.*

Failure is the inability of a software system or component to perform its required functions within specified performance requirements. Failure happens when defects that lurk in software are executed. Some defects or failures are not detected during a normal development process *mainly due to the limitation of a user simulation.*

> *S-Company proudly launched a new infotainment system, which can link a phone to the car navigation and audio system. Dynamic road conditions and the stream of music service can be received via smartphone into the car. Unfortunately, the phone link function to the car infotainment system failed only in particular cars. After a lengthy investigation in the market, they found that the link fails only in the case of particular users' phones—those which contain special characters in the phone's name other than English. The cause of this problem was in the car infotainment system software because the system expected to have an English name as an input function variable for the phone link function.*

TABLE 10.4

Cost of Software Quality

1. Prevention (Internal Cost)	2. Error and Defects (Internal Cost)	3. Failures (External Cost)	4. Crisis Cost (External Cost)
• Design • Training • Review of algorithm, design, and coding • Process improvement • Audit	• Debugging and recoding • Repeated reviews • Extra management • Reconfirmation tests • Penalty for late delivery to the client	• New versions, patches and their management • Penalties (Warranty, Goodwill) • Loss of sales opportunity • Loss of reputation	• Damage compensation to the clients • Recall • Penalties (Legal, PL, Lawsuit)

10.3.3.1 Cost of Poor Quality

The basic mechanism of software quality cost is similar to the cost of hardware products as discussed in Chapter 2. Most of software quality cost comes from man-hour costs for recovery activities—detailed contents are summarized in Table 10.4. Outflowed software problems can lead to potential safety or crisis issues, and its economic scale can be astronomical.

Numerous observations and historical data demonstrates that the internal cost is much cheaper than external recovery cost, at scales similar to that of hardware product as we explained in Figure 2.12. The *"10-Times Scale Rule" also applies to* the downstream scale of software quality cost: *the downstream recovery cost is 10 times higher than the cost of an earlier stream activity* (*see* Figure 10.13).[17] *In other words, the equation of (2.8) is still valid concerning the impact of software quality: downstream recovery cost is 10 × (Upstream countermeasure cost).*

Preventive activities should be the core of software quality management. For instance, reviews of algorithm, design, and coding can catch more than half of a product's defects regardless of the domain or the maturity level of an organization.[18] To minimize the errors, defects, and failures, it is also critical for us to understand the sources of the problems and reflect the concerns into software quality planning as the core activity of software development.

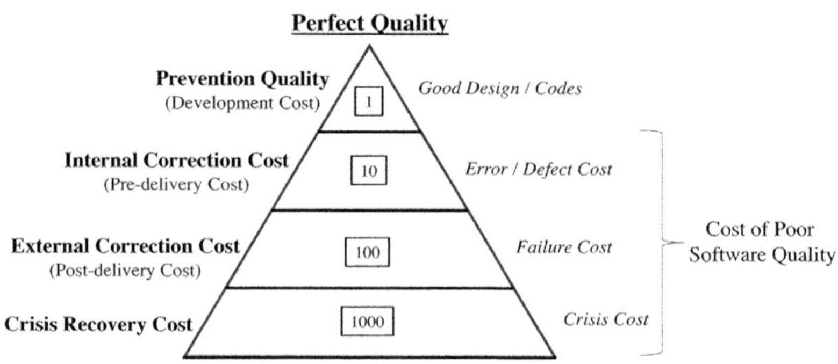

FIGURE 10.13
Cost impact of software quality (10-Times Scale Rule).

10.3.4 Sources of Software Problems and Remedies

Defect sources for software problems are numerous, such as the lack of programming skills of the developer, weak engineering planning and design, poor communication between teams and the client, and weak system and processes.[19]

The root causes of such problems can be broken down into three fundamental categories: engineering capability of a developer, system and process, and management. These are the "Quality Pillars" that we discussed in Chapter 3 (see Figure 3.1).

- *Engineering Quality:* Software design is a creative process; therefore, individual skills, engineering experiences, and technical capability significantly influence the outcome of software quality. These factors are reflected in the category of *software engineering quality.*
- *System and Process Quality:* Systems and processes are important pillars for quality activities inside an organization. For example, software specifications, development, testing, and improvement activity follow specific systems and processes, which we can call "QC (Quality Control) and QA (Quality Assurance) processes." Standardized systems and processes for product development are a critical foundation for project planning, development, and validation for good-quality software.
- *Management Quality:* W. Edwards Deming said, "management is responsible for building the system, process, and quality culture."[20] Good standards or processes that people do not follow generate inefficiencies and confusion. The hiring of a perfectly skilled workforce is impossible. Principle-based management and leadership practices are essential to motivate people to respect systems and work together to create a high-quality level of a software product (refer to Chapter 4). It is also management's responsibility to improve systems and workforce skills continually—*good engineers are not born, but educated, trained, and managed.*

10.3.5 Software Quality Management

As we discussed in Chapter 3, quality management is composed of four steps: QP (quality planning) + QC (quality control) + QA (quality assurance) + QI (quality improvement). Figure 10.14 represents the four steps of the software quality management model.

1. *Quality Planning:* As we discussed in Chapter 7, scientific quality planning is the most critical step in quality management. As Figure 10.14 shows, the first step starts with the analysis of the customer requirements and the technical and business feasibility study of a project undertaken. If the study result is positive, we can

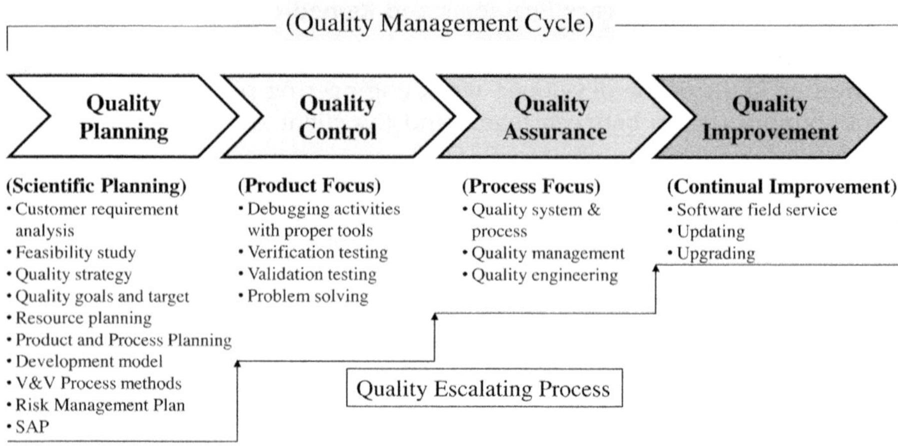

FIGURE 10.14
Software quality management model.

move to set up strategies, tactics, and quality targets. Then we need to prepare necessary resources, processes, and methods as well as specific action plan (SAP) to achieve the quality target. The SAP should include the client's main events, to which intermediate versions of software need to be applied and confirmed. When we plan for required resources, such as workforce and budget, we should consider the life-cycle cost of a software quality model.

The principal items of quality targets should be composed of several parameters: the quality levels of each attribute in Table 10.3, errors and faults targets at each stage, and the final failure target in the market as in Figure 10.15. Moreover, the software's test plans should be linked with the client's confirmation activities via a joint test plan as illustrated in Figure 10.15.

2. *Quality Control:* The fundamental concept of quality control is to not ship any non-conformed products to customers. Thus the focus on quality control is not on processes but products, i.e., software products. *Quality control activity is to discover and debug any errors or faults that might exist in the software.* In other words, software debugging, verification, and validation are critical activities, which should take place during software development and its delivery approval process. During a quality control process, a variety of methods and tools need to be utilized, such as simulation, emulation, bench test, and real product test. Discovering problems and resolving them in an early stage is a critical part of quality control activities regarding customer satisfaction and cost management, as discussed in Figure 10.13.

A systematic breakdown of the types of errors, faults, and failures by category is a method that should be in place and taught to the

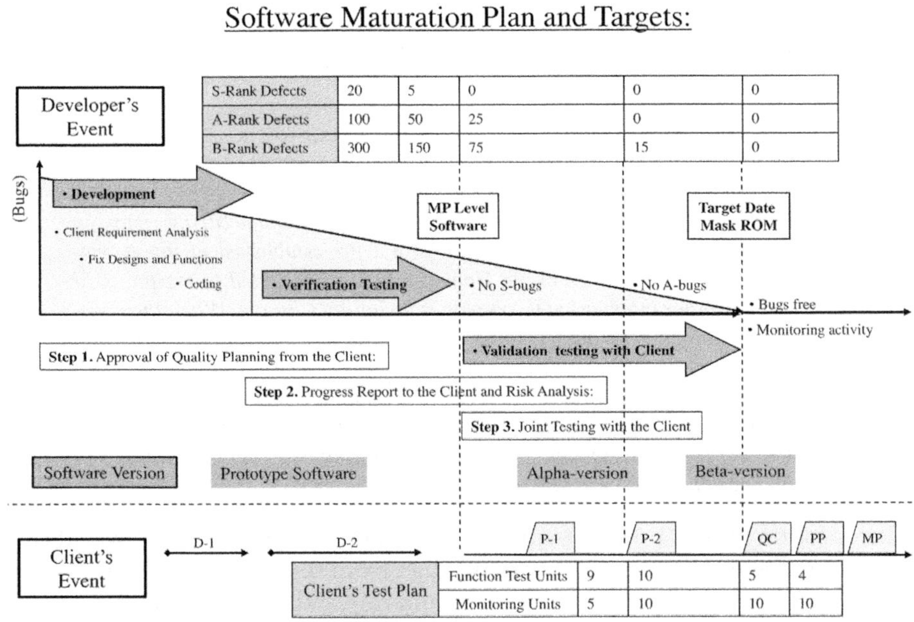

FIGURE 10.15
Software quality management plan.

software developers. Moreover, the scientific problem management techniques introduced in Chapter 6 can be used effectively to solve problems found, discover causes, and reflect them into the upstream part of the software development. Implementing quality management in such a way can improve the current project as well as future projects' software quality.

3. *Quality Assurance and Quality Improvement:* The original concept of QC (Quality Control) is focused on testing and inspection activity to ship only the right quality products meeting their design specifications. However, when problems continue to be made upstream, checking and filtering out 100% is impossible. As such, the quality of software cannot be warranted by QC activity alone. There should be systems and processes to prevent and eliminate errors from the beginning—this is the QA (quality assurance) approach.

   Software Quality Assurance is a set of quality activities to ensure quality in the first place through software quality engineering, system, and management processes; a quality product is the natural outcome of quality assurance activities. QA activities establish and evaluate systems and processes that produce software products—this is a process-focused action.

   QA needs to evolve through QI (Quality Improvement) activities, through which QA continually improves in product design, systems, processes, and management. QI put an organization in a competitive position in the market, increasing customer satisfaction.

In the 21st-century economy, sourcing electronic control parts happens globally. The control units are an embedded system; in other words, it has hardware and software.

J-German supplier of HVAC controller set up a joint manufacturing company in China. Its initial business was stable with a pre-established business relationship with German automakers. However, they also wanted to expand their businesses to other international companies including Chinese. When they contacted other automakers, a question arose regarding their technical and quality capabilities not only of hardware but also software to meet the client company's requirements. One of the main concerns from the potential clients was about their quality assurance system of software such that they can recognize the robustness of their systems and products when the supplier designs a new system and develop software for them.

The question here is how potential clients on the other side of the world can recognize the quality of a software product from unknown suppliers?

From this perspective, when building and improving QA and QI processes, software companies should follow standard, international approaches. They need to prove that the overall maturity level of the software development process in their organization is the right level that the client can trust. They also need to identify weak processes and make improvement plans so that their process maturity level can continually increase to satisfy their potential clients.

The most popularly used models are CMMI (Capability Maturity Model Integration) and SPICE (Software Process Improvement and Capability dEtermination)—ISO/ICE 33001:2015.

10.3.5.1 CMMI (Capability Maturity Model Integration)

CMMI was developed for the U.S. federal government from 1987 until 1997 by the Software Engineering Institute (SEI) of Carnegie Melon University. CMMI integrated many different models into one framework, and version 1.1 was released in 2002. In the beginning, CMMI was used to improve the capabilities of the American software industry, and later the model came to be used worldwide.

CMMI is an institutional approach to ensure that plans are established and processes are defined for all software development in the organization. This model provides a way of rating 22 process areas on a scale from 0 to 5; the 22 processes are grouped into four categories: *process management, project management, engineering*, and *support categories*.[21]

For example, in the process management category, there are five process areas: organizational process definition, organizational process focus, organizational training, organizational process performance, and organizational innovation and deployment.

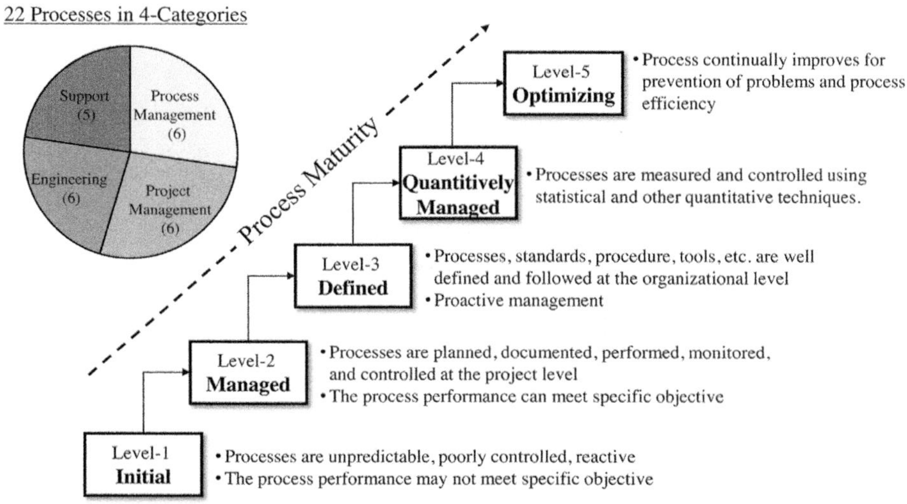

FIGURE 10.16
The CMMI—5 maturity level.

CMMI model provides a way of evaluating these 22 processes by five levels: *Level-1 Initial, Level-2 Managed, Level-3 Defined, Level-4 Quantitatively Managed, and Level-5 Optimizing* (see Figure 10.16), which determines the organization's overall development capability, including software quality assurance system and process.

10.3.5.2 ISO 33001 (SPICE)

SPICE (Software Process Improvement and Capability dEtermination (ISO/IEC 15504) was developed by the International Organization for Standardization (ISO). The first version of SPICE was released in 1995, and a major revision was completed in 2004.[22] Later, in 2015, SPICE has been revised by ISO/IEC 33001:2015.

The purpose of SPICE (ISO/IEC 33001) is to motivate an organization to strengthen its overall development process for high-quality software. It starts from the assessment of the current process—the workflow of SPICE is summarized in Figure 10.17.

First, a current software development process is examined through an assessment, which leads to process capability determination and process improvement strategy. Capability determination identifies the capabilities and risks of a process, and improvement strategy identifies the changes necessary with appropriate strategies. The recommended changes should be applied to the existing processes. *Software capability determination generally motivates an organization to make process improvements continually.*[23]

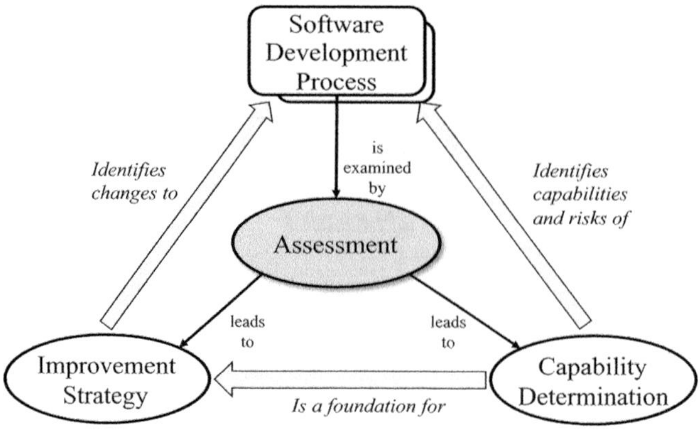

FIGURE 10.17
SPICE workflow.

The concept of SPICE is based on a two-dimensional framework: a process dimension and a capability dimension. Figure 10.18 illustrates the concept image of the capability determination of 65 processes in 5 categories, such as customer-supplier, engineering, supporting, management, and organization. Each process is evaluated using capability levels numbered from 0 to 5, where 0 describes an incomplete process, 1 a performed process, and so on, with 5 equating to a continually improving process.[24] Through the assessment, an organization would be able to detect strengths and weaknesses in its software development process such that they would be motivated to make improvements in their processes through self-improvement activity.

From the original SPICE, *Automotive SPICE (ASPICE)* was *originally* developed in 2001 as a variant of ISO 15504 (SPICE) by the AUTOSIG (Automotive Special Interest Group)[26]. ASPICE extracted key elements from SPICE for automotive

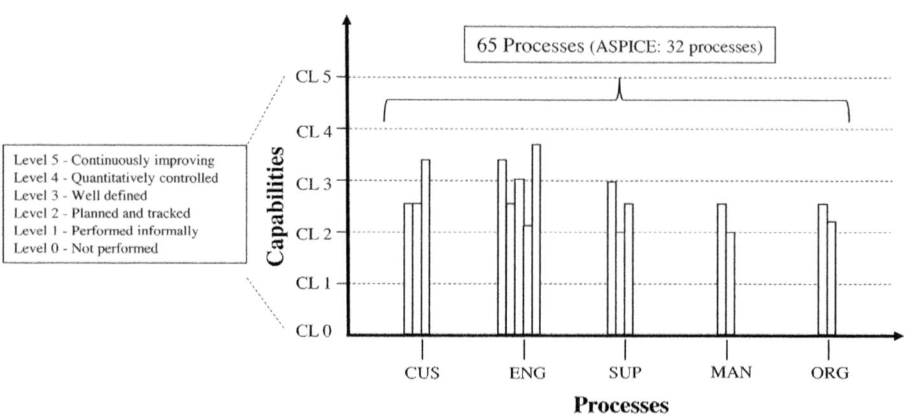

FIGURE 10.18
Framework of SPICE.[25]

application; it has 32 processes out of the original 65 processes of SPICE. In 2017, a new version of ASPICE-ver. 3.1 has been issued based on ISO 33001.

For example, the *software engineering process group* has six process areas to check capabilities: software requirements analysis, software architectural design, software detailed design and unit construction, software unit verification, software integration and integration test, and software qualification test.

More and more automotive manufacturers, especially European companies, request their suppliers to get ASPICE-certified even though the requested target level is different. If their process maturity level does not meet the client's requirements, suppliers are requested to improve the development processes, or in the case of high risk/low capability levels, the suppliers are excluded from sourcing.

CMMI and SPICE are comparable, and both are focused on overall process improvement in an organization. CMMI is generic process improvements for all organizations regarding software development while in SPICE there are some special reference models for some specific clients like the automobile industry—and SPICE is currently popular in Europe, especially for automobile software.

As discussed in Chapter 9, certification for certification has no meaning. The organization should take advantage of using such high international standards to improve its true process capability by motivating and educating the organization to continually improve software development quality.

10.4 Software Quality Confirmation

The purpose of software quality management is to make sure good-quality software is created consistently and repeatedly through sound and systematic management. Development for a large software project requires lengthy and complex processes. Development progress needs to be confirmed at each stage; otherwise, errors and faults (defects) accumulate and it becomes more difficult to find them at a later stage of development.

The core method of software quality confirmation is through robust testing. Generally, there are two stages of processes for software quality confirmation: *verification and validation testing (V&V testing). Verification* is a quality confirmation process to make sure the software meets the intended design specifications based on customer requirements. In contrast, *validation* is a final confirmation process to make sure the software meets end-users' expectations. Therefore, validation confirmation needs a broader range of testing with various testing scenarios.

V&V confirmation testing should go through a "Multi-Tier Testing Strategy"—checking the program codes, testing a unit program, integration testing, system level testing in a target embedded system, and field testing in the host machine (product) environment (see Figure 10.19). Each stage of

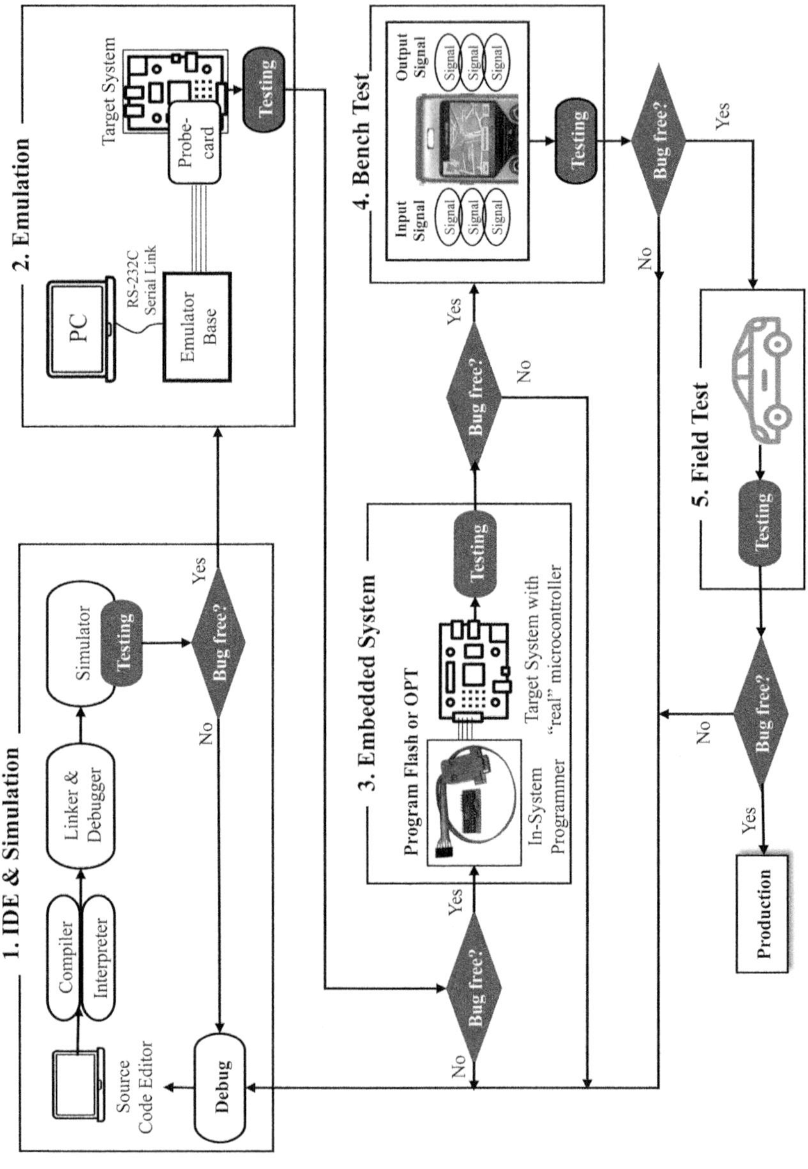

FIGURE 10.19
Software quality confirmation in development.[28]

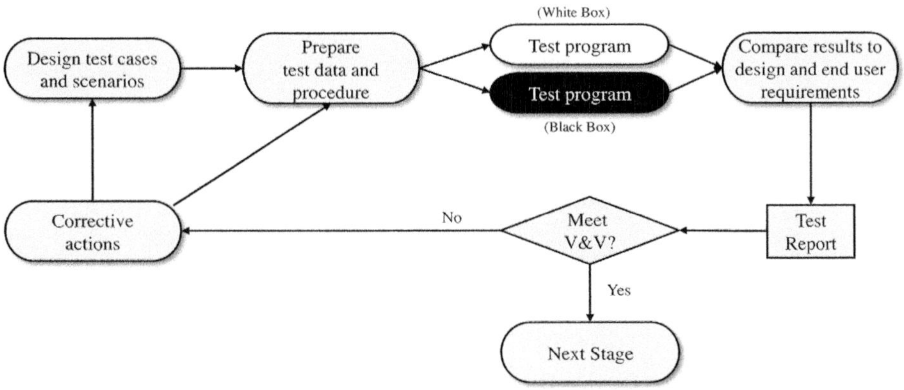

FIGURE 10.20
Macro model of the software testing workflow.

testing activity should go through a specific testing procedure and reconfirmation testing after corrected actions. In some cases, clients and selected end users can participate in the testing process through *alpha* and *beta testing* to help find errors or faults (defects) before the final release of the software into the market.[27]

Detailed testing cases and scenarios need to be prepared in advance with detailed test data and procedures like the V-model shown in Figure 10.9. Two kinds of testing procedures can be made: white box testing and black box testing, depending on the testing purpose and scope. The testing workflow can be summarized as in Figure 10.20,[29] and the flow needs to be used appropriately in conjunction with Figure 10.19.

10.4.1 Coding and Debugging through IDE and Simulation

Programmers, on average, spend half of their development time fixing bugs—debugging; investigating programming errors and faults (defects), and making the code work.[30] Creating bug-free software in the first place is the most critical step for stable quality development in later stages. However, being perfect the first time in coding or programming is impossible. There have been many surveys and research concerning errors inadvertently made by programmers. The number of errors varies depending on the process and capability maturity of software engineering as well as the training and competency of software developers.[31] *A typical developer makes 50 to 250 errors per 1000 lines of code.*[32,33]

10.4.1.1 Confirmation of Clean Codes in Programs

Computer code is the smallest element in a computer program. Computer code, or program code, is the set of special instructions forming a computer program.

A

```
for (i = 0; i < 200, i++) {
        printf("i = %d", i);
}
```

B

```
for (i = 0; i < 200, i++) {
printf("i = %d", i);
}
```

C

```
for (i = 0; i < 200, i++) { printf("i = %d", i);}
```

FIGURE 10.21
Practice of clean coding.

Bjarne Stroustrup, the inventor of C++, emphasized the simple and clean programming principle, stating "the logic should be straightforward and make it hard for bugs to hide…."

Clean code is easy to read, analyze, and manage. For example, in the case of C-language programming, each instruction should be in a new line with a proper indent for easy reading. For example, Figure 10.21 shows an example for a loop routine in C-language—all three programs work, but we know "A" coding is the easiest to read.[34]

A clean coding policy should be respected and practiced to the standard of a programming language and an organization's guidelines. During the review of code in a development stage, senior programmers need to coach junior programmers to practice clean coding. *Good code makes a difference.*

10.4.1.2 Debugging in IDE (Integrated Development Environment)

A traditional manual debugging process by running a compiler each time is not efficient in complex programming. In a manual debugging method, typically a programmer puts in "stop and go" breaking points in the code to pinpoint the bugs. Modern debugging management relies on advanced debugging tools built in IDE (integrated development environment) systems.

A debugging process by using IDE is explained in the first step of Figure 10.19. IDE is a single integrated development environment including code editor, compiler/interpreter, and debugger in one package. IDE provides convenient visual and graphical debugging methods. For example, when there are syntax or typing errors during coding, problematic code or lines are automatically highlighted with blinking or red colors. With IDE, dynamic debugging is also possible, e.g., a programmer can monitor the running of each line of code by breaking the program running conveniently without a need to insert separate codes into the program and can change variable values while the program is running to see how the program runs.

A firmware designer should confirm the supporting capability of an IDE system for a particular microcontroller of its target embedded system. When choosing a microcontroller, the development ecosystem needs to be confirmed with respect to whether it is supported by a capable IDE. In industry, there are popular IDEs supporting many different types of microcontrollers: the most popular ones are Eclipse, AVR Studio, and Keil uVision.[35] Most of the IDEs provide built-in simulation functions as a debugging tool as well.

10.4.1.3 Debugging in Simulation

Embedded software—firmware—operates under the interactions with its peripheral hardware, where it processes input signals and sends out output signals to actuators or other ECUs. Any software needs to be confirmed under such real-world situations. However, since software and hardware are being developed concurrently in the engineering development process, it is impractical to confirm the whole software operation in the actual target embedded system in an early stage of development.

Simulators are a software approach to efficiently debug potential concerns by simulating signals as if they are from a real-world embedded system. Simulation testing can provide various stimulus inputs, such as pushbutton types connected to I/O and regular clock inputs for simulating A/D conversion data or serial communication inputs. Many embedded systems can be effectively debugged using proper peripheral stimulus methods.

While simulators offer a cost and time-effective development environment, *real-time system simulations are difficult*. In the real world, input and processing speeds by the processor are 100 to 1,000 times faster, with a variety of wide-range signals.[36]

10.4.2 Confirmation by Emulation

While simulation is useful for algorithm development, an embedded system by nature requires its full speed operation and access capability to peripherals such as I/O ports, timers, A/Ds, PWMs, and so on. In other words, *a program that runs perfectly in simulation mode can crash in a host embedded system when it operates in real-time.*

Emulation is a process to mimic the operation of software on a target embedded hardware system. As Step 2 of Figure 10.19 illustrates, an actual MCU (microcontroller unit) on the target board is replaced with either a bond-out or a regular production chip installed on the pod. There are emulators for different types of processors. In general, the higher the level of integration with the end-product hardware, the greater the benefit of a tool in improving the debugging process, but the higher the cost as well.

For example, an in-circuit emulator supports many functionalities including software and hardware breakpoints and real-time tracing capability. By downloading the code into the emulated chip, a microcontroller's functions are emulated in real-time while using external static RAM as an emulated micro's program memory. RAM allows for code to be changed quickly and an indefinite number of times, with access to on-chip registers and data memory. Any crash points can be easily traced, debugged, and can be rerun, allowing effective debugging.

10.4.3 Confirmation at a Target Embedded System

After debugging through the emulation process, the program still needs to be confirmed at the actual target embedded system. The embedded system has inputs and outputs in digital or analog forms. Therefore, this step of the process requires the actual target system with a "real" microcontroller—see Step 3 of Figure 10.19.

The microcontroller is flashed or OTP (one time programmable) with a final version of the software from Step 2 of Figure 10.19. Then different test cases (scenarios) are run through the embedded system as shown in Figure 10.20. The primary purpose is to confirm that the program in an actual microcontroller runs without crashing inside the embedded target system and with real-time operations.

10.4.4 Confirmation through Bench Test

After software testing in a target embedded system, a major remaining concern is interactions between microcontrollers, hardware, and peripherals in a host embedded system. Inputs from the peripherals are converted into digital signals via A/D converter in a PCB board and processed through a microcontroller. If actual inputs from sensors or other controllers do not match the particular types or ranges of signals that the microcontroller expects, there will be system errors.

Such concerns need to be checked through a bench test. As Step 4 of Figure 10.19 shows, bench testing should have real signal inputs from the peripherals of a host system (an end-user hardware product). However, for the difficult cases, the actual input signals can be substituted with simulated signals. Then, testing can be performed to the plan designed in an early stage as shown in Figure 10.20. The bench test execution can be automated, and the test results need to be confirmed with the expected results either by automatic or manual process.

A typical automotive navigation unit processes many different types of input information, e.g., GPS signals, real-time traffic information, and operator's inputs. It needs to display not only the map for a driver's location but also route guidance to the driver's target destination. The test bench should mimic a real traffic situation by inputting artificially generated signals to the navigation system. By utilizing image processing technology, the operation function, speed, and the accuracy of an actual map display and route guidance can be checked automatically with a video camera installed in front of the navigation screen in the test bench.

10.4.5 Confirmation through Field Testing

Software in the actual usage environment of the target system may not function or perform to the customer's requirements or the end-user's expectations. For instance, the end user can input two operation buttons at the same time, or some input signals may contain electronic magnetic interference (EMI) noises.

As Step 5 of Figure 10.19 shows, the best confirmation of such concerns is through field operation testing, which is the final stage of release testing of a software product. The field testing should typically include the clients and selected end users; however, special expert groups in the market can represent the end users as well.

Different testing scenarios need to be planned in advance to the testing procedure in Figure 10.20. Any concerns identified in the field testing may require the client's or software developer's corrective actions following scientific problem solving procedures discussed in Chapter 6.

There can be two different strategies in the final testing of software as shown in Figure 10.22. Model-A represents a conventional way, where the client focuses on their own product's validation testing with software provided by suppliers, and suppliers focus on the final validation of their software by themselves. In contrast, Model-B represents the client and suppliers working together, i.e., joint testing to confirm the quality of the final software. The advantage of Model-B is that if any abnormalities occur during the joint testing, their causes can be promptly analyzed and counter

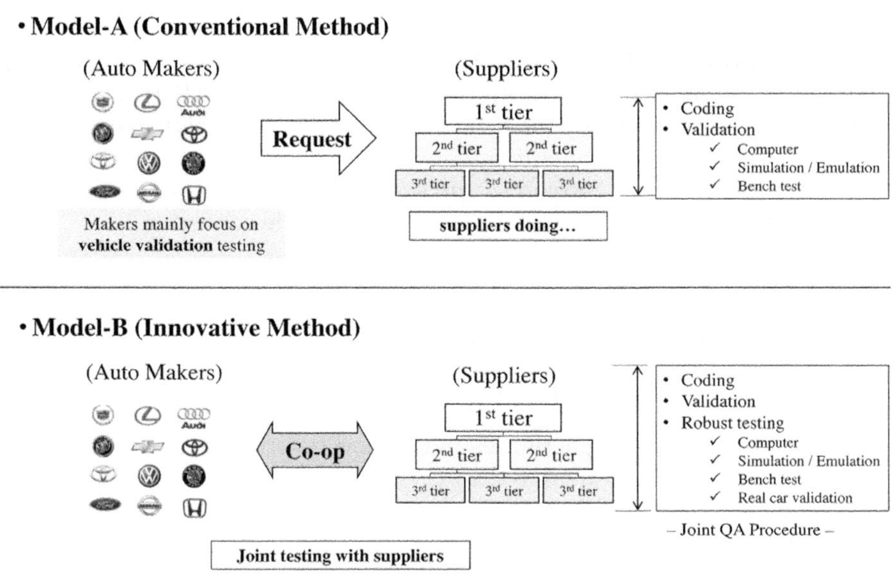

FIGURE 10.22
Software field testing strategy.

measured either by the client or suppliers. The client's action may require changes to their hardware, and the software developer's actions may require debugging or updating their software.

10.4.6 Release Testing

10.4.6.1 Alpha Testing

Alpha testing occurs when software development is completed. Usually, this type of testing can be performed with the client. The client is more knowledgeable about the usage conditions and environment of their system. The client can invite a selected group of end users to the testing process. Depending on the severity and the causes of the problems, the client or/and the developer should take corrective actions.

10.4.6.2 Beta Testing

Typically, the final version of the software is released to a larger group of end users to allow them to experiment, use, and monitor the system. Issues arising during the testing of the system should be thoroughly documented including relevant photos and video clips. The issues should be repeated, analyzed, and corrected by the software developers or/and the client.

10.4.6.3 Acceptance Testing

The final version of the software is submitted to the client as a part of the PSW (Parts Submission Warrant) process required in PPAP as discussed in Chapter 5. *This is testing by the client's development and quality assurance team using their own methods and criteria.* The team should also *reconfirm the previously identified concerns during their product's tes*ting including potential side effects of the countermeasures applied to resolve the concerns.

10.5 Software Quality Engineering

In spite of the debugging activities performed through the *5-Steps of quality confirmation process* explained in the previous section, some bugs may remain not completely fixed and lurk in the final software product, causing failures in the market. Therefore, scientific management of the bugs and failures with cause analysis is critical both for current problem management and for improving the software quality process in the organization as a part of preventive quality management.

Quality engineering should play a critical role in this perspective. A scientific method of improving software quality by using engineering methods, discipline and process is "software quality engineering." Many of the tools that we introduced in Chapter 5 can be adapted to use for software quality engineering activities.

10.5.1 Scientific Software Problem Management

Most software problems are found during the internal verification and validation process. Statistically speaking, the more errors and defects occur during the development stage, the more potential failures occur in the market. Therefore, a thorough investigation of the problems identified at an early development stage is the most critical maturation strategy once a project has been started.

The scientific problem management method introduced in Chapter 6 and quality engineering tools introduced in Chapter 5 can be fully utilized to solve these software problems, and recurrence prevention activity should follow.

10.5.1.1 Pareto Chart

As explained in Chapter 5, 80% of phenomena are attributed to 20% of causes, which is the 80/20 principle observed in statistics, economics, sociology, and quality. A Pareto chart is a handy graphical method to identify 80% of the problems by focusing on 20% of the causes.

80% of the errors and defects that occurred during software development may have 20% of common root causes. By eliminating this 20% of common causes, 80% of the problems can be eliminated or prevented in advance.

Table 10.5 shows software errors found during A-Model Project development. The first column is the error types, the second column shows the

TABLE 10.5

Software Errors and Causes (Example: A-Project)

Error Types	No. of Errors/ 1000 lines	Major Cause of Errors
Syntax Error	23	Skills
Wrong Variable Name	13	Skills
Memory Allocation Mistake	10	Design
Data Problem	7	Communication
Algorithm Mistake	5	Design
System Integration	4	Design
Calculation Error	4	Requirements
Run Time Error	3	Requirements
Other	5	Other

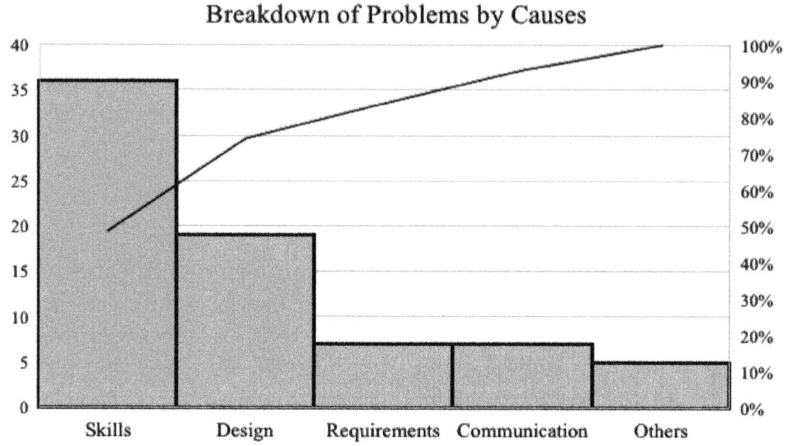

FIGURE 10.23
Pareto chart of development errors.

number of errors per 1000 lines of code, and the third column explains the most plausible root causes of the errors identified through a cause and effect diagram.

At first glance, the problems seem scattered. However, when we draw a Pareto chart (Figure 10.23), we discover that the causes of the problems are concentrated in two major causes: the developers' skills and design issues.

10.5.1.2 Cause and Effect Diagram

As we discussed in Chapter 5, a "cause and effect diagram" can be used widely across various industries to analyze the causes of a specific event. The same "cause and effect diagram" can also be used in software quality engineering to identify the causes of software errors, defects, and failures.

Figure 10.24 shows how we can logically use the diagram by using a "why" analysis in a fishbone approach. From this diagram, we can identify that the primary causes of coding errors in Table 10.5 are hiring and training problems.

10.5.2 Scientific Quality Engineering Planning

Software errors, defects, and failures need to be resolved promptly during the quality control process. However, prevention is better than cure. Most quality features and functions regarding Big Q are determined at the beginning of the software design process. The quality engineering and methods discussed in Chapter 5 can be applied to software design at the quality planning stage as well.

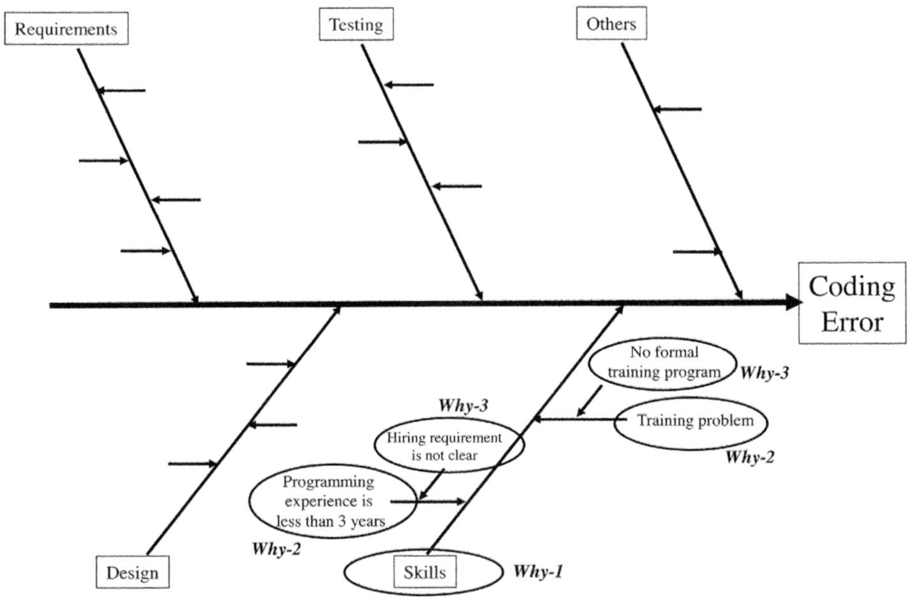

FIGURE 10.24
Cause and effect diagram. (Example: A-project.)

10.5.2.1 QFD (Quality Function Deployment)

As we discussed in Chapter 5, QFD is an essential tool for quality planning of a new product. It can analyze the relationship between customers' viewpoints and engineering design parameters. Figure 10.25 shows an example of how QFD can be used for the software design of a "car navigation system." The matrix explains the relationship between customers' requirements and software design parameters. The numbers inside the matrix are calculated and show design priorities at the bottom as a "Total" number: fast processing of GPS signal, fast route calculation algorithm, processing of dynamic traffic conditions, high accuracy map data, and so on. The original design concept of "X-Model" is also checked, weak points are identified, and improvement plans can be made for each design priority item, such as "Apply the latest published new algorithm" for "fast processing of GPS signal."

10.5.2.2 FMEA (Failure Mode and Effects Analysis)

FMEA was developed in the 1960s by NASA. Since then, this quality engineering methodology has been widely used in a variety of industries from manufacturing to software development. FMEA carried out during the software design phase helps in identifying structural weaknesses in the software design.[37]

QFD MATRIX QFD: Quality Function Deployment		Software Design Parameters								Benchmarking			New Model
Customers' viewpoints	Weight	High Accuracy Map Data	Fast Processing of GPS Signal	Fast Route Calculation Algorithm	Processing of Dynamic Traffic Condition	Voice Recognition Algorithm	Supporting High Resolution Screen	Fast Processing of Touch Screen Command	Reset Function	V-company T-Model	F-company K-Model	H-company E-Model	X-Model Concept
Accurate Display of Map	10	◎	◎	○	○		▽		▽	9.0	9.0	8.0	9.0
Dynamic Route Guidance	9	○	◎	◎	◎					7.5	8.0	7.0	6.5
Speed of Navigation Guide	8		◎	◎	◎			▽	◎	8.0	8.0	7.0	7.0
High Resolution Display	7	○					◎			7.5	7.5	8.0	7.5
Touch Screen Function	6							◎		7.0	8.0	7.0	9.0
Voice Command Function	5					◎				9.0	8.5	7.0	8.0
POI Information	4	○								9.0	9.0	7.5	8.0
Total		110	135	115	115	25	45	38	50	397	405	362	383
		Add a function of map update remotely	**Apply the latest published Algorithm**	**Apply the latest published Algorithm**	**Reconfirmation of the current function accuracy and speed**	Keep the current design concept	Keep the current design concept	Keep the current design concept	Planned already				**X-model Design (Improvement Areas)**

	Relevance	Weight
◎	Strong	5
○	Normal	3
▽	A little	1

FIGURE 10.25
QFD. (Example: Navigation system.)

Figure 10.26 shows an example of FMEA for "navigation system design" according to the procedure discussed in Chapter 5. A primary concern identified is "map data," which is a necessary input for a route guidance program. The navigation can freeze when proper data are not inputted. Through FMEA analysis we can identify such risks and deploy countermeasures in the design, such as detailed map data in rural areas and installation of a reset function in the software in case of system freeze.

10.5.2.3 FTA (Fault Tree Analysis)

FTA is a deductive procedure—a top-down approach. It starts with a top failure event rather than starting from a software system or function level like FMEA. Thus software FTA is applied to software design to locate potential problems in the development stage, and the root causes of potential failures can be fixed before the development is finished.

Failure Mode and Effects Analysis (Design-FMEA)

Item / Function	Potential Failure Mode	Potential Effect(s) of Failure	SEV	CLASS	Potential Cause(s)/ Mechanism(s) Of Failure	OCCUR	Current Design Controls Prevention	Current Design Controls Detection	DETEC	RPN	Recommended Action(s)	Responsibility & Target Completion Date	Action Results				
													Actions Taken	SEV	OCC	DET	RPN
Read map data of the current location matching GPS location information	Cannot find map information	Navigation System freeze	10		Map data is not available	6	Only city map data is available	Testing in city and urban areas	9	**540**	1. Tiger map data are required in rural areas 2. Add a reset function	Map supplier 'Google', Jan. 2020	Testing in rural areas using proto map data	10	2	5	**100**

FIGURE 10.26
FMEA. (Example: Navigation system.)

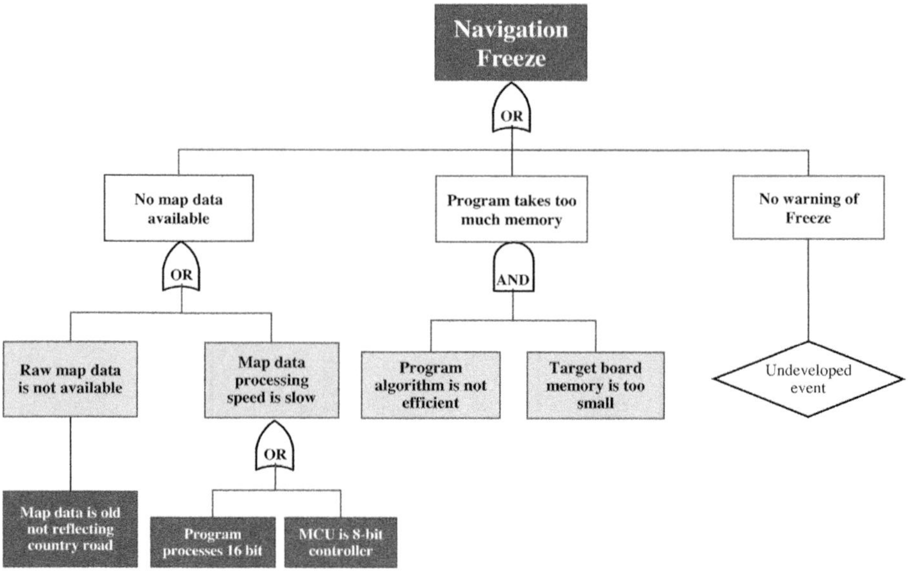

FIGURE 10.27
FTA. (Example: Navigation system.)

As Figure 10.27 shows, to carry out the software FTA we first identify the undesired top events that concern the system functionality or critical attributes, such as the navigation system freeze as in this example. Taking the undesired event as the topmost event, we logically identify lower events, such as intermediate and basic events that contribute to the top failure event.

The FTA approach can be a complementary tool to FMEA because FTA can identify multiple causes of the top failure event. In the example in Figure 10.27, we have identified the potential failure events: raw map data are too old—not reflecting new roads; map data processing speed is slow; program algorithm is not efficient; and target board memory is too small. Some basic events can be further broken down into root causes, such as the program processing 16-bit while the MCC (micro control unit) has an 8-bit controller. In such a case, the FTA revealed not only potential software problems but also hardware problems that are interlinked with the software.

Through this type of analysis, preventive actions need to be taken in advance in software or/and hardware design when they are interlinked together.

Quality engineering principles and methods have not yet fully penetrated the software industry. Some companies do FMEA or FTA just as a routine procedure to show the process is done. However, viewpoints and contents are more critical to reap the true value of such powerful tools. Many catastrophic failures in software industry history happened because engineers and managers did not deploy the process in a structured way of quality management—the contents are more important than just following the predefined process; if someone skips the process, many people would be able to recognize them; however, one big mistake lurking in software is difficult to find out. As we discussed in Chapters 4 and 8, thorough confirmation of the detailed contents prior to event evaluation should be a critical quality management viewpoint. One missed content kills hundreds of people.

Notes

1. Software quality management, https://en.wikipedia.org/wiki/Software_quality_management (last modified February 19, 2018).
2. Wilmshurst, T. *An Introduction to the Design of Small-Scale Embedded Systems*, Palgrave, 2001.
3. Wilmshurst, T. *Designing Embedded Systems with PIC Microcontrollers*, Elsevier, 2010.
4. Firmware, https://en.wikipedia.org/wiki/Firmware#cite_note-3 (last modified May 29, 2019).
5. Wilmshurst, T., *Designing Embedded Systems with PIC Microcontrollers*, Elsevier, 2010.
6. Laporte, Claude Y, *IEEE Computer Society*, Wiley, Inc., 2018.
7. Burkacky, Ondrej, Johannes Deichmann, Georg Doll, and Christian Knochenhauer, "Rethinking car software and electronics architecture," Mckinsey&Company, February, 2018.
8. https://blog.udemy.com/embedded-c-tutorial/ (accessed February 19, 2019).
9. May, Chelsie. July 6, 2018, "Top 10 Embedded Software Outsourcing Companies for Automotive," https://hackernoon.com/top-10-embedded-software-outsourcing-companies-for-automotive-1c609b25057c (accessed February 16, 2019).
10. Boeing to Make Key Change in 737 MAX Cockpit Software, March 18, 2019, www.wsj.com/articles/boeing-to-make-key-change-in-max-cockpit-software-11552413489?emailToken=2a95a4ae43a376b010d2324010ccd806eGjZfLpNXe3p7ku//MOMteXCOwp+mpkEsXyQLXnX6WHSB65Zi7AO23CWqQZ6UoZZkBAVEoGV/VNXVJOEGAiwCzToXUtabmRMymTgYUDcFO4%3D&reflink=article_email_share (March 18, 2019).
11. V-Model (software development), https://en.wikipedia.org/wiki/V-Model_(software_development) (last modified March 16, 2019).
12. What Is Incremental model—advantages, disadvantages and when to use it?, http://tryqa.com/what-is-incremental-model-advantages-disadvantages-and-when-to-use-it/ (accessed February 17, 2019).
13. Sommerville, Ian. *Software Engineering*, 9th edition, Addison-Wesley, 2011.
14. Software Quality Engineering, 2018, www.wagmob.com/ (accessed February 19, 2019).
15. Software Quality Engineering, 2018, www.wagmob.com/ (accessed September 10, 2018).
16. Burnstein, Ilene. *Practical Software Testing*, Springer, 2003.
17. Ebert, C. and R. Dumke. *Software Measurement*. Springer, Heidelberg, 2007.
18. Shull, F. et al: "What we have learned about fighting defects," *Proceedings of the 8th International Symposium on Software Metrics*. IEEE, Los Alamitos, USA, pp. 249–258, 2002.

19. Burnstein, Ilene. *Practical Software Testing*, Springer, 2003.
20. Deming, W. Edwards, *Out of Crisis*, The MIT Press, 1982.
21. Sommerville, Ian, *Software Engineering, 9th edition*, Addison-Wesley, 2011.
22. ISO/IEC 15504, https://en.wikipedia.org/wiki/ISO/IEC_15504 (last modified April 1, 2019).
23. Pyhajarvi, Marko. "SPICE International Standard for Software Process Assessment," Helsinki, 31st November 2004, Seminar on Quality Models for Software Engineering Department of Computer Science, University of Helsinki.
24. VDA QMC, *Quality Management in the Automotive Industry, Automotive SPICE® Process Reference Model*, Process Assessment Model Version 3.0.
25. Pyhajarvi, Marko. "SPICE International Standard for Software Process Assessment," Helsinki, 31st November 2004, Seminar on Quality Models for Software Engineering Department of Computer Science, University of Helsinki.
26. This group consists of the SPICE User Group, the Procurement Forum, and German Automotive Companies and some other international automotive manufacturers like Fiat, Ford, Jaguar, Land Rover, and Volvo, www.plays-in-business.com/automotivespice/#History_ofAutomotive_SPICE, (accessed February 19, 2019).
27. Solis Tech, *Quality Assurance: Software Quality Assurance Made Easy*, 2015.
28. Developing Embedded Systems—A Tools Introduction, www.embedded.com/design/programming-languages-and-tools/4017874/Developing-Embedded-Systems-151-A-Tools-Introduction (accessed February 19, 2019).
29. Sommerville, Ian, *Software Engineering*, 9th edition, Addison-Wesley, 2011.
30. "Reversible Debugging Software," The Cambridge MBA, University of Cambridge, 2012.
31. Laporte, Claude Y. and Alain April. *Software Quality Assurance*, IEEE Computer Society, Inc., 2018.
32. https://coralogix.com/log-analytics-blog/this-is-what-your-developers-are-doing-75-of-the-time-and-this-is-the-cost-you-pay/ (accessed February 19, 2019).
33. Humphrey, W.S. "The software quality challenge. Cross Talk," *The Journal of Defense Software Engineering*, June 2008, pp. 4–9.
34. www.khanacademy.org/computing/computer-programming/programming/writing-clean-code/e/quiz–clean-code (accessed February 19, 2019).
35. Microcontroller IDE Comparison, https://blog.mbedded.ninja/programming/general/microcontroller-ide-comparison (accessed February 19, 2019).
36. Debugging tools for embedded systems, www.electronicproducts.com/Software/Debugging_tools_for_embedded_systems.aspx (February 19, 2019).
37. Rajan, Chitra. "Software FMEA and Software FTA—An Effective Tool for Embedded Software Quality Assurance," White Paper, Mahidra Satyam.

11

Quality 4.0 in Smart Manufacturing

Since the first industrial revolution began in the 18th century, machine power began to replace animal and human labor forces. Work efficiency has significantly improved in terms of labor usage, processing capability, and productivity. The second industrial revolution in the 19th century made mass production possible with the help of electrical energy, and manufactured goods became much cheaper. The third industrial revolution in the 20th century started with the power of computers and IT technology, which made automated production possible by controlling equipment and machines. Increasingly, computers became the central means of improving productivity and managing complex products. Now the fourth industrial revolution has started with smart manufacturing harnessing the power of IoT, Big Data, and AI (see Figure 11.1).

Smart Manufacturing[2] or *Industry 4.0*[3] is defined as a "fully-integrated collaborative manufacturing system that responds in real time to meet the changing demands and conditions in the factory, in the supply networks, and customers' needs." The physical world is integrated with cyber technology, including control of the equipment, machines, and the flow of materials in plants. It connects people, machines, and data in new and smart ways.

Quality should take the same view of smart manufacturing. The definition of quality remains the same; however, the implementation of quality management should ride the wave of smart manufacturing technology. For instance, there should be communication and cooperation between processing machines regarding measurement and corrective actions. The quality actions should be taken automatically in module processes under *centralized and decentralized management*.

In such a new paradigm of quality management, all the processes and product quality data should be connected from the product design, supplier's manufacturing, and in-house manufacturing to the market. Big data should be processed and analyzed to manage quality at each process by the integrated power of machines, humans, and AI. *A new way of quality management, which is possible by smart manufacturing, will bring a new level of sophisticated products to customers.*

1. Industry 4.0
2. Quality 4.0
3. Future-Oriented Quality Management

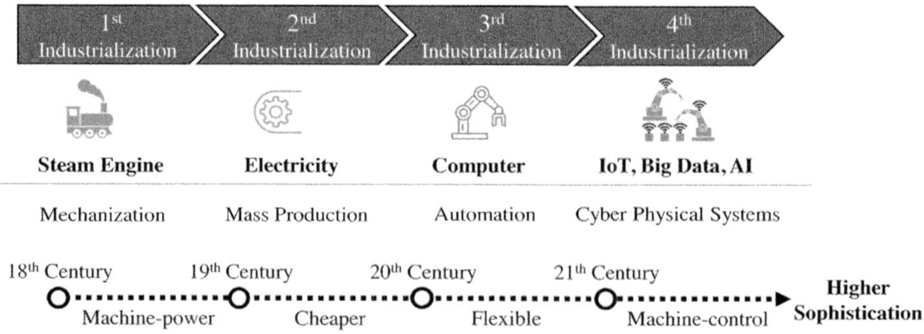

FIGURE 11.1
Industrial revolutions and future view. (Modified from POSCO.)[1]

11.1 Industry 4.0

Previous industrial revolutions sought more quantity: more materials, more goods, and more services. These outputs were not attained for free. We created environmental and social problems. Global warming and climate change now threaten the future well-being of our next generation. The disparity of wealth created a social struggle between owners and employees. Factory workers are overloaded with numerous working hours and health issues.

Now, in a new industrial era, significant changes come from technology, and the impact will be much greater than the accumulated power of the past industrial revolutions.

Our technological world has already progressed in the handling of laborious, dangerous, and complicated work. For example, industrial robots are prevalent in many industries, such as automobile plants, logistics warehouses, power plants, and medical industries. Manufacturing electronic parts has already become highly automated. The hardware power of equipment and machines is at a maturity stage in its S-curve in the technology life cycle.

As we discussed in Chapter 10, the real industrial revolution in the 21st century came with the power of software. It has already brought about huge changes in communication, receiving information, developing and producing things, and performing various economic activities. Now a new era of software is arriving in the power of Big Data processing and AI, which makes a new industrial revolution, Industry 4.0, possible.

However, software alone has less power and impact in creating value than when it is combined with hardware. More and more hardware products have software embedded systems as discussed in Chapter 10. However, the trend of the future is more than embedded systems. More smart products are being developed with decentralized, centralized, IoT, cloud computing, 5G, and mobile control capabilities. For instance, we can control automobiles, home appliances, and intelligent robots via 5G and smartphones. Apple, Amazon, and Google are all developing hardware interlinked with software. In the case of automobiles, electronics and software combined already constitute up to 40% of the cost of a vehicle.[4]

Now the future is clear. The future of industry is not just about a separate advancement of hardware or software technology. The future is a combination of hardware and software power; however, unlike traditional embedded systems, the future integrations *require a network of interacting elements with physical inputs and outputs* instead of standalone devices, i.e., the integration of the physical and the cyber world. That is the approach of smart manufacturing, or Industry 4.0.

Software will develop further with an entirely new platform, that is, AI. AI platforms will eventually replace intellectual tasks that human beings are doing today. We will integrate Big Data processing, AI, and versatile machine power to change the landscape of future industries.

Currently manufacturing is a hierarchically managed process with the power of the computer and human intervention, which is the concept of Industry 3.0. However, the first step of future manufacturing will be managed and controlled with the power of AI not only centrally but also locally by equipment and machines at each module. Figure 11.2 shows a contrasting control method between Industry 3.0 and 4.0.

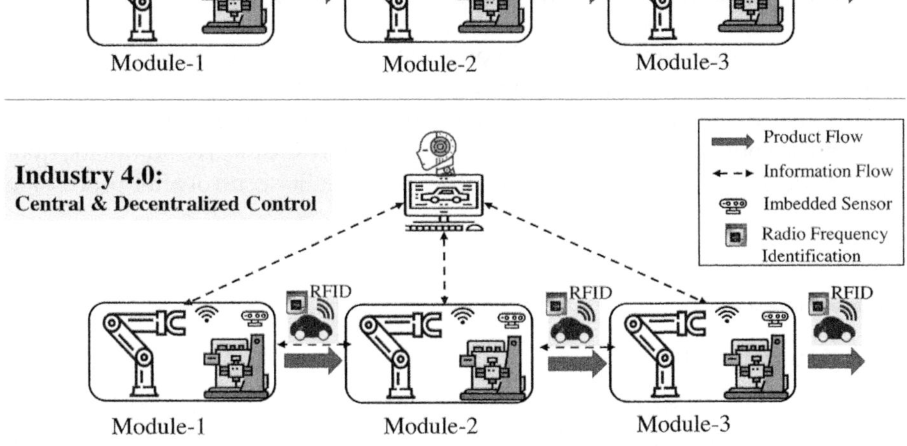

FIGURE 11.2
Manufacturing control difference between Industry 3.0 vs. 4.0.

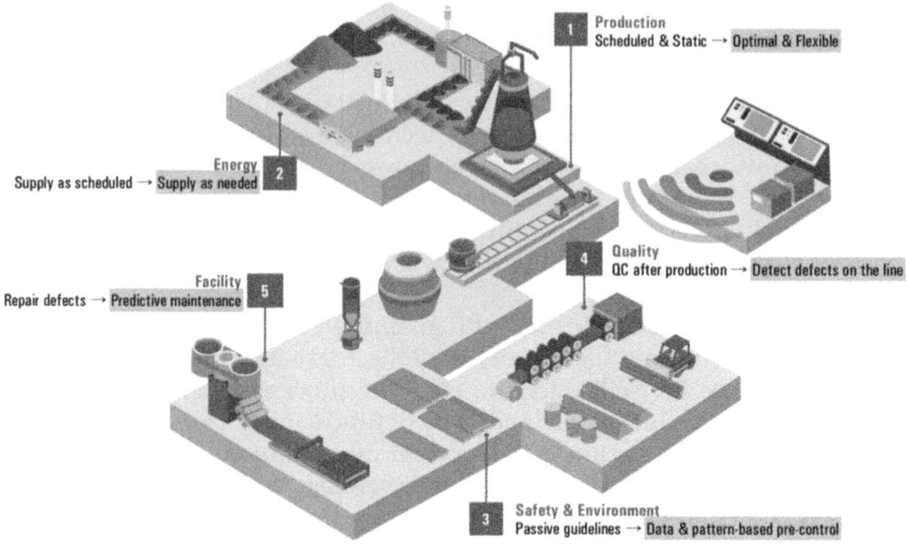

FIGURE 11.3
Industry 4.0 and steel industry. (Image permission from POSCO.)[5]

In Industry 4.0, each process module reads production information from an RF (Radio Frequency) ID tag attached to a part or product. Then each machine processes the product based on processing instructions from its own and central cyber locations. Afterward, the processed information is shared with other machines and humans for appropriate actions.

Figure 11.3 shows an actual example of Industry 4.0 applied in the steel industry. A primary role of humans is to devise a standardized and optimized plant design at the beginning, such as the plant floor and module layouts, equipment and machine allocation, and communication and control systems. Once the platform has been created, it will be self-operational in a maximum optimal condition. That is the ideal goal of Industry 4.0.

The distinctive concept of Industry 4.0 regarding quality is, as No. 4 in Figure 11.3 shows, quality is assured on the line in real-time, i.e., automatic quality control and assurance on the line rather than quality control after processing.

11.2 Quality 4.0

We assume *the first stage of Industry 4.0 and Quality 4.0 are the integration of cyber systems and machines through human interventions.* Industry 4.0 is not aimed at replacing humans, but helping them work more efficiently. A successful digital transformation will not be possible without good collaboration between

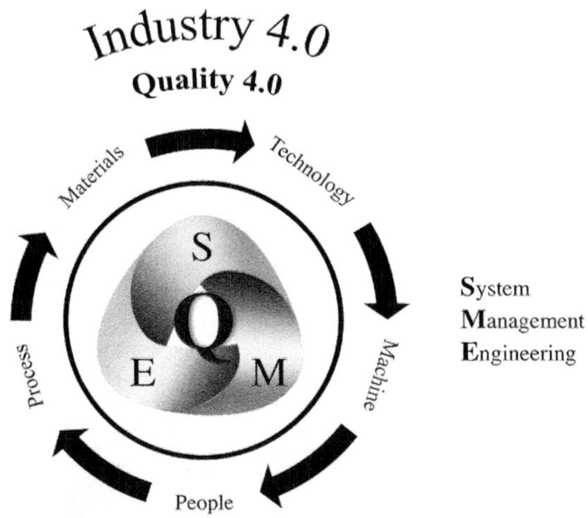

FIGURE 11.4
Evolution of traditional quality integrated with Quality 4.0.

humans and machines.[6] Furthermore, human intervention promotes efficient, robust, and reliable control of cyber-physical systems via top-level decisions when required. *Therefore, in this chapter, our scope of Quality 4.0 is on this first stage. As we discussed earlier, intelligent human intervention is a big part of Industry 4.0.*

Quality 4.0 is a reference to Industry 4.0. As Industry 4.0 is integration with IT, so is Quality 4.0. Determining the optimal implementation of quality management in the Industry 4.0 framework is the focus of Quality 4.0.

Technologies constitute a significant scope of quality management because they enable the transformation of culture, collaboration, communication, and compliance between processes, people, products, machines, etc.[7] *Quality 4.0 is not about technology itself, but how a company creates and maximizes quality value through quality activities in the framework of Industry 4.0.* Figure 11.4 explains how Quality 4.0 integrates the traditional quality elements (System, Management, Engineering) with the power of Industry 4.0, *interlinking technology, machines, people, processes, and materials.*

With the aid of advanced technologies, Quality 4.0 is a combination of IT (Information Technology) and OT (Operation Technology) through a digital transformation with human intervention as represented in Figure 11.5.

As discussed in Chapter 2, conventional quality management is not performed in real time, so there is always a time lag between the product and its quality data. For instance, when the quality data are available, many parts have already been made or shipped.

However, with digital transformation technology, innovative quality management methods are possible: each machine can be controlled in *real time* at each workstation by self-checking its own work, *quality Big Data* from various sources can be collected and interpreted, *cyber-aided communication* between machines and people can occur in real time, and *AI can* control the entire system.

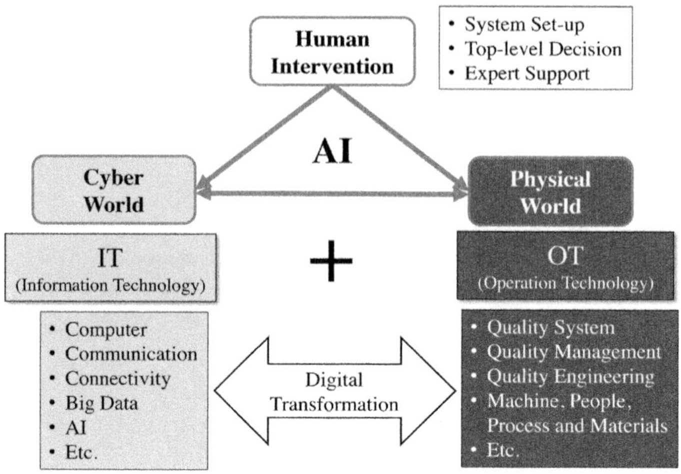

FIGURE 11.5
Quality 4.0 combining IT and OT through digital transformation.

The practical quality innovations and examples are summarized in Figure 11.6. Other practical extensions of Quality 4.0 can be further developed as Industry 4.0 becomes widespread in the industry.

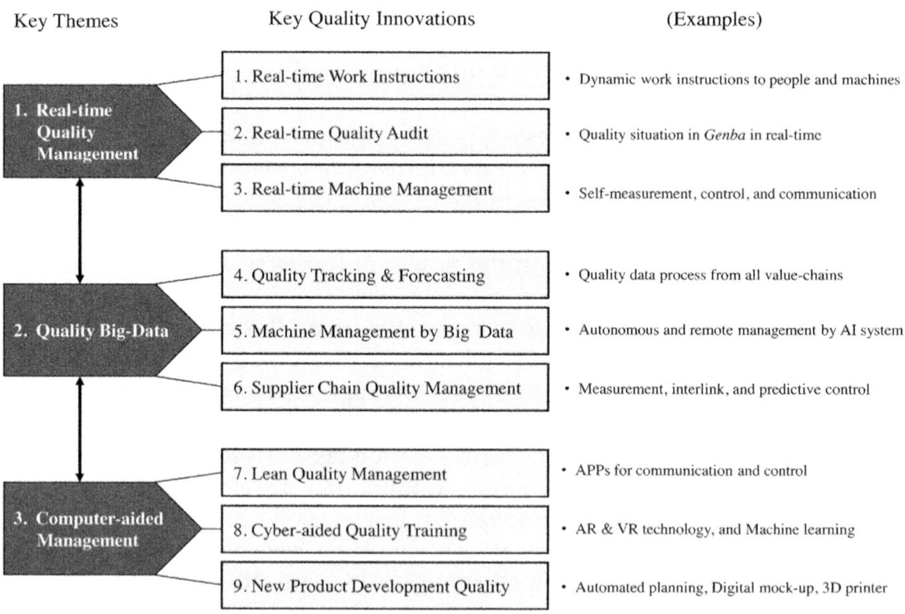

FIGURE 11.6
Application of Quality 4.0 in the manufacturing industry.

11.2.1 Real-Time Work Instructions

Work instruction is an essential part of quality management to make employees or machines understand and follow the right procedures to make things correctly. However, work instructions are always a nightmare for every organization, especially when many mixed models are produced through one production line. The work instructions contain complex information depending on model types: different parts required, different machining processes, different assembly sequences, different critical control points, and so on. Whenever even small changes occur in a product model, we have to adjust or issue new work instructions.

Nowadays, many electronic displays are easily available on a production line or in a pocket. When a product model passes through an assembly line, an RFID tag can be read automatically, uploading visual work instructions in real time. In the case of equipment or machining processes, the machines can detect specific model types and adjust their processes automatically according to the real-time updated instructions.

We can also communicate with our employees or machines with real-time quality data and information found from the downstream production processes or inspection stations so that any repeated mistakes can be prevented in advance. We can dynamically update work instructions through real-time quality data to adjust relevant processes in their processing locations.

11.2.2 Real-Time Quality Audit

A normal quality audit process requires many resources like human workforce and time. During the audit, we also need to interview people and check documents and actual work operations. This kind of typical audit process bothers people and interferes in the normal operation of the plant or business.

Now, with IT and production information combined, we can check a real-time situation much more efficiently and can have an accurate quality audit report generated by the system. For example, Big Data analysis combined with an AI system can process real-time quality data from the market, enabling auditors to see the specific problematic process. It can determine the exact location of the problem source much more quickly and accurately. The system can also check how the relevant work is done in each station through automatic data collection and situation analysis.

Since real-time quality audit can be deployed, we can see the bigger picture of current or potential quality concerns in advance rather than merely reacting to past or current concerns.

We can minimize or eliminate regular quality meetings or quality reports from functional teams by substituting concrete, reliable data and information for the meetings. This kind of quality audit can be an effective tool for preventive and lean quality management by moving people and machines for countermeasures in advance.

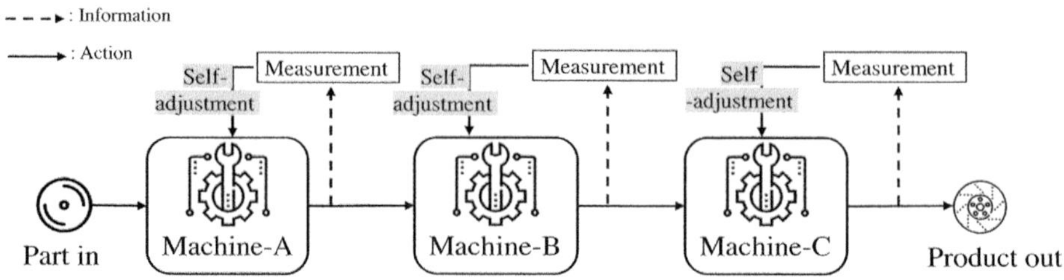

FIGURE 11.7
Real-time machine management.

11.2.3 Real-Time Machine Management

It is a myth that once machines or equipment have been set, they will produce a homogeneous quality of products, e.g., uniform appearance, same properties, and equal dimensions. Using equipment, machines, or robots does not guarantee perfect quality. They experience wear and tear, and their motions may jitter depending on the environment, such as variations in input voltage and humidity.

Machines, robots, and equipment are excellent tools to produce things with better productivity, for they do not get tired and excel at dangerous or repetitive work. However, guaranteeing quality consistently without variation from product to product is an entirely different matter. They need calibration and regular maintenance and settings to produce things with consistency and accuracy.

The traditional way to control the quality of machines and equipment was to stop them regularly for human intervention such as for adjustments or repairs based on the regular measurement of produced parts. During this process, statistical methods and tools have been used for trend analysis of the data, such as using a control chart.

However, as a part of Industry 4.0, many machines are becoming smart enough to measure their produced parts' dimensions or properties automatically, either by sampling or 100% checking. The data are being collected as part of the manufacturing process at each machine as shown in Figure 11.7. Depending on the data analysis result, each machine at each module or work station helped with smart sensors will do self-calibration and self-adjustment, and intelligently communicate with the system and humans when they need helps.

11.2.4 Quality Tracking & Forecasting

While quality data are interconnected through the life cycle of a product, from the development of the new product, to supply chains, manufacturing, and service, in the past, the proper tools to interconnect and correlate the data were not available. Each process and function had data in a computer system, but the data still needed to be interpreted and analyzed with human intervention. As a result, real-time data analysis and subsequent action plans took a long time or were too late for proactive countermeasures.

Preventive or risk management now becomes much more possible with real-time communication and analysis between each function group regardless of physical location. In the future, many manufacturing processes need to be more than JIT (Just In Time) and JIS (Just In Sequence) operations to improve productivity and quality in the stream of Industry 4.0. With the power of the digital transformation of IT plus OT, quality data can be easily traceable from parts to a completed end product.

Figure 11.8 shows an example of Quality 4.0 that demonstrates how steel defects can be traced to a root cause.

> *Traditional quality control involves manually collecting and analyzing the thousands of data sets generated during each process to identify the causes of defects. However, in the framework of Quality 4.0, all the relevant structured and unstructured data from connected processes are automatically analyzed to identify the causes of defects.*
>
> For example, if the surface defect detector (SDD) detects any cracks in a plate, the operation, quality control, and maintenance data from the previous processes are analyzed in 100 mm increments of the product in each phase, tracing back to previous processes to the mother plate and slab, as shown in Figure 11.8. Doing this allows engineers to determine the precise cause of the defect, allowing preventive action to be taken immediately.
>
> Such real-time analysis capability can minimize unnecessary quality control in the upstream process, e.g., scarfing of a whole lot of slabs when a defect is found in a sample slab.[8]

Another application can be improving the traceability of defective products. In industry, traceability is a big headache because not all parts are recorded with one-to-one matching to an end product.

> *When we attach an RFID tag to a parts bin, the problem range can be traced to a much narrower extent. Market warranty data can be processed automatically to identify the exact range of the concerned products within 10 to 50 parts instead of 1,000 to 5,000. For example, if the customers find any abnormal initial or durability quality issues, the system can automatically retrieve the quality data from the service network and define the problem range precisely to deploy a special service or recall campaign.*

11.2.5 Machine Management by Big Data

The ultimate purpose of quality management is to guarantee high quality in the market. There should be a connection between the market quality data and the in-house manufacturing process. When the machines process parts *within their control specifications, a reasonable variance is allowed* as long as they are within the specifications—"Little q" approach (see Table 5.1).

However, this kind of criteria may be not enough when the end product's performance is significantly influenced by critical processing parameters.

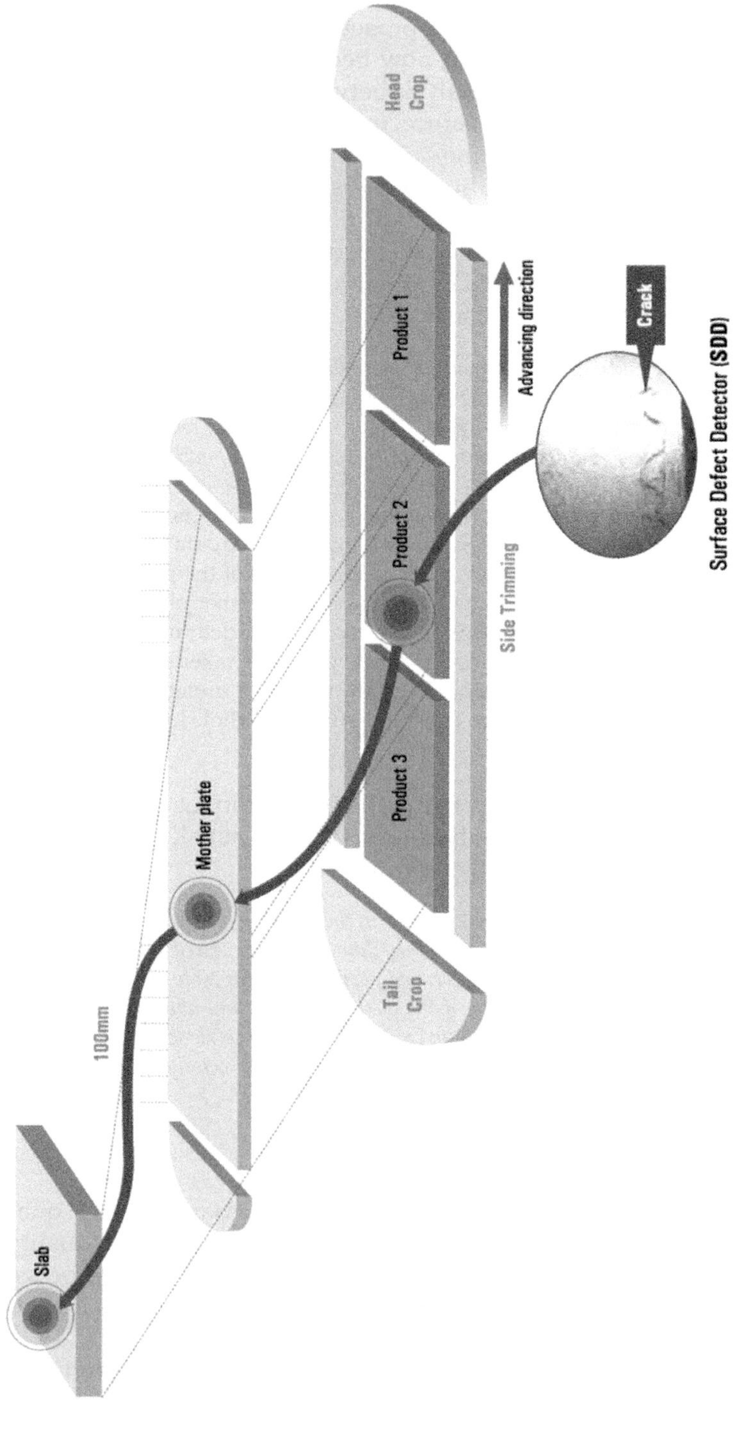

FIGURE 11.8
Quality tracking by surface defect traceability. (Image permission from POSCO.)[9]

Sometimes it is difficult to determine which parameters sensitively influence the customers' perceptions in the market. We need to analyze the data and determine critical concerns in the market and connect them to specific manufacturing control processes and parameters.

During this process, the market warranty data alone is not enough. The end product's quality data comes in the form of Big Data, as the forms of customers' complaints are very complicated. Some data come in a clear warranty format, as hard data, but some are vague complaints like VOC (voice of customers) from a call center or social media. In other words, warranty information are in structured data form while VOC in unstructured one. Such a collection of big data sets is so large and complex that it becomes difficult to process using conventional database management tools or traditional data processing applications.[10]

H-Company's quality organization has a very competitive MI (market information) team with a well-established system and experienced members. Their mission is to receive data (Warranty, VOC) from the market, analyze them, and give the problem information to the quality engineering team for them to tackle the issue (root cause analysis and countermeasure of the problem). However, the market data processing by MI team is done manually—handling of a huge amount of data in Excel sheet, such as sorting the data by the system, part number, and problem phenomenon. Such manual processing demands much overtime working, not to mention that some VOC data are not fully utilized due to its unstructured form.

Big Data processing requires a powerful computing power for data acquisition, storage, and analysis accompanied by advanced data processing method. In this scope, the biggest challenge of Big Data is in data velocity, volume and variety. Now with the introduction of Big Data processing technology, we are able to analyze the big data efficiently. For example, Google, Yahoo, and Amazon have developed very powerful Big Data processing algorithm and tools, such as MapReduce, Hadoop, Elastic MapReduce.[11] Later, Greenplum from EMC and HANA from SAP have also developed real-time analysis architectures.[12]

Utilizing Big Data processing power, we can provide the foundation for manufacturers moving from an Excel filing system to a Big Data analytics server.[13] With the Big Data processing, AI can play a critical role in communicating with equipment or machines that are responsible for the market warranty or complaints so that corrective actions can be taken by self-adjusting of key processing parameters in real-time (see Figure 11.9).

Brake disk quality influences many quality issues in the market, such as noise, judder, vehicle pulling while braking, and braking performance. Many quality parameters in machining, such as external dimensions, runout, and the disk surface, can influence those market quality concerns. Controlling every parameter to the nominal value of the design specification is impossible.

When the big data from the market is processed automatically, it may reveal that a critical quality issue is "brake judder." Then with the analysis of the big data from each machine, the system can instantaneously point out "Lateral runout of

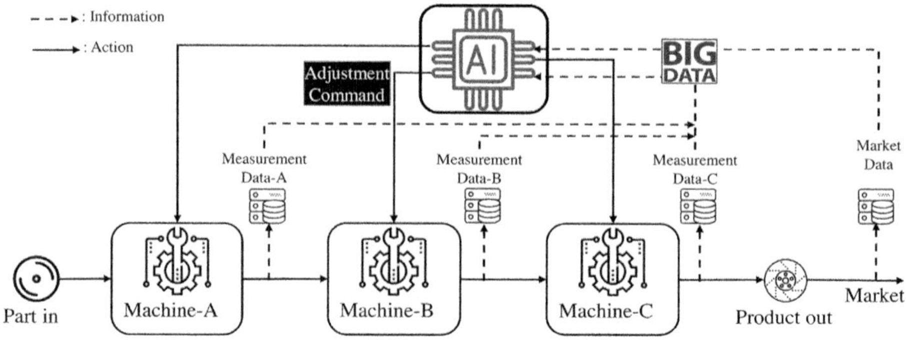

FIGURE 11.9
Real-time equipment management (decentralized and centralized control).

> *the disk surface" as a major cause—lateral runout is the side to side movement of the brake rotor as it rotates. The nominal design spec is 20μm, but the actual part has a variance of 20+/–5μm, which is still within the design spec. The AI system can then request the specific machining process to be further adjusted to improve the machining accuracy to a narrower variance of 20+/–2μm.*

11.2.6 Supply Chain Quality Management

In advanced supply chain quality management, predictions can be made based on supplier performance records and incident data.[14] With the power of connectivity the client can monitor suppliers' in-line parts' quality data and can check for any potential impacts on the client's plant regarding manufacturability or quality concerns.

> *A seat frame's mounting hole data sensitively influences the reading value of the seat weight sensor, part of the OCS (Occupant Classification Systems) sensor, which is used for the airbag suppression system to protect a child by measuring the weight of a passenger.[15] If the mounting holes of a seat frame and car body are aligned in the opposite direction, a stacked up tolerance issue can occur to distort the accuracy of the reading value of the weight sensor, even though each part is within the specified tolerance.*
>
> *The quality impact of such a concern can be pre-checked by virtually connecting the data with in-house body parts, and then we can take appropriate actions immediately: correction of the frame mounting jig in the supplier location or adjustment of welding jigs in the in-house welding line.*

Having the supplier's data in advance can give us many options to improve quality proactively. Appropriate actions can be pre-studied and applied immediately depending on the suppliers' live quality data.

11.2.7 Lean Quality Management

At the end of the 20th century, Japanese manufacturing power impressed the Western manufacturing industry. The first wave was "high quality" and the second wave was "high productivity." As we discussed in Chapter 9, lean manufacturing was led by Toyota, and the lean principle applies to all business scopes to improve productivity and business efficiency.

A principle alone has difficulty making outputs in a dynamically changing business environment. When we have a variety of product models, the complexity increases, and it is difficult to see both the macro and micro pictures for us to focus on waste elimination and reduction of non-value-added steps.

We need to have proper tools to measure our productivity in each cell and module in real time and communicate with relevant people and machines interactively. The seven wastes in manufacturing and quality management, as discussed in Chapter 9, must be analyzed, monitored, and controlled in real time.

To process the production information, we need a tool to compare the current production situation with an *ideal theoretical model* generated by the computer depending on product models, target quantity, etc. For this purpose, a MES (Manufacturing Execution System)[16] can be utilized, but more importantly, the MES should be connected with the overall plant and quality operation system, i.e., the EQMS (Enterprise Quality Management System).

Seeing is believing. The information can be sent to each station operator, office staff, engineer, and manager via mobile device by internet communication. Many user-friendly apps can be developed for information collection and actions to take. The real-time information is shared with the relevant people and suppliers, machine operation parameters are adjusted remotely, and a shipping blocking process can be initiated when quality issues occur.

Our wastes in transportation, inventory, motion, waiting, over-processing, over-production, inspection, repair, data collection, documentation, meetings, and surveys can all be monitored and controlled by computer-aided management. With the power of AI system connected with all production systems, such as MES and EQMS, we can identify the waste concerns in real time and implement more accurate corrective actions much quicker to improve our productivity and quality by using the lean approach.

11.2.8 Cyber-Aided Quality Training

Training is a big part of the quality assurance process in the manufacturing industry. In a human organization, the most significant stagnation point can be cultural stagnation. Unless we change the people, the culture of the organization cannot be improved.

We can maximize our human capability by changing our thinking and attitude through education and training. Traditional quality training can be complemented with a cyber method of training. With Industry 4.0 technology we can utilize a variety of IT tools to make training resemble real situations on the manufacturing floor. For example, we can use AR (Augmented Reality) and simulation technology to teach people to perform properly on the job. We can avoid building an expensive pilot production center. Instead, employees can have simulated training on their expected work by using real-time sensors attached to their bodies and tools in a *virtual factory environment*. We can operate and control machines better. Moreover, machines can also train themselves through self-learning or machine learning processes.

As for quality assurance procedures, workers can also have a virtual experience of what could go wrong when they do not process things properly, by simulating the effects of potential failures. They can also virtually learn how to confirm quality issues and trace them to the right places to identify the root causes. This kind of training strategy also applies to the equipment and machines through a machine learning process—AI.

11.2.9 New Product Development Quality

Traditional new product development requires much manual processing and control of the project from the concept stage to making drawings, verification and validation, and industrialization. It is typical for a company to develop multiple products with limited resources and a short development cycle. Conventional quality management has difficulty controlling the whole development process of a new product—it would take a tremendous amount of workforce and resources.

Now, most of the management processes mentioned above can be deployed intelligently through innovative technology combined with Industry 4.0. For example, for new product planning for a competitive position in the market, an advanced quality planning system can search the history of previous projects and reveal the lessons to reflect onto a new project. Moreover, it can also compare the concept of a new product with the best performing product in the market regarding sales, function, performance, and features. Then it will point out key modifications, new features, or functions necessary to meet a new project's goals and strategy. Basically, QFD (Quality Function Deployment) in Chapter 5 can be deployed automatically in Quality 4.0.

> *By analyzing Big Data of VOC, an advanced quality planning system may recommend an innovative camera design for a new smartphone, such as performing better in dark environments, but not recommend other changes from the previous designs because they are already well accepted in the market. Then engineers can focus only on the recommended area for the new phone, instead of expending their energy and resources on other features.*

With respect to verification and validation processes, designers or human operators use current digital technology (CAD or CAM) to review quality areas of concern *manually*. Some advanced tools like XVL Design Studio of Corel CAD or CAE of CATIA are popularly used. However, in a new design process, the manual checking process can be substituted with an AI system. The AI helps to creates digital mock-up models and runs potential concern areas *automatically* for the entire design system based on the previously known database, the information of a new manufacturing process layout, and new equipment and machines virtually installed. Virtual factories can be built in a cyber environment and simulated to identify any concerns that we may experience during the industrialization process. The system then creates a complete report on potential concerns of design functions, performance, and manufacturability or usability of a future product on behalf of customers in the production process and end-users.

3D printing technology can also be extensively used to deploy real concurrent engineering. For example, a rapid working prototype can be made promptly and cost-effectively for early design verification and validation physically. With this functional physical mock-up part, designers, suppliers, manufacturing engineers, and marketing staff can check the feasibility of the new design concept in terms of manufacturability and marketability. A customer panel can be given the physical mock-up part to check the new concept of a new product at a very early design stage for function, performance, and usability.

All the identified concerns can be compared and analyzed with the Big Data from the manufacturing processes and the users' experiences from the market, and appropriate quality actions can be effectively deployed in the product concept and design stages.

11.3 Future-Oriented Quality Management

Future plants will be run by more equipment, machines, and robots. Future quality management will work more with systems, computers, and machines than with people. Our viewpoints need to be more aligned with Big Q, focusing on advanced features, functions, and performance rather than Little q, reducing defects and variance. *Quality management methodology will completely transform from corrective and preventive actions to predictive management.*

Historically in the manufacturing industry, people ran equipment until it broke and then fixed it, a very poor practice. "Preventive maintenance" and a calibration system were devised to regularly check the equipment and change oil, filters, and parts or adjust the components to prevent sudden breakdown or misreading. Now, with smart sensors and communication technology, we are able to predict the future concerns of equipment or machines so that we can fix future concerns in advance, which is called "predictive maintenance."

Using the same principles and methodology, we can check the function and performance of equipment and machines wirelessly and much more accurately. To make this happen, we need to build equipment and machines with built-in smart sensors, data measurement, analysis, self-control, and information sharing capabilities. Figure 11.10 explains how we can set up a system with such *real-time quality management capabilities* inside the machine, from machine to machine, from machine to human, and from human to machine.

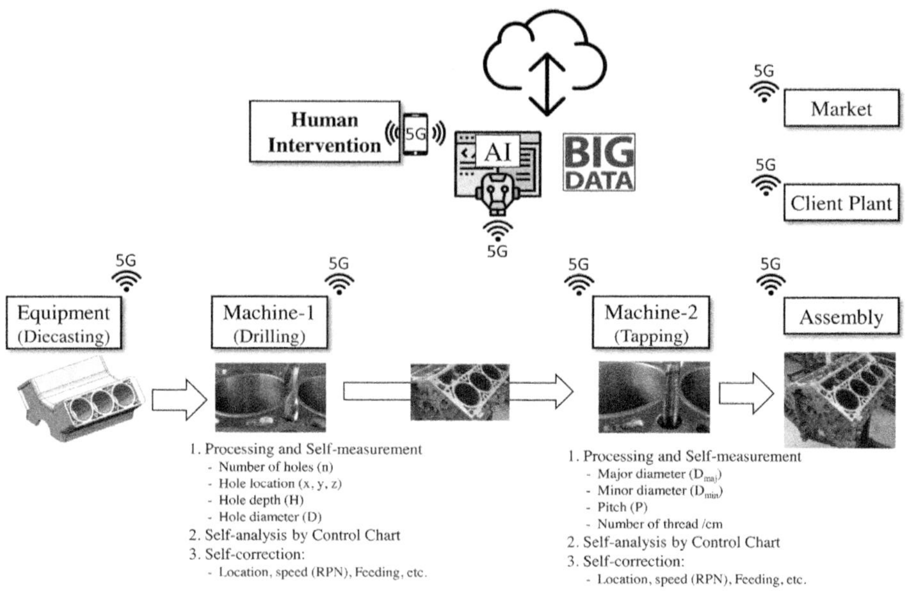

FIGURE 11.10
Real-time measurement, self-analysis, self-control, and communication. (Example: Engine manufacturing process.)

Through the big data analysis from each equipment and machine, we can pinpoint which machines can generate potential quality problems to the next processes or the market in the future.

First, each machine checks its own status, as explained in Figure 11.9. For example, if higher processing temperature or higher dimension is a trend in a particular machine, this information is processed, analyzed, and self-adjusted by the machine itself first. At the same time, each measurement data is exchanged with others in the production line to take appropriate actions, such as tighter control of the specific process in a particular machine or equipment. Throughout each different process, different data are gathered and processed through big data analysis, including the market data. The data is uploaded into a cloud computer via a 5G network for Big Data processing and AI system. AI system can point out a particular process, equipment, or machine as a concern in the trend, and send a command to the machine or equipment to tune up or to warn information to human to take appropriate interventions for predictive quality management.

We say, "Knowledge is power." Why?

Because knowledge has the power to predict the future. From a mathematician, statistician, scientist, economist, psychologist, sociologist to a businessman, everybody's common interest is to know the future. By knowing the future, we can have the power to prepare for the future.

Man of quality management should be able to answer the following questions for quality management:

- Is our quality planning robust enough to design and produce excellent products?
- Is this level of quality satisfactory to the client?
- What are the remaining risks we may have to resolve before the launch of a new product?
- If this problem goes on, what would be a risk in the market?
- Should we stop making the product and block the shipping of parts already made?

Mathematically, any reasonable prediction or forecasting requires a minimum of two sets of data: the past and the present. With the power of Big Data processing, we have the power of predicting the future more accurately and faster than a conventional methodology. Now with the power of AI, we have power managing the future.

"Quality Management in Engineering" should continue to evolve with the philosophy of future-oriented quality management.

This kind of new thinking and methods, combined with new technologies in Industry 4.0, can be applied to future quality management in engineering by scientific and systematic approaches—this is the new paradigm of quality for the future.

Notes

1. Cheong, Je-Ho. "The Fourth Industrial Revolution: The Winds of Change are Blowing in the Steel Industry," *Asian Steel Watch*, Vol. 2, POSCO Research Ins, October 2016.
2. "Definition of NIST (The National Institute of Standards and Technology)," January 16, 2017, www.manufacturingtomorrow.com/article/2017/02/what-is-smart-manufacturing–the-smart-factory/9166 (accessed February 10, 2019).
3. Jacinto, Joan. "Smart Manufacturing? Industry 4.0? What's It All About?," July 31, 2014, www.totallyintegratedautomation.com/2014/07/smart-manufacturing-industry-4-0-whats/ (accessed February 10, 2019).
4. Sachseneder, Christine, Violetka Dirlea, Ram Kidambi, and Stephan Krubasik. "Quality 4.0 Preventive, Holistic, Future-Proof," www.atkearney.com/industrial-goods-services/article?/a/quality-4-0-preventive-holistic-future-proof (accessed February 10, 2019).
5. Cheong, Je-Ho. "The Fourth Industrial Revolution: The Winds of Change are Blowing in the Steel Industry," *Asian Steel Watch*, Vol. 2, POSCO Research Ins, October 2016.

6. Roy, Clementine. Industry 4.0: The Role of Humans in applying New Technologies, https://insights.bridgr.co/industry-4-0-the-role-of-humans-in-applying-new-technologies/ (March 20, 2019).
7. Quality 4.0 Impact and Strategy Handbook, LNS research.
8. Jeong, Duk-Kyoon. "Accelerating Digital Transformation with Smart Factory to Unlock New Value: Case of POSCO," *Asian Steel Watch*, Vol. 2, POSCO Research Institute, October 2016.
9. Jeong, Duk-Kyoon. "Accelerating Digital Transformation with Smart Factory to Unlock New Value: Case of POSCO," *Asian Steel Watch*, Vol. 2, POSCO Research Institute, October 2016.
10. Padgavankar, M.H., Gupta S.R., and et al, "Big Data Storage and Challenges," *International Journal of Computer Science and Information Technologies,* Vol. 5 (2), 2014, 2218–2223.
11. O'Reilly Radar Team, Big Data Now, O'REILLY, 2012.
12. Padgavankar, M.H., Gupta S.R., and et al, "Big Data Storage and Challenges," *International Journal of Computer Science and Information Technologies,* Vol. 5 (2), 2014, 2218–2223.
13. Roy, Clementine. Industry 4.0: The Role of Humans in applying New Technologies, https://insights.bridgr.co/industry-4-0-the-role-of-humans-in-applying-new-technologies/ (March 20, 2019).
14. Sachseneder, Christine, Violetka Dirlea, Ram Kidambi, and Stephan Krubasik, "Quality 4.0 Preventive, Holistic, Future-Proof," www.atkearney.com/industrial-goods-services/article?/a/quality-4-0-preventive-holistic-future-proof (March 20, 2019).
15. How Occupant Classification Systems Work, OCS Sensor, https://auto.howstuffworks.com/car-driving-safety/safety-regulatory-devices/ocs1.htm (accessed February 11, 2019). If you're driving an American car or truck made after 2006, airbag suppression has been applied. Once the onboard computer knows the passenger's size and weight, the car's dual stage airbags come into play. Based on the occupants' size, these types of airbags can deploy at full speed, partial speed, or not at all.
16. Manufacturing Execution System, https://en.wikipedia.org/wiki/Manufacturing_execution_system, (last modified June 21, 2019). Manufacturing execution systems (MES) are computerized systems used in manufacturing to track and document the transformation of raw materials to finished goods. MES provides information that helps manufacturing decision makers understand how current conditions on the plant floor can be optimized to improve production output. MES works in real time to enable the control of multiple elements of the production process (e.g., inputs, personnel, machines, and support services).

Index

Note: Locators in *italics* represent figures and **bold** indicate tables in the text.

Q